Walton Cubley
Surveyor General's
Office

Saint Paul
Minnesota

BURT'S SOLAR COMPASS.

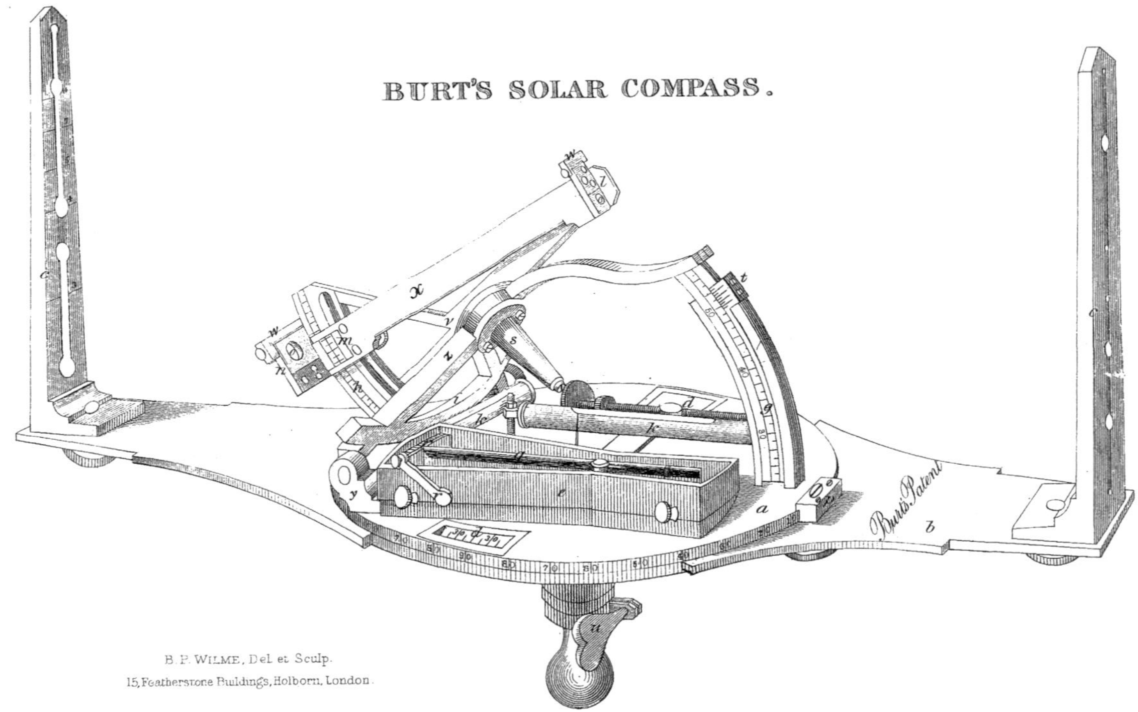

B. P. WILME, Del et Sculp.
15, Featherstone Buildings, Holborn, London.

A KEY

TO THE

SOLAR COMPASS,

AND

SURVEYOR'S COMPANION;

COMPRISING

All the Rules necessary for Use in the Field.

ALSO,

DESCRIPTION OF THE LINEAR SURVEYS, AND PUBLIC LAND SYSTEM OF THE UNITED STATES; NOTES ON THE BAROMETER, SUGGESTIONS FOR AN OUTFIT FOR A SURVEY OF FOUR MONTHS, ETC., ETC.

BY WILLIAM A. BURT,

U. S. DEPUTY SURVEYOR.

PHILADELPHIA:
PRINTED FOR THE AUTHOR BY WILLIAM S. YOUNG,
No. 50, North Sixth Street.
1858.

PREFACE.

Much perplexity and difficulty has been felt by surveyors in the use of the Magnetic Compass, in consequence of its variations from the true meridian, at various localities or stations, and also its almost constant diurnal changes as well as aberrations, caused by local attraction. A more perfect guide for the surveyor than the Magnetic Needle was, therefore, very desirable. The long continued efforts made by the author to accomplish this object, resulted in the invention of the Solar or Astronomical Compass. A model of this instrument was made in the year 1835, by the inventor, in order to test its principles, and in the latter part of the same year, the first Solar Compass was made, under his direction and supervision, by William J. Young, of Philadelphia, Pa. The instrument was then submitted to a committee of the Franklin Institute, of the State of Pennsylvania, who after a full examination of its principles and merits, awarded the inventor a premium of twenty dollars and a "Scott's Legacy" medal. The Solar Compass as then made, like most newly invented instruments, was soon found susceptible of improvement and of greater usefulness than at first anticipated. Accordingly the inventor made several alterations and improvements suggested by experience, and in December, 1840, again submitted the instrument, as improved, to a committee of the same Institute, who reported a decided improvement, in point of accuracy, and the simplicity of its adjustments and use. The inventor has since continued to improve this instrument as more experience in the use of it seemed to suggest. And in 1851 exhibited it, as improved, at the World's Fair, in the city of London, where a premium medal was awarded the exhibitor by the jurors on Astronomical Instruments.

Since its invention in 1835, and during its progressive improvements, the inventor has been called upon, personally or by letter, from a large portion of the surveyors of the public lands, for information how to adjust and use it. Such inquiries could be but imperfectly answered by letter, or a few hours' conversation, and the author could not, without being discourteous, avoid replying in some manner to such necessary inquiries, though a serious tax sometimes on his business. To prevent this the inventor published a few pages of instructions, showing how to adjust and use this instrument, and distributed them among the surveyors; but soon after this, new discoveries were made in the construction and adjustments of the Solar Compass, consequently what had been done only supplied their wants in part, and the inventor was solicited by many of the surveyors of the public lands for full instructions on this subject, and a treatise on surveying adapted to their wants in the field of survey. The foregoing remarks constitute the apology of the author for assuming a task so foreign to his habits of life, and to which duty seemed to impel him in the absence of any prospects of this much needed work being soon accomplished by any other person. This treatise contains much original matter, mostly derived from experience in practical surveying. The elements of surveying as published and taught in the schools, are purposely omitted to lessen the size of this work, the object of which is to furnish the practical surveyor with a convenient pocket companion suited to his business while engaged in his field work. The inexperienced surveyor in this branch of the public service has need of all necessary information to enable him to accomplish his arduous duties in a proper manner. The frequent failures in part, or in whole, by many Deputy Surveyors, have done much injury to the public surveys, and ruined their hopes and reputation.

This is a sufficient reason for introducing into this work the necessary outfit and preparations for a large survey in the wilderness, the want of which has been one of the principal causes of these failures.

The author does not presume that this treatise is without defects; he indulges the hope, however, that it will answer the purpose for which it is designed, until further experience shall furnish a better. The author has availed himself of the experience of several practical surveyors, in preparing this work, and has also consulted the best authorities that appeared to throw light upon the subjects treated of.

The tables of Natural Sines and Tangents, at the end of the work, have been carefully compared with different standard works, and are offered to the surveyor with a confidence that he will find them accurate. The table of chords has been added to supply a want, frequently experienced, in draughting, where a reliable protractor is not at hand. The majority of protractors accompanying draughting instruments are either so small or so inaccurate as to be productive of sensible errors in large draughts.

CONTENTS.

PART II.

A KEY

TO THE

SOLAR COMPASS, AND SURVEYOR'S COMPANION.

THE SOLAR COMPASS DESCRIBED.

SEE PLATE I.

The Solar Compass works astronomically in determining latitude, and in measuring horizontal angles from the true meridian, and in determining the declination, and hour arcs, of celestial objects within the Zodiac; and is further used as a magnetic compass. This instrument is used on a tripod, with a ball and socket, in order to adjust it readily to an approximate level by the hand, after which it is adjusted to a true level by means of four thumb-screws at the lower end of the socket, by which it is attached to the tripod. No part of these are seen in the plate, except the ball, clamp and screw, at *u*. This clamp fastens the instrument on the tripod in any required position. The Solar Compass has two main plates, seen at *a.* and *b.*—*a.* is the upper and *b.* the under plate, the latter is that on which the compass sights *cc.* are attached by screws and steady pins. This plate revolves underneath the upper plate on a conical centre piece, and may be clamped to it at any required angle by two clamps, one of which is seen at *p.* There is, also, an inlaid silver ring on the under plate, divided into half degrees, which is covered by the upper plate, except at two openings at opposite points, with a vernier attached to each, *d. d.* Upon the upper plate is attached a needle-box, *e*, by a conical centre piece below the cap of the needle *q.* This needle has an arc of about 36°, divided into halves, for its north end only. A lever, *r*, is to raise the needle from its pivot when not in use.

The needle-box has a limb extending at right angles from its centre, which is not seen in the plate. At the end of this limb is a vernier and arc to set off the needle's variation; the tangent screw to this limb and vernier is seen between *k* and *d*. In consequence of the imperfection of magnetic needles, the arc is attached to the upper plate by two screws, and made adjustable, so that all instruments of this kind can be made to read the same magnetic variation. On the upper plate two adjustable spirit levels are placed at right angles to each other, for the purpose of adjusting the instrument to a true level, when an observation is made on any celestial object. The edge of the upper plate is divided to every five degrees of a circle; and in its centre is placed a brass pin, rising a little above the needle-box; by this arrangement, the surveyor can readily see the approximate course of any object in view, without turning the sights in its direction.

Together with the foregoing described parts, on the upper plate is placed the solar apparatus, which is attached to it by two small blocks, fastened by screws and steady pins, one of which is seen at *y*. Into each of these blocks one axis of the latitude arc *g* enters. These axes are connected by the hour arc *i* and two radial arms *z z* from its centre at *s*. From this centre of the hour arc, a curvilinear arm extends to the latitude arc *g*. The latitude arc moves in a grooved arc to which its vernier *t* is attached. The grooved arc is fastened to the compass plate by a flange at its base, and two screws. The latitude arc *g* has a radius of about five inches, and is divided into quarter degrees, and its vernier *t* reads these divisions to minutes. The latitude arc is clamped at any required latitude, by a clamp screw on the back side, not seen on the plate. The hour arc *i*, as above stated, lies between, and connects the axes of the latitude arc: it is only a portion of the hour circle, and is divided to half degrees. This arc gives the hour angle of celestial objects within the Zodiac of about 55° or 60° east and west of the meridian.

The revolving limb *v*, with its declination arc *h*, is mounted on the centre of the hour arc, and has a free motion on its conical spindle or axis, within the conical socket *s*, at the lower end of which is a collar and screw, for the purpose of giving a suitable tenseness to its movement. This is called the polar axis.

In connexion with the revolving limb is another moveable limb *x*, attached to it by a short conical centre at *l*; the other end with its vernier *m*, moves over the declination arc *h*, and is clamped to it at any required declination, by a clamp screw on the back side of the arc. This arc has about the same radius, and the same divisions and vernier as the latitude arc.

A small brass plate is attached by screws to each end of the limb *x*, standing out at right angles from the limb; and into the upper half of one plate, and the lower half of the other, is set a small convex lens, as seen at *o o*, called the solar lenses; and on the opposite brass plate to each lens, is attached a small adjustable silver plate by means of three screws. On each of these silver plates two sets of parallel lines are drawn, crossing each other at right angles, at a suitable distance apart to embrace the sun's image, which falls between them from the lens.

The set of lines which are parallel to the hour arc are called the equatorial lines, and the set which are vertical to the hour arc are called the hour lines. On the upper edge of each brass plate above named, is placed an equatorial sight *w w*, which can be attached or detached at pleasure, by means of small thumb screws.

There is also another limb (not seen in the plate) called an adjuster, which can be substituted in the place occupied by the limb *x*, for the purpose of adjusting to a parallelism with the lenses, the equatorial lines on the silver plates. It is a brass bar about six inches long, and one-fourth of an inch thick, with a plane surface, and three small pins at each end. The pins are for the purpose of keeping the limb *x*, when on the adjuster, in its place.

The adjuster when used must be attached to the same place occupied by the limb *x*, with the same centre and screws that held the latter. (See second adjustment.)

PRINCIPLES OF THE SOLAR COMPASS BRIEFLY EXPLAINED.

Where a solar compass is correctly adjusted in all its parts, and also to the latitude and meridian of the place of observation, with its vernier *m*, of the declination arc clamped at 0, or zero, then the polar axis *s*, of the instrument, will be parallel to the axis of the earth, and the moveable limb *x*, with its lenses and equatorial sights, will consequently be at right angles to the polar axis, and will revolve on this axis parallel to the plane of the equator; therefore, it is clear that this motion coincides with the diurnal motion of any heavenly body that has no declination, and it is equally clear, that this coincidence holds good when a celestial object has north or south declination, if its declination be set off on the declination arc of the instrument; for, the diurnal motion of the heavenly object will be like the motion of the moveable limb *x*, parallel to the equator and equidistant from it. Now if the instrument be turned horizontally out of the meridian, the polar axis will not be parallel to the axis of the earth, nor will the moveable limb *x* revolve parallel to the equator; consequently it will not follow the diurnal motion of any heavenly body; therefore, if the sun's declination be set off on the declination arc, the sun's image from the lens will not fall between the equatorial lines on the silver plate, but will fall above or below them, and will not fall between them until the compass is turned again into the true meridian.

It is from these principles of the solar compass that the true meridian is obtained, and the variation of the needle determined, etc.

ADJUSTMENTS OF THE SOLAR COMPASS.

Before using the solar compass it must be correctly adjusted. This consists in bringing its different parts to their proper place, and in determining the index errors of the instrument in its graduated arcs, which is chiefly done by reversals and adjusting screws.

FIRST ADJUSTMENT.

To adjust the two spirit levels k k, to a horizontal movement of the instrument on its lower axis.

Place the compass on the tripod, and level it, or nearly so, with the hand, then by means of the levelling screws at the lower end of the ball and socket, bring the bubble in each level to the middle of its opening. If the bubbles do not move while the compass is turned horizontally around on its lower axis, this adjustment is right; but if they move, the levels must be adjusted by the screws at the end of each for that purpose, until the bubbles will remain stationary while the instrument is turned horizontally around.

SECOND ADJUSTMENT.

To make the solar lenses and the equatorial lines on their opposite plates parallel to each other.

Detach the limb x, by taking out its fastening screws, and attach the adjuster in its place, with the same screws that held the limb; then clamp it at the moveable end, to the sun's declination as near as practicable. Now let the compass be placed on the tripod where the sun shines, and level it, with the sights north and south, or nearly so; then place the limb x on the adjuster, between the pins, the same side up that was upon the compass, and then bring it to bear on the sun as in other observations, and turn the compass horizontally, if necessary to bring the sun's image precisely between the equatorial lines on the silver plate; now, without moving the compass in the least out of level, or otherwise, take the limb x from the adjuster and turn the upper side down, without changing ends, and place it on the adjuster again; then see if the sun's image falls between the equatorial lines as before. If it does, this plate is in adjustment; but if it does not, loosen the three small screws which hold the silver plate, (having oblong holes under their heads,) and move this plate one-half of the observed difference, up or down as the case requires, and lightly tighten the screws again. Repeat these observations and adjustments, as above described, until the sun's image falls precisely between the equatorial lines, either side up. This plate then will be in correct adjustment.

Now reverse the ends of the limb x, and adjust the other silver plate in the same manner as the first. When this is done, the parallelism of the lenses and equatorial lines are as perfect as reversals will make them, and the equatorial sights are also parallel to these. The adjuster may now be taken off and the limb x returned to its place. It will not be necessary to repeat this adjustment unless the silver plates get moved by accident or otherwise. The best time to make these adjustments, is between the hours of 10 A. M. and 2 P. M.

In making this adjustment the limb x should fit accurately on the adjuster, and the brass plates in which the lenses are set must be precisely of the same breadth; if they are not, this adjustment cannot be correctly made. Therefore, these plates should be carefully tried with a gauge, and any difference in size corrected.

THIRD ADJUSTMENT.

To find the index error of the declination arc.

FIRST METHOD.

Set the vernier *m* of the declination arc *h* at 0, or zero, place the compass on the tripod, and incline it north or south, as the sun may have north or south declination, until the sun's image falls precisely between the equatorial lines on the silver plate; then reverse the lenses by turning the revolving limb half way around, and see if the sun's image falls precisely between the equatorial lines on the other silver plate; if it does, there is no index error in this arc; but if it does not, move the limb *x* up or down, as the case requires, on the declination arc one-half of the observed difference, and try the reversals again, and so repeat them, if necessary, until the sun's image falls precisely between the equatorial lines on both silver plates. The amount of index error in this arc can now be read by its vernier *m*. If the index error is *below* the graduated zero point on the declination arc, its amount must be *subtracted* from the declination of the celestial object, before it is set off on the declination arc; but if *above*, it must be *added.*

SECOND METHOD.

Set the vernier *m* of the declination arc *h* at zero, as before, and bring the equatorial sights to bear on some distant object; then, without moving the compass in the least, reverse the revolving limb *v*, and see if the line of sight is the same as before; if it is, there is no index error; but if not, proceed as described, by reversals on the sun, until the equatorial sights will bear on the same objects when reversed.

FOURTH ADJUSTMENT.

To bring the polar axis to a right angle with the axis of the latitude arc.

This adjustment generally is, and always should be made by the instrument maker, but the surveyor should test his instrument in all of its parts. First detach the solar apparatus from the upper plate, by taking out the clamp screw of the latitude arc, and the screws that fasten its axis and blocks to the upper plate; then take a piece of board about four inches wide and a foot long, with smooth edges, and nail one edge to another board about one foot square, so that it will be at right angles to its surface. Place this on a stand or table, in a convenient place to view some distant object, then take the blocks that hold the axes of the latitude arc, and place them on their axes, and fasten them by their screws to the upper edge of the narrow board; by this arrangement the polar axis *s* can be brought to a perpendicular, and then reversed, by giving motion to the axes of the latitude arc of 180°.

The moveable limb *x* must now be clamped to its true zero point, as found by the third adjustment, and the polar axis *s* brought to a perpendicular; the revolving limb *v* must now be turned parallel to the axis of the latitude arc; then observe some distant object through the equatorial sights; now reverse the polar axis as above directed, and see if the equatorial sights bear on the same object as before reversing the polar axis; if they do, the polar axis is at right angles to the axis of the latitude arc; but if not, the face of the flange, or the seat of the conical socket *s*, must be ground on one side enough to correct this error, so that the equatorial sights will bear on the same object when reversed as above stated. If the error be small, it may be corrected by placing a thin piece of tin foil, or some other firm substance, under one side of the flange of the conical socket *s*.

FIFTH ADJUSTMENT.

To make the compass sights coincide with the true meridian, when an observation is made with the solar compass.

Place the compass on the tripod, and clamp the sights to an east, west course; then take out the clamp screw to the latitude arc, and raise this arc until the polar axis *s* is horizontal, or nearly so, and fasten it in this position, which can be easily done by placing a small wedging piece of wood between the edge of the hour arc and the upper plate of the compass, and a small brace of wood between the brass centre pin and the conical centre *s* of the hour arc. Then clamp the vernier *m* of the declination arc at its true zero point, as found by the third adjustment. Now bring the equatorial sights to bear on some distant object in or near the horizon; then unclamp the main plates *a* and *b*, and bring the compass sights to bear on the same distant object; (it is well to reverse the equatorial sights and make the same observation again;) if both sights still coincide, read at the verniers *d d*, the amount of the index error, if any, between these plates.

This adjustment should always be made by the instrument maker, and *cleared* of *index error*, by a proper adjustment of the compass sights on the lower plate. But if any index error is found in the instrument, while in the hands of the surveyor, it should be allowed for in all courses run by him, or he may correct it by removing one of the compass sights the required amount so as to make the line of sight to coincide with the meridian. This can be done by enlarging, with a small round file, the holes on one side of the steady pins and screw that hold the compass sight to the lower plate, enough to correct the index error. The vacancy on the side of the steady pins may be filled with tin foil, or some other substance that is not magnetic.

SIXTH ADJUSTMENT.

To find the index error of the latitude arc.

This is most correctly done by determining the latitude of any station by north and south stars, or, determine the latitude by the sun, and again by the pole star; (see article, "Latitude by the Solar Compass;") one-half of the difference of latitude thus found, if any, is the index error of this arc. If the latitude determined by an observation on the sun, or star within the zodiac, be less than the latitude by the north star, the half difference must be added, to the zodiacal observation, to obtain the true latitude of the station; but if greater, it must be subtracted. But this index error is not used for any other purpose than to find the true latitude, for the latitude given by an observation on a celestial object within the zodiac, is the latitude to be used for all other purposes.

SEVENTH ADJUSTMENT.

To find the index error of the hour arc.

Adjust and clamp the compass sights to the true meridian, as directed in the remarks to find the meridian, variation of the needle, &c.; also, set a stake in the meridian, four or five chains south of the instrument, and keep the compass sights directed to it. Then at the distance of ten or twelve feet south of the instrument, suspend a plumb line from the top of a suitably inclined pole set in the ground, and firmly supported with crotches, and of a sufficient height to observe, near the top of the line, the meridian passage of the sun. Then with the aid of a suitable dark glass, observe through the north sight vane, the meridian contact of the sun's west limb with the line, while an assistant has kept the sun's image accurately between the hour lines on the silver plate. At this point, read on the graduated side of the declination arc, at either end of the revolving limb, its distance from the graduated zero point, and the same again with the last contact of the sun's east limb: half the difference on the hour arc, between these two observations, will be its true zero point; from which read the index error.

It should be remarked here, that the principles of the solar compass have been applied in various ways to surveying instruments, to suit the views of mathematical instrument makers, or surveyors for whom they were made; but the solar compass described in the foregoing pages, and for which the adjustments are given, has been found, after much experience in its use, to be the best adapted to surveying the public lands, and for this purpose it is generally used; for the reason that it is more safely and conveniently carried and used through all the exposures which are unavoidable in the wilderness. Some change, however, may be made in its mechanical construction, for the purpose of city surveying, and for running the lines and curves of railroads, etc. But in whatever form they may be made, it is important to a good solar apparatus, that the latitude and declination arcs have a radius not less than five inches, so that their divisions may be sufficiently large to be easily read, and the arcs readily and accurately adjusted for use. The importance of this will be understood by considering the frequency of these adjustments, and the circumstances under which they are made while running lines in the field. So far as known to the author, but few surveyors have qualified themselves to use the solar compass on any other celestial object than the sun; and, perhaps, as few have fully understood its principles and adjustments. The reason of this is found in the fact, that no work has been published before this, sufficiently elucidating its principles, adjustments and use. The sun is the principal celestial object used in surveying lines with this instrument, which only requires a knowledge of the true declination of the sun for each hour of the day, in the longitude where the survey is to be made. Therefore, with the instructions here given, no accomplished surveyor with the magnetic compass, need hesitate to use the solar compass on the sun; and he will soon acquire the further knowledge of using it on other heavenly bodies at night, to determine the variation of the needle, and for other purposes treated of in this work.

If the solar compass has been truly adjusted in all of its parts, previous to its being used in the field, the surveyor may feel the fullest confidence in the true course of his lines run with it.

ASTRONOMY.

Though merely a knowledge of the apparent diurnal motion of the sun in the heavens, will serve for the single purpose of using the solar compass on that luminary; yet, for all the purposes for which this instrument can be employed by night on the planets and fixed stars, a more extended knowledge of astronomy is required.

Therefore, the following brief notice of astronomical facts and phenomena is deemed necessary to be understood by all surveyors, to enable them to use the solar compass to the best advantage.

SOLAR SYSTEM.

The sun is the centre of the solar system, around which all the planets revolve in elliptical orbits, from west to east,* with diminished velocities as their distances increase from the sun: the planes of their orbits are nearly coincident with the plane of the ecliptic; therefore, their greatest declinations will be sometimes more or less than the sun's greatest declination, by the amount of the angle of inclination of each of their orbits to the plane of the ecliptic. See the following table.

Planet's names.	Mean diameter in English miles.	Mean distance in English miles from the Sun.	Mean sidereal period in mean Solar days.	Inclination of orbit to the ecliptic.	Hourly motion in orbit in miles.
The Sun, . . .	883.246				
Mercury, . . .	3.224	37.000.000	87.969.225	7° 0' 9".1	109.400
Venus, ...	7.687	68.000.000	224.700.787	3°23'28".5	80.060
The Earth, . .	7.912	95.000.000	365.256.361		68.080
The Moon, . .	2.160	95.000.000	27.321.661	5° 8'47".9	2.290
Mars, ...	4.189	142.000.000	686.979.646	1°51' 6".2	55.000
Jupiter, . . .	89.170	495.000.000	4.332.584.821	1°18'51".3	28.000
Saturn, ...	79.042	906.000.000	10.759.219.817	2°29'35".7	20.000
Uranus, . . .	35.112	1.820.000.000	30.686.820.830	0°46'28".4	15.000
Neptune, . . .	35.000	3.600.000.000	60.128.000.000		

* East and west are relative, or local terms. It is meant here, that they move in their orbits around the sun, in the same direction as the opposite side of the earth from the sun moves around its axis.

THE EARTH.

The earth is an oblate spheroid, whose equatorial diameter exceeds its polar diameter about 26 miles; the cause of this difference is supposed to be the centrifugal force of the earth's rotary motion around its axis.

The north and south poles of the earth are two points on its surface, opposite to each other; and a straight line between these two points is called the axis of the earth, around which the earth revolves, from west to east, once in a sidereal day.

The axis of the earth is always inclined from a perpendicular to the plane of its orbit; in other words, the axis of the earth has an angle to the axis of the ecliptic, of about 23° 28′. Therefore, the axis of the earth is always in the same direction in regard to the heavens, in every part of its orbit.

This angle of inclination causes the declination of the sun north and south of the celestial equator, during each revolution of the earth around the sun. It is, also, the principal cause of the declinations of the planets; the different seasons of the year; and the different length of days and nights.

EQUATOR.

The Equator encircles the earth at right angles to the axis, and is equidistant, or 90° from its poles; its plane divides the earth into two equal parts, called northern and southern hemispheres.

The plane of the equator, if extended to the heavens, is called the celestial equator, which has an angle to the plane of the ecliptic, (like the angle between their axes) of about 23° 28′.

The motion of the earth around its axis is uniform; but the velocity of the earth in its orbit around the sun is unequal, the mean of which is 59′ 8″ each day. The sun will therefore return to any given meridian each day in unequal times; hence the difference between apparent and mean time, called the equation of time.

A tropical year is 365 d., 5 h., 48 m., 49 s. A sidereal year, reckoned in mean solar time, is 365 d., 6 h., 9 m., 9. 6s., and reckoned in sidereal time, is 366 d., 6 h., 9 m., 9. 6s.

The reason of this difference is; the earth has moved once around the sun in its orbit the same way the equator moves around its own axis. The earth must therefore complete one revolution and 59′ 8″ on its axis each day, to bring the sun to the same meridian This is called solar time.

The earth has precisely one revolution on its axis from the transit of a fixed star to the next transit of the same star, which is a sidereal day of 24 hours; but, if reckoned in mean solar time, it is 23 h., 56 m., 4 s., 9'''.

An astronomical day commences at noon, and is reckoned from one to 24 hours successively; the civil day commences at the preceding midnight, and is reckoned from 1 to 12 hours, twice in a civi day: therefore the last 12 hours of the civil day correspond to the first 12 hours of the astronomical day. All astronomical calculations are computed in astronomical time.

LATITUDE.

Latitude on the earth is reckoned north and south of the equator in degrees, etc., of the meridian, to the poles (or 90°.) Difference of latitude is an arc of the meridian, between any two parallels of latitude.

LONGITUDE.

Longitude on the earth is reckoned east and west from any prime meridian, in arc or time to 180° or 12 hours. Difference of longitude is the difference in arc or time, between any two meridians, reckoned on any parallel of latitude.

ECLIPTIC.

The Ecliptic is a great circle of the heavens, and its plane is the extension of the plane of the earth's orbit, indefinitely, into space, or the starry heavens.

The sun is always in the ecliptic, and the orbits of all the planets cut or intersect the ecliptic at opposite points, called their nodes, in which only eclipses occur.

ZODIAC.

The Zodiac is an imaginary belt or circle of the heavens, and occupies a space of 8° on each side of the ecliptic; within which all the planets appear to perform their revolutions around the sun.

DECLINATION.

Declination of a heavenly body is reckoned north and south of the equatorial plane. The complement of the declination of a celestial object is its nearest polar distance.

RIGHT ASCENSION.

The right ascension of heavenly bodies is reckoned in time from the first point of Aries, or the vernal equinox, around in the order of the signs, on the equator, to the same point again. The longitude of heavenly bodies is reckoned from the same point, and in the same order on the ecliptic, in degrees, etc., as right ascension is reckoned in time on the equator.

ALTITUDE AND ZENITH DISTANCE.

The altitude of a celestial object is the angle in which it is observed above the horizon. The zenith distance of a heavenly body is its angular distance from the zenith, or point directly over head of the observer.

HORIZON.

An observer has two horizons, the sensible and rational. The sensible horizon is a circle at the extent of view in all directions, on a horizontal plain, or on the ocean. The plane of the rational horizon divides the earth into two equal parts through its centre, parallel to the sensible horizon; it is, therefore, the semi-diameter of the earth below the sensible horizon.

REFRACTION AND PARALLAX.

The atmospheric refraction causes a heavenly body to appear above its true place in the heavens, except it be in the zenith. The parallax of a celestial object is the difference in altitude that would appear between an observation made from any point on the earth's surface and from its centre. Therefore, parallax causes heavenly bodies to appear below their true place in the heavens, except they are in the zenith; hence the corrections for parallax and refraction of instrumental observations on celestial objects.

AZIMUTH.

The azimuth of a heavenly body is reckoned on the horizon of the observer, between a vertical plane of the meridian, and another vertical plane passing through the centre of the celestial object, to the zenith of the observer. In other words, it is the true bearing of a heavenly body referred to the horizon from the meridian.

Azimuths are generally reckoned from the north in north latitude, and from the south in south latitude.

The amplitude of a heavenly body is its true course or bearing at rising or setting, from the east or west points of the horizon.

NAUTICAL ALMANAC.

Blunt's Nautical Almanac and Astronomical Ephemeris, (on account of its size) is the most convenient that has yet been published for the surveyor to take data from, for the use of the solar compass. The heading of each page and column is a sufficient explanation of its contents and use.

This almanac is adapted to mean noon at Greenwich, England, except the sun's declination, which is more properly given for apparent noon.

It will be seen that the quantities in the columns are continually varying from day to day; therefore some reduction is necessary to adapt them to any other time or longitude, than that for which they were registered. This is accomplished by applying the hourly differences, where they are given, according to their sign or precept; and where the hourly differences are not given, take the required proportional part of the difference between the preceding and succeeding noon at Greenwich, and add to or subtract from the registered quantities, according as they are increasing or decreasing, as the case requires.

FIXED STARS.

The following table of the mean places of 35 fixed stars has been selected from the Nautical Almanac, for January 1st, 1854, for the purpose of night observation with the solar compass. The sign + prefixed to an annual variation is to be *added to*, and the sign — is to be *subtracted from* the right ascension: also, for stars having *north* declination, + signifies *add*, and — *subtract;* but for stars of *south* declination + denotes that the variation is to be *subtracted from*, and — that it is to be *added to* the declination.

FIXED STARS.

MEAN PLACES OF THIRTY-FIVE PRINCIPAL FIXED STARS FOR JANUARY 1ST, 1854.

STAR'S NAME.	MAG.	RIGHT ASCENSION.	ANNUAL VAR.	DECLINATION.	ANNUAL VAR.
		H. M. S.	S.	0. 1/. "	"
β Ceti, ...	2	0 36 15·414	+ 3·0127	S. 18 47 20·28	+ 19·832
α Urs. Min. (Polaris,)	2	1 6 11·891	18·0600	N. 88 31 52·32	19·241
θ' Ceti, ...	3	1 16 43·553	2·9997	S. 8 56 17·08	18·740
α Arietis, ...	2	1 58 57·027	+ 3·3634	N. 22 46 11·30	+ 17·274
α Ceti, ...	2·3	2 54 39·087	3·1269	N. 3 30 50·23	14·399
η Tauri, ...	3	3 38 48·759	+ 3·5527	N. 23 38 59·65	+ 11·536
α Tauri (Aldebaran,)	1	4 27 32·809	3·4336	N. 16 12 42·11	7·702
β Orionis (Reigel,) .	1	5 7 31·330	+ 2·8803	S. 8 22 27·30	+ 4·540
δ Orionis, ...	2	5 24 32·994	3·0663	S. 0 24 40·87	3·048
ε Orionis, ...	2	5 28 48·357	+ 3·0436	S. 1 17 57·31	+ 2·709
α Orionis, ...	var.	5 47 16·094	3·2469	N. 7 22 31·37	+ 1·112
μ Geminorum, ...	3	6 14 7·640	3·6357	N. 22 35 1·21	— 1·367
α Canis Maj. (Sirius,)	1	6 38 42·914	2·6447	S. 16 31 11·02	4·602
α Can. Min. (Procyon,)	1	7 31 39·317	3·1459	N. 5 35 43·82	8·859
α Hydr. Æ., ...	2	9 20 24·674	+ 2·9480	S. 8 1 41·32	—15·343
α Leonis (Regulus,) .	1·2	10 0 35·517	+ 3·2026	N. 12 40 43·82	—17·380
δ Leonis, ...	2·3	11 6 20·253	3·2064	N. 21 19 22·28	19·645
β Leonis, ...	2	11 41 36·517	+ 3·0656	N. 15 23 16·95	—20·084
α Virginis (Spicæ,) .	1	13 17 30·330	3·1495	S. 10 23 52·41	18·946
α Bootis (Arcturus,) .	1	14 9 0·134	2·7332	N. 19 56 40·39	18·919
α² Libræ, ...	2·3	14 42 48·491	+ 3·3070	S. 15 25 55·72	15·234
β Ursæ Minoris, . .	2	14 51 10·977	— 0·2687	N. 74 45 7·18	14·760
β Libræ, ...	2	15 9 9·271	+ 3·2202	S. 8 50 27·49	13·601
α Serpentis, ...	2·3	15 37 4·683	+ 2·9514	N. 6 53 17.13	11·643
β' Scorpii, ...	2	15 56 57·182	+ 3·4784	S. 19 24 6·72	10·275
α Herculis, ...	var.	17 7 59·418	+ 2·7322	N. 14 33 37·00	4·449
α Ophiuchi, ...	2	17 28 9·421	+ 2·7796	N. 12 40 11·94	2·967
ζ Aquilæ, ...	3	18 58 41·894	2·7546	N. 13 39 0·00	+ 5·022
γ Aquilæ, ...	3	19 39 19·055	+ 2·8553	N. 10 15 38·89	+ 8·434
α Aquilæ (Altair,) .	1·2	19 43 39·502	2·9286	N. 8 29 10·27	9·154
β Aquarii, ...	3	21 23 52·159	3·1673	S. 6 12 39·37	15·609
β Cephei, ...	3	21 26 45·508	0.8045	N. 69 55 12·83	+ 15·686
ε Pegasi, ...	2·3	21 37 0·915	+ 2.9510	N. 9 12 28·01	+ 16·297
α Aquarii, ...	3	21 58 16·921	3·0826	S. 1 1 38·53	17·300
α Pegasi (Markab,) .	2	22 57 29·402	2·9834	N. 14 25 14·62	19·310

LATITUDE BY THE SOLAR COMPASS.

After the solar compass has been correctly adjusted in all of its parts, its future usefulness depends upon finding the latitude as given by the instrument, at the place where it is used.

That it may not be repeated again, hereafter, it should be remarked, that in all observations with the solar compass, it must be placed on the tripod, and accurately levelled, with the latitude arc turned toward the equator; except, that when making an observation on the pole-star, it must be turned in that direction. This can be done approximately by the magnetic needle.

Thus prepared, set off the sun's declination for noon on the declination arc, allowing for its index error, if any, and the sun's meridional refraction, also, adjust the latitude arc approximately to the latitude of the place, and the revolving limb *v.* at its true zero point on the hour arc *i.*: in other words, for noon.

Commence the observation for latitude about fifteen minutes before the sun culminates, by turning the instrument horizontally on its lower axis, so that the sun's image will fall between the hour lines on the silver plate, and raise or lower the latitude arc, if necessary, to bring the sun's image between the equatorial lines. Then follow the motion of the sun, by turning the compass horizontally, at short intervals of time, and adjust the latitude arc, to keep the sun's image between the equatorial lines, until he culminates. The latitude of the station can then be read at the vernier of the latitude arc.

The same method may be pursued by night to determine the latitude by an observation on any celestial object within the zodiac, viewed through the equatorial sights. In making these observations, it will sometimes be necessary for an assistant to hold a lighted candle a little behind and above the head of the observer, in such a manner that the equatorial sights can be seen; but not so bright as to obscure the star.

LATITUDE BY THE POLE-STAR.

It should be remarked, that the latitude given by an observation on any heavenly body within the zodiac, is read direct on the latitude arc; but when the latitude arc is turned to the north for an observation on the pole-star, or some other star near the pole, the latitude arc will read the co-latitude of the station; it must, therefore, be subtracted from 90° to obtain the true latitude. In these latter observations, the polar distance of the star must be set off on the declination arc instead of its declination, and if the upper meridian passage of the star be observed, the declination *arc* must be turned toward it; but, if the lower meridian passage of the star be observed, the declination arc must be turned from the star.

See sixth adjustment to find the index error of the latitude arc.

EASTERN ELONGATIONS OF POLARIS.

Days.	April.	May.	June.	July.	August.	Sept.
	H. M.	H. M.	H. M.	H. M.	H. M.	H. M.
1	18·18	16·26	14·24	12·20	10·16	8·20
7	17·56	16·03	14·00	11·55	9·53	7·58
13	17·34	15·40	13·35	11·31	9·30	7·36
19	17·12	15·17	13·10	11·07	9·08	7·15
25	16·49	14·53	12·45	10·43	8·45	6·53

WESTERN ELONGATIONS OF POLARIS.

Days.	Oct.	Nov.	Dec.	Jan.	Feb.	March.
	H. M.	H. M.	H. M.	H. M.	H. M.	H. M.
1	18·18	16·22	14·19	12·02	9·50	8·01
7	17·56	15·59	13·53	11·36	9·26	7·38
13	17·34	15·35	13·27	11·10	9·02	7·16
19	17·12	15·10	13·00	10·44	8·39	6·54
25	16·49	14·45	12·34	10·18	8·16	6·33

To find the time of the *meridian passages* of the pole star, add 5 hr. 59 min. to the time of its elongation.

TO FIND THE TRUE MERIDIAN, AND HORIZONTAL ANGLES FROM IT; ALSO, THE VARIATION OF THE NEEDLE.

Clamp the sight of the compass at 0 or zero, and adjust the latitude arc to the latitude of the place; also, set off the sun's declination for the time of day, allowing for index error, if any, and the sun's meridional refraction; then bring the sights of the compass approximately into the meridian by the needle, and the solar lenses into the direction of the sun; if the sun's image does not fall between the equatorial lines, turn the instrument horizontally, and the revolving limb *v.* on its axis, in a manner to bring the sun's image between the equatorial lines, allowing for refraction, if required; then the compass sights will be in the true meridian. Now if the needle *q.* be lowered on to its pivot by the lever *r.*, its variation from the true meridian can be read, and set off on the arc for that purpose; the tangent screw of the vernier limb is seen at *k.* and *d.* (See Plate 1.)

To set the sights of the compass to any other course or angle from the meridian it is only necessary to unclamp the under plate from the upper, and turn the sights to the course required, the angle of which can be read at the verniers *d. d.*

Observations for the same purpose can be made in the night, on any celestial object within the zodiac, by the use of the equatorial sights, instead of the lenses; and by observing two stars, one east, and the other west of the meridian, the variation of the needle, or the course of a line, may be more accurately defined.

ZENITH DISTANCE AND ALTITUDE.

Clamp the compass sights to 90°, or for an E. and W. course, also, set off on the declination arc 23 degrees, and bring the revolving limb *v* to zero or noon on the hour arc. Then by turning the instrument horizontally on its lower axis, bring the solar lenses and equatorial sights into the direction of the sun or star to be observed, and raise or lower the latitude arc as the case requires, until the sun's image falls between the equatorial lines, or the star is seen through the equatorial sights. If the observation be made with the declination arc turned from the object, 23 degrees must be added to the reading of the latitude arc, to obtain the zenith distance of the object observed; but if the declination arc is turned toward the object, 23 degrees must be subtracted from the reading of the latitude arc, to obtain the zenith distance.

If the zenith distance be subtracted from 90 degrees, the altitude of the object will be had.

TIME OF DAY BY THE SUN.

After an observation is made to determine the variation of the needle, or the course of a line by the sun, bring the revolving limb to one division on the hour arc in advance of the sun, then observe the movement of the sun's image to the instant it arrives between the hour lines, and correct for index error of the hour arc, and the effects of refraction, and the hour angle from the meridian at that time, expressed in degrees will be had, which may be converted into time by allowing 15 degrees for an hour, and for each degree four minutes of time. If mean time is required, add or subtract the equation of time according to its precept, and mean time will be had.

DIURNAL VARIATION OF THE NEEDLE.

It has been found by numerous observations, that the diurnal variation of the needle is more in summer than in winter months, and the amount of these aberrations is more or less on different days of the same season of the year, and is probably caused by heat and cold.

But the order in which these diurnal changes take place, can be a little more clearly defined. The north end of the needle will arrive at its most easterly declination between one and two hours after sunrise. It will soon after gradually decline westerly until one or two o'clock, P. M., soon after which it will decline eastward, and at sunset it will have returned half way back to where it was in the morning. Its daily movement may be better understood by an examination of the following table:—

The following observations were made by the author in latitude 42 degrees 42 minutes North, near Detroit, in July, 1839.

1839.	THERMOMETER.			WEATHER.	WEATHER.	WIND.	MAGNETIC VARIATION.		
July.	5½ A. M.	1 P. M.	6½ P. M.	A. M.	P. M.		5½ A. M.	1 P. M.	6½ P. M.
13	60	79	62	clear,	light showers,	W. S. W.	1° 42′	1° 28′	1° 42′
14	59	72	67	clear,	flying clouds,	N. W.	1 42	1 26	1 33
15	56	73	64	cloudy,	light showers,	N. W.	1 32	1 28	1 28
16	55	71	66	cloudy,	some. cloudy,	West.	1 38	1 28	1 30
17	52	80	69	clear,	clear,	W. N. W.	1 30	1 28	1 30
18	55	85½	83	clear,	clear,	West.	1 41	1 28	1 35
19	56	89	82	clear,	flying clouds,	S. W.	1 40	1 28	1 35
20	63	80	74	clear,	cloudy,	S. S. W.	1 40	1 25	1 35
21	70	82	77	clear,	cloudy,	South.	1 42	1 28	1 30
22	72	86	75	cloudy,	some. cloudy,	West.	1 40	1 28	1 35
23	65	88	77	clear,	clear,	East.	1 41	1 23	1 36
24	72	86	77	rain,	clear,	W. S. W.	1 43	1 25	1 35
25	69	83	80	clear,	clear,	N. W.	1 41	1 15	1 32
26	66	88	79	clear,	cloudy,	West.	1 40	1 23	1 35
27	69	80	76	clear,	shower,	West.	1 41	1 30	1 37
28	64	86	80	clear,	clear,	West.	1 42	1 24	1 30
29	66	87	78	cloudy,	clear,	West.	1 41	1 21	1 30
30	69	90	79	clear,	showers,	West.	1 41	1 25	1 33

It will be seen that the average variation for eighteen days
at 5h., 30m., A. M., is 1° 39′ 50″, E.
at 1h., 00m., P. M., is 1° 25′ 37″, E.
at 6h., 30m., P. M., is 1° 33′ 23″, E.

The difference of these numbers gives the diurnal variation as follows:—

Between morning and evening—6′ 27″,
Between morning and noon—14′ 13″,
Between noon and evening—7′ 46″.

From these facts it may be seen, that the variation of the needle, as found at one time, cannot be safely relied upon in running lines at any length of time subsequently. Hence the importance of finding its variation at the time the line is being run.

To guard against errors occurring on account of the variation, the surveyor should at the end of each line, or at the point where the variation of the needle is found, for the purpose of running a line from it at some future time, take the bearing of some distant object, and make a note of the same. On resuming the work, if the sun should be obscured by clouds so as to prevent finding the variation of the needle, he can observe the course of the same object again, and the difference in its course, if any, is the change of variation, and must be allowed for to correct the variation previously determined.

Local attraction, also, so frequently changes the direction of the needle, that the surveyor cannot safely extend his line far without an observation to find its variation; and it will be frequently found that a little delay for this purpose, will more than compensate for all the supposed advantages of running the line without it.

TO FIND THE MERIDIAN PASSAGE OF A FIXED STAR, AND ITS HOUR ANGLE AT ANY HOUR OF THE DAY.

Subtract the sun's Right Ascension for the day and hour of observation, from the star's Right Ascension, borrowing 24 hours for the latter when necessary, and the difference will give the star's meridian passage in solar time; if mean time be required, add to or subtract from the solar time, the equation of time, according to its precept, and the meridian passage of the star will be given sufficiently near for that purpose. Then, if the hour of observation, (astronomical time) can be subtracted from the time of the star's meridian passage, the star's hour angle, east of the meridian will be given; but if the meridian passage of the star be subtracted from the hour of observation, it will give its hour angle, west of the meridian. And thus it may be determined what stars are most favourably situated, for the purpose of finding the variation of the needle, at any time of night.

If any one of the fixed stars named in the preceding table are not truly known to the observer by the geography of the heavens, it is necessary to find the time of meridian passage in order to know the star's hour angle at the time of the proposed observation.

This being known, set the instrument to the star's declination and the equatorial sights to the hour angle of the star, on the hour arc, then bring the sights of the compass into the meridian as near as may be by the needle; the equatorial sights will then direct the eye, nearly, to the star sought for, and by a little movement of the instrument horizontally on its lower axis, bring the line of sight to bear directly on the star, and the observation is complete.

THE EFFECT OF REFRACTION AND PARALLAX IN THE USE OF THE SOLAR COMPASS EXPLAINED.

The equatorial and hour lines of the solar compass will vary their angles from the horizon, as the object observed by the instrument recedes from, or approaches to the meridian of the observer; and when at 90°, or six hours from the meridian, the equatorial lines will have an angle to the horizon, equal to the co-latitude, and the hour lines equal to the latitude of the place of observation. Now if the equatorial lines were at all times in a vertical plane, passing through the centre of the celestial object, refraction would not produce any effect in the course of lines run with the solar compass; but as they will have an angle, as above stated, at different hours of the day, a proportion of the whole amount of refraction, according to the angle, must be allowed for, when large enough to produce a sensible effect in the course of the lines. The equatorial lines are parallel to the horizon when observing a celestial object on the meridian; therefore, the whole amount of the meridional refraction must be allowed for, in setting off its declination. The hour lines are only affected by the whole amount of the refraction, or parallax, when on the equator, or latitude 0°.

The effect of parallax of the sun and large planets, is too small to be regarded, except in the most refined observations. But the parallax of the moon is too large to be neglected in any; for this reason, a table of refraction in altitude is given in this work.

Refraction does not decrease in regular proportion to the altitude of the object. When a celestial object is in the zenith, it has no refraction or parallax; but when it is in the horizon, its refraction is 33′ 51″, and at an altitude of 45° about one minute, (more exactly 58″;) the natural co-tangent of the altitude of a heavenly body, express nearly its refraction.

For the purpose of determining with facility the whole amount of refraction in altitude of a celestial object, the compass sights have lines drawn across them at various distances from the top; at each of these lines are figures, which indicate, in minutes of a degree, the amount of refraction in altitude of a celestial object, as seen from each line in range with the top of the other sight.

From the amount of refraction thus found, subtract the meridional refraction, then the following table will give the proportion of the remainder, expressed in hundredths, to be added to its declination, when the latitude is of the same name; or subtracted from it, when of a contrary name, from one to six hours in time, east and west of the meridian; also, the proportion of the whole amount of the sun's refraction, to be subtracted in time from his hour arc, in the forenoon, and added to it in the afternoon, to obtain the true apparent time. This table will also be useful in observations on the Moon; for the same proportion of the moon's parallax in altitude, must be allowed for on the declination arc, in a reversed order from that of refraction; in other words, the same proportion of the moon's parallax in altitude, corrected for refraction, (see table for that purpose) must be subtracted from her declination, when the latitude is of the same name, and added to it, when of a contrary name.

For the purpose of making corrections for refraction expeditiously, while running lines by the sun, there are three lines drawn below the equatorial lines, 5′ apart, by which to estimate the proportion of refraction to be allowed, by bringing the lower limb of the sun's image the number of minutes below the lower equatorial line on the silver plate, instead of setting it off with the sun's declination. When the surveyor becomes familiarly acquainted with making these allowances for refraction, in using the solar compass, he will seldom need to refer to the tables, or to mathematical calculations, to enable him to make a proper allowance for refraction at all hours of the day, except when the sun is within 5° of the horizon. But for an observation by night on a star, its refraction should be set off with its declination, in the manner before stated.

PROPORTION OF REFRACTION TO BE ALLOWED IN HUNDREDTHS OF THE WHOLE.

	ON THE EQUATORIAL LINES.						ON THE HOUR ARC.					
	HOURS FROM THE MERIDIAN.						HOURS FROM THE MERIDIAN.					
Lat.	1 H.	2 H.	3 H.	4 H.	5 H.	6 H.	1 H.	2 H.	3 H.	4 H.	5 H.	6 H.
10°	97	87	72	52	31	17	26	49	70	85	95	98
12°	97	87	72	53	33	21	25	49	69	85	95	98
14°	97	87	73	53	35	24	25	48	69	85	94	97
16°	97	88	73	55	36	28	25	48	68	83	93	96
18°	97	88	74	57	39	31	25	48	67	82	92	95
20°	97	88	75	58	42	34	24	47	67	81	91	94
22°	97	89	75	59	45	37	24	46	66	80	89	93
24°	97	89	76	61	47	41	23	46	65	79	88	91
26°	97	89	77	63	50	44	23	45	64	78	87	90
28°	97	90	78	65	52	47	23	44	63	76	85	88
30°	97	90	79	66	55	50	22	43	61	75	84	87
32°	98	91	80	68	57	53	22	42	60	73	82	85
34°	98	91	81	71	60	56	22	42	59	71	80	83
36°	98	92	82	71	62	59	21	40	57	70	78	81
38°	98	92	83	73	65	62	20	39	56	68	76	79
40°	98	92	84	75	67	64	20	38	54	66	74	77
42°	98	93	85	77	69	67	19	37	53	64	72	74
44°	98	93	86	78	72	69	19	36	51	62	69	72
46°	98	93	87	80	74	72	18	36	49	60	67	69
48°	98	94	89	81	76	74	18	33	46	58	65	67
50°	99	95	89	83	78	77	17	32	45	56	62	64
52°	99	95	90	85	80	79	16	31	43	53	59	62
54°	99	96	91	86	82	81	15	29	42	51	57	59
56°	99	96	92	87	84	83	14	28	39	48	54	56
58°	99	96	93	89	86	85	14	26	37	47	51	53
60°	99	97	94	90	87	87	13	25	35	43	48	50

DR. YOUNG'S REFRACTIONS.

The Barometer being at 30 inches, and the *internal* Thermometer at 50, or the *external* at 47 degrees, with the correction for + 1 inch in the Barometer, and for — 1 degree in the Thermometer of Farenheit.

App. Alt.	Refr. B. 30. Th. 50°.	Diff. for + 1 B.	Diff. for — 1° Fa.	App. Alt.	Refr. B. 30. Th. 50°.	Diff. for + 1 B.	Diff. for — 1° Fa.	App. Alt.	Refr. B. 30. Th. 50°.	Diff. for + 1 B.	Diff. for — 1° Fa.
° ′	′ ″	″	″	° ′	′ ″	″	″	° ′	′ ″	″	″
0· 0	33·51	74	8·1	3· 0	14·35	30	2·3	8· 0	6·35	13·3	·85
5	32·53	71	7·6	5	14·19	29	2·2	10	6·28	13·1	·83
10	31·58	69	7·3	10	14· 4	29	2·2	20	6·21	12·8	·82
15	31· 5	67	7·0	15	13·50	28	2·1	30	6·14	12·6	·80
20	30·13	65	6·7	20	13·35	28	2·1	40	6· 7	12·3	·79
25	29·24	63	6·4	25	13·21	27	2·0	50	6· 0	12·1	·77
30	28·37	61	6·1	30	13· 7	27	2·0	9· 0	5·54	11·9	·76
35	27·51	59	5·9	35	12·53	26	2·0	10	5·47	11·7	·74
40	27· 6	58	5·6	40	12·41	26	1·9	20	5·41	11·5	·73
45	26·24	56	5·4	45	12·28	25	1·9	30	5·36	11·3	·72
50	25·43	55	5·1	50	12·16	25	1·9	40	5·30	11·1	·71
55	25· 3	53	4·9	55	12· 3	25	1·8	50	5·25	11·0	·70
1. 0	24·25	52	4·7	4· 0	11·52	24·1	1·70	10· 0	5·20	10·8	·69
5	23·48	50	4·6	10	11·30	23·4	1·64	10	5·15	10·6	·67
10	23·13	49	4·6	20	11·10	22·7	1·58	20	5·10	10·4	·65
15	22·40	48	4·4	30	10·50	22·0	1·53	30	5· 5	10·2	·64
20	22· 8	46	4·2	40	10·32	21·3	1·48	40	5· 0	10·1	·63
25	21·37	45	4·0	50	10·15	20·7	1·43	50	4·56	9·9	·62
30	21· 7	44	3·9	5· 0	9·58	20·1	1·38	11. 0	4·51	9·8	·60
35	20·38	43	3·8	10	9·42	19·6	1·34	10	4·47	9·6	·59
40	20·10	42	3·6	20	9·27	19·1	1·30	20	4·43	9·5	·58
45	19·43	40	3·5	30	9·11	18·6	1·26	30	4·39	9·4	·57
50	19·17	39	3·4	40	8·58	18·1	1·22	40	4·35	9·2	·56
55	18·52	39	3·3	50	8·45	17·6	1·19	50	4·31	9·1	·55
2· 0	18·29	38	3·2	6· 0	8·32	17·2	1·15	12· 0	4·28·1	9·00	·556
5	18· 5	37	3·1	10	8·20	16·8	1·11	10	4·24·4	8·86	·548
10	17·43	36	3·0	20	8· 9	16·4	1·09	20	4·20·8	8·74	·541
15	17·21	36	2·9	30	7·58	16·0	1·06	30	4·17·3	8·63	·533
20	17· 0	35	2·8	40	7·47	15·7	1·03	40	4·13·9	8·51	·524
25	16·40	34	2·8	50	7·37	15·3	1·00	50	4·10·7	8·41	·517
30	16·21	33	2·7	7· 0	7·27	15·0	·98	13· 0	4· 7·5	8·30	·509
35	16· 2	33	2·7	10	7·17	14·6	·95	10	4· 4·4	8·20	·503
40	15·43	32	2·6	20	7· 8	14·3	·93	20	4· 1·4	8·10	·496
45	15·25	32	2·5	30	6·59	14·1	·91	30	3·58·4	8·00	·490
50	15· 8	31	2·4	40	6·51	13·8	·89	40	3·55·5	7·89	·482
55	14·51	30	3·3	50	6·43	13·5	·87	50	3·52·6	7·79	·476

TABLE OF REFRACTIONS—*continued.*

App. Alt.	Refr. B. 30. Th. 50°.	Diff. for +1 B.	Diff. for —1° Fa.	App. Alt.	Refr. B. 30. Th. 50°.	Diff. for +1 B.	Diff. for —1° Fa.	App. Alt.	Refr. B. 30. Th. 50°.	Diff. for +1 B.	Diff. for —1° Fa.
° ′	′ ″	″	″	°	′ ″	″	″	°	′ ″	″	″
14·0	3·49·9	7·70	·469	36	1·20·0	2·68	·161	66	25·9	·87	·052
10	3·47·1	7·61	·464	37	1·17·1	2·58	·155	67	24·7	·83	·050
20	3·44·4	7·52	·458	38	1·14·4	2·49	·149	68	23·5	·79	·047
30	3·41·8	7·43	·453	39	1·11·8	2·40	·144	69	22·4	·75	·045
40	3·39·2	7·34	·448	40	1· 9·3	2·32	·139	70	21·2	·71	·043
50	3·36·7	7·26	·444	41	1· 6·9	2·24	·134	71	19·9	·67	·040
15·0	3·34·3	7·18	·439	42	1· 4·6	2·16	·130	72	18·8	·63	·038
30	3·27·3	6·95	·424	43	1· 2·4	2·09	·125	73	17·7	·59	·036
16·0	3·20·6	6·73	·411	44	1· 0·3	2·02	·120	74	16·6	·56	·033
30	3·14·4	6·51	·399	45	58·1	1·95	·116	75	15·5	·52	·031
17·0	3· 8·5	6·31	·386	46	56·1	1·88	·112	76	14·4	·48	·029
30	3· 2·9	6·12	·374	47	54·2	1·81	·108	77	13·4	·45	·027
18·0	2·57·6	5·94	·362	48	52·3	1·75	·104	78	12·3	·41	·025
19	2·47·7	5·61	·340	49	50·5	1·69	·101	79	11·2	·38	·023
20	2·38·7	5·31	·322	50	48·8	1·63	·097	80	10·2	·34	·021
21	2·30·5	5·04	·305	51	47·1	1·58	·094	81	9·2	·31	·018
22	2·23·2	4·79	·290	52	45·4	1·52	·090	82	8·2	·27	·016
23	2·16·5	4·57	·276	53	43·8	1·47	·088	83	7·1	·24	·014
24	2·10·1	4·35	·264	54	42·2	1·41	·085	84	6·1	·20	·012
25	2· 4·2	4·16	·252	55	40·8	1·36	·082	85	5·1	·17	·010
26	1·58·8	3·97	·241	56	39·3	1·31	·079	86	4·1	·14	·008
27	1·53·8	3·81	·230	57	37·8	1·26	·076	87	3·1	·10	·006
28	1·49·1	3·65	·219	58	36·4	1·22	·073	88	2·0	·07	·004
29	1·44·7	3·50	·209	59	35·0	1·17	·070	89	1·0	·03	·002
30	1·40·5	3·36	·201	60	33·6	1·12	·067	90	0·0	·00	·000
31	1·36·6	3·23	·193	61	32·3	1·08	·065				
32	1·33·0	3·11	·186	62	31·0	1·04	·062				
33	1·29·5	2·99	·179	63	29·7	·99	·060				
34	1·26·1	2·88	·173	64	28·4	·95	·057				
35	1·23·0	2·78	·167	65	27·2	·91	·055				

The correction for an increase of altitude of one inch in the Barometer, or for depression of one degree in the Thermometer, is to be *added* to the tabular refraction; but when the Barometer is lower than *thirty* inches, or the Thermometer higher than 47 degrees, the correction becomes *subtractive.*

When great accuracy is required, 0·003 inch should be deducted from the observed height of the Barometer, for each degree that the Thermometer near it is above *fifty* degrees, and the same quantity added for an equal depression.

CORRECTION OF MOON'S APPARENT ALTITUDE FOR PARALLAX AND MEAN REFRACTION.

Moon's apparent altitude.	Moon's Horizontal Parallax. Barom. 30 in. Therm. 50°.								Moon's apparent altitude.
	54′	55′	56′	57′	58′	59′	60′	61′	
°	′ ″	′ ″	′ ″	′ ″	′ ″	′ ″	′ ″	′ ″	°
8	46 59	47 58	48 58	49 57	50 57	51·56	52 56	53 55	8
10	47 56	48 55	49 54	50 53	51 52	52 51	53 50	54 49	10
12	48 26	49 25	50 23	51 22	52 21	53 19	54 18	55 17	12
15	48 39	49 37	50 35	51 33	52 31	53 29	54 27	55 25	15
20	48 7	49 3	50 0	50 56	51 53	52 49	53 45	54 42	20
24	47 9	48 4	48 59	49 54	50 49	51 44	52 38	53 33	24
27	46 12	47 6	47 59	48 53	49 46	50 40	51 33	52 27	27
30	45 3	45 55	46 47	47 35	48 31	49 23	50 15	51 7	30
32	44 12	45 3	45 54	46 45	47 35	48 26	49 17	50 8	32
34	43 17	44 7	44 56	45 46	46 36	47 25	48 15	49 5	34
36	42 18	43 6	43 55	44 44	45 32	46 21	47 9	47 58	36
38	41 14	42 2	42 49	43 36	44 23	45 11	45 58	46 45	38
40	40 8	40 54	41 40	42 26	43 12	43 58	44 44	45 30	40
42	38 59	39 43	40 28	41 12	41 57	42 41	43 26	44 11	42
44	37 44	38 28	39 11	39 54	40 37	41 21	42 4	42 46	44
45	37 7	37 50	38 32	39 14	39 57	40 39	41 22	42 4	45
46	36 29	37 10	37 52	38 34	39 15	39 57	40 39	41 20	46
47	35 50	36 31	37 11	37 52	38 33	38 14	39 55	40 36	47
48	35 10	35 50	36 30	36 10	37 50	37 30	38 11	39 51	48
49	34 29	35 8	35 48	36 27	36 7	37 46	38 25	39 5	49
50	33 48	34 26	35 5	35 44	36 22	37 1	37 39	38 18	50
51	33 6	33 44	34 21	34 59	35 37	36 15	36 52	37 30	51
52	32 22	32 59	33 36	34 13	34 50	35 27	36 4	36 41	52
53	31 39	32 15	32 51	33 27	34 3	34 40	35 16	35 52	53
54	30 55	31 30	32 5	32 41	33 16	33 51	34 27	35 2	54
55	30 11	30 45	31 19	31 54	32 28	33 3	33 37	34 11	55
56	29 25	29 59	30 32	31 6	31 40	32 13	32 47	33 20	56
57	28 40	29 12	29 45	30 18	30 50	31 23	31 56	32 28	57
58	27 53	28 25	28 57	29 29	30 10	30 32	31 4	31 36	58
59	27 7	27 37	28 8	28 39	29 10	29 41	30 12	30 43	59
60	26 19	26 49	27 19	27 49	28 19	28 49	29 19	29 49	60

MEASURING LINES.

In the surveys of the United States lands it is required, that the measuring chain should be two poles, or thirty-three feet in length, and containing fifty links, which must be compared with, and adjusted to the length of the *standard chain* in the Surveyor General's Office, and afterwards to be frequently compared with a standard chain kept by the surveyor for that purpose. But all the measurements, and calculations, are kept, and entered in the field book, in four pole chains, of one hundred links.

The surveyor is required to use eleven tally pins; they should be made of steel, and not more than about one foot in length, and large enough near the points, to cause them to drop perpendicularly; at the top end of each pin, a loop or eye should be made, in which a piece of red cloth may be fixed, that they may be more readily found, when stuck among weeds, grass, &c.

In all measurements the level or horizontal length is to be taken; for this purpose, in ascending hills, banks, &c., the chain-men must let down one end of the chain to the ground, and raise the other end to a level therewith, at the *elevated end* of which a tally pin should be plumbed and let fall, to ascertain the spot for setting it; and, when the surface of the ground is very steep, it may be necessary to take so much of the length of the chain as can be raised to a level, so as to obtain the true horizontal measurement.

In measuring lines, one of the eleven tally pins must be set at the starting-point, and when the remaining ten are set, it is called a tally or out, (five chains) and the forward chain-man cries "Tally," and each chain-man registers the distance by slipping a thimble or loop on a tally belt worn for that purpose. The back chain-man then comes up, and having counted, in the presence of his fellow, the tally pins which he had taken up, so that both may be assured that none have been lost, takes the forward end of the chain and proceeds to set them. Thus the chain-men alternately change places, each setting the pins that he had taken up, so that one is forward in all the odd, and the other in all the even tallies; which contributes to the accuracy of the measurement, facilitates the recollection of the distances to notable objects on the line, and renders a mis-tally almost impossible.

Measurements with the chain and tally-pins are often very imperfectly performed by the chain-men, and much more error is made than is generally supposed. It has been found by many trials, with as good men as can generally be obtained, that with two sets of chain-men, instructed alike in the proper manner of keeping their chain level and straight on the line, and of setting the tally-pins plumb, as well as holding the ends of the chain to them, a difference has sometimes been made of 36 links, and an average difference of 15 or 16 links to a mile, in common timbered land. But repeated measurements over the same mile, by the same chain-men, and near the same time, will generally agree within five links; yet after several months' employment in the field, a measurement of this line may not agree so nearly. Again, the same chain-men will make a different measurement to some extent, over swamps, marshes, windfalls and thickets, when there is snow on the ground and when there is none, in cold and in warm weather, effecting a change in the length of the chain, and by measuring fast or slow the amount of error to each would be difficult to estimate. Therefore the surveyor should keep a vigilant watch over his chain-men, and see that their duties are performed in the best manner, to counteract all these sources of error as far as practicable.

TELESCOPIC MEASUREMENT.

This method of measuring, when properly conducted, is more uniformly the same, and therefore correct, than measurements made by the chain by various chainmen. It is well adapted to measure along the shores of lakes and rivers where obstacles are frequently found of a character to prevent a good measurement with a chain, also for measuring short distances over streams, ponds, &c.

The following arrangement and method of measuring with a telescope and rod will be found very convenient for meandering rivers, lakes, &c. A good telescope must be provided, of about 16 or 18 inches in length when adjusted for use, with two parallel lines correctly set in its principal focus, forming between them, in the field view, not less than 45′ of a degree. This telescope is attached to the sight of the compass with a suitable fixture for that purpose, when wanted for use. Provide a sliding-rod, such as are commonly used for taking levels for canals, railroads, &c., with two targets, one stationary at the top of the rod, the other moveable, with a vernier for the usual readings, on the lower part.

When measurements with the telescope and rod are to be made, the telescope must be attached to the compass sights and adjusted for an observation; then measure four chains from it very accurately, and place the rod at that point, with the targets facing the compass, then bring the upper line in the telescope to bear correctly on the upper target by means of the levelling screws, and adjust the moveable target to range with the lower line, then by observing accurately the distance the targets are apart on the rod, when they measure the angle formed by the parallel lines in the telescope at the given distance from the compass, the observer will have data from which a table may be readily constructed for all other distances, of which the telescope will enable the observer to view the distance between the targets accurately. It may conduce to the correctness of this method of measuring to make observations at various distances, to test the accuracy of the table thus formed; after this, the surveyor may feel a confidence in the correctness of his measurements with the telescope and rod.

Lines run and measured by this arrangement along the shores of lakes and navigable streams are most conveniently and expeditiously done with two skiffs or canoes, or even with two light rafts, with the compass in one and the rod in the other, which can be landed at suitable points and distances apart on their shores; then, after the bearing and distance between them has been taken, the compass can be moved, with the skiff or canoe, to the position occupied by the rod, and the latter again stationed at the next suitable point, and its course and distance taken as before, and so on to the close of the survey.

In all observations, care should be taken to hold the rod at right angles to the line between it and the compass; but it is often necessary to lean the rod at right angles to this line, sometimes even to a level with the horizon; in all such cases, the telescope must be rolled in the y's to bring the parallel line at right angles to the rod.

By this method, the shores of lakes and rivers, however difficult to be measured with the chain, may be correctly meandered by course and distance, without encountering the obstacles on shore with the compass and chain.

To prevent confusion or mistake in the locality of the different stations and notable objects on the land or off the shore, a temporary map should be fully kept up with the survey, on which each object must be represented in order to furnish data for the construction of a good and correct map.

No surveyor, however, should presume to meander important surveys by this method, except he has previously made the necessary preparations, and has qualified himself by some practical experience beforehand.

TABLE

CHAINS TO FEET.—FEET TO CHAINS.

Links, 7·92 inches.—Chain, 66 feet, = 792 inches.

CHAINS INTO FEET.				FEET INTO CHAINS.			
Chains. Links.	Feet.	Chains. Links.	Feet.	Feet.	Links.	Feet.	Links.
0·1	0·66	3·0	198	0·10	0·15	10·0	15·1
0·2	1·32	4·0	264	0·20	0·30	15·0	22·7
0·3	1·98	5·0	330	0·25	0·38	20·	30·3
0·4	2·64	6·0	396	0·30	0·45	24·	36·3
0·5	3·30	7·0	462	0·40	0·60	27·	40·9
0·6	3·96	8·0	528	0·50	0·76	30·	45·4
0·7	4·62	9·0	594	0·60	0·91	33·	50·0
0·8	5·28	10·0	660	0·70	1·06	36·	54·5
0·9	5·94	20·	1320	0·75	1·13	39·	59·1
0·10	6·60	30·	1980	0·80	1·21	40·	60·6
0·20	13·20	35·	2310	0·90	1·36	42·	63·3
0·30	19·80	40·	2640	1·00	1·51	45·	68·2
0·40	26·40	45·	2970	2·0	3·0	48·	72·7
0·50	33·00	50·	3300	3·0	4·5	50·	75·7
0·60	39·60	55·	3630	4·0	6·0	51·	77·3
0·70	46·20	60·	3960	5·0	7·5	54·	81·8
0·80	52·80	65·	4290	6·0	9·1	57·	86·3
0·90	59·40	70·	4620	7·0	10·6	60·	90·9
1·00	66·00	75·	4950	8·0	12·1	63·	95·4
2·00	132·	80·	5280	9·0	13·6	66·	100·

CONVENIENT METHODS FOR MEASURING DISTANCES OVER RIVERS, LAKES, MIRY-MARSHES, ETC.; WHICH CANNOT BE MEASURED DIRECTLY WITH THE CHAIN.

It may be remarked here, that in surveying large districts of new country, many obstacles of this kind are to be expected, and are met with sometimes under many difficulties, such as the direction and swampy or thickety character of their shores, also, the annoyance felt by the presence of increasing swarms of blood-thirsty flies and moschetoes, which largely infest such shores in summer; hence the importance of the best management, and correct and expeditious methods of passing such obstacles.

The following illustrations will assist the inexperienced surveyor in the accomplishment of this object. They are given on the principle of reducing the base, whatever may be its course or courses, to a right-angled base to the course of the line to be measured. This can be readily done if care be taken to run and measure the base, at such angles that their latitude and departure can be taken from the traverse table.

FIGURE 1.

Distance required over lake from *A* to *C*, course East,—right-angled base,—from *A* to *B* 690 links. Angle at *C* 20° 20′

Natural co-tangent of the angle at *C*, ……………	=2.698525
Multiplied by base *A*. B.	690
	242867250
	16191150
Over lake 1862 links	1861.982250

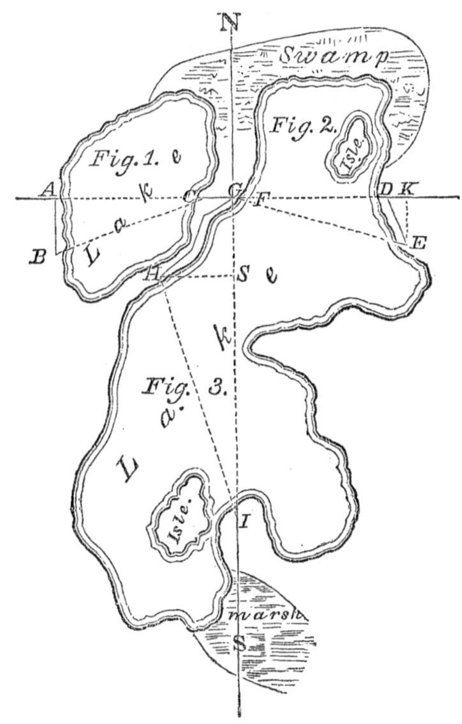

FIGURE 2.

Distance required over lake from *D* to *F*, course West.—From *D* to *E*, S. 20° E 752 links—gives 707 links southing, which is the right-angled base *K G*, and 257 links easting from *D* to *K*. Angle at *F* 15¾°.

Natural co-tangent of the angle at *F*.	=3.545732
Multiplied by the base *K. E*,	707
(Nat. co-tan. *F*×*K E*.)—*K D*=*D F*.	24820124
(3.545.702×707)=2507—257=2250.	24820124
	2506.832524
Subtract distance from *D* to *K*,	257
Distance from *D* to *F*,	2250 links nearly.

FIGURE 3

Distance required over lake from *G* to *I*, course South.

To obtain a base in this example, we run

	Southing.	Westing.
S. 55¾ degrees, W. 400	225	331
S. 19¼ do W. 440	415	145
S. 50 do W. 548	352	420
Distance from *G* to *S*	992 .... *H* to *S*	896 links.

Co-tangent of the angle 16° 38′ at *I*,	3.347319
Multiply by the base *H S*,	896
	20083914
	30125871
	26778552
Distance from *S* to *I*	2999.197824
Add distance from *G* to *S*	992
Distance over lake from *G* to *I*	3991 links.

DISTANCE OVER A RIVER BY "OFF-SET."

EXAMPLE.

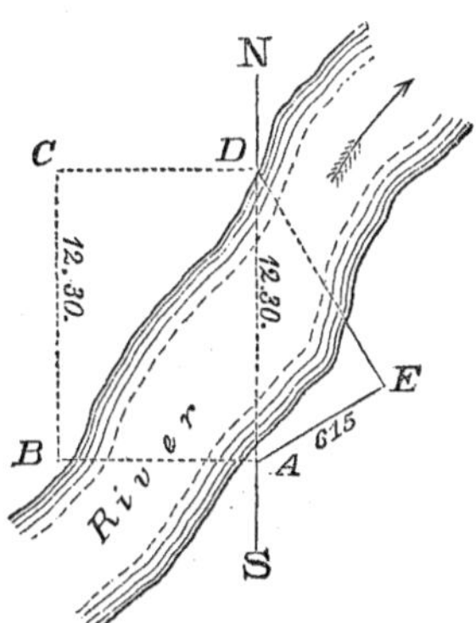

FIGURE 4.

In running a line north, intersect the right bank of a river at *A*, (course N. N. E.,) and erect an object, turn the compass sights to west, to an object at *B*, and pass over the river to it, then run and measure a line north to *C*, and "off-set" east into line at *D*, the distance between *A* and *D* will be equal to the distance between *B* and *C*. Or, if a line be run and measured from *A*, N. 60°, E. until an object in line at *D* bears N. 30° W., the distance *A. D.* will be twice that of *A. E.*, for the reason that the triangle thus formed is one-half of an equilateral triangle.

Frequently off-sets are made in passing small lakes, bends of rivers, etc.: sometimes the distances can be advantageously taken over such obstacles, with the telescope and rod, (see article, Telescopic Measurement.) Also, it often happens that a suitable angle can be taken, and the base to that angle measured afterwards; in such cases the distance can be taken from the traverse table; but if no traverse, or other proper tables are at hand, the following angles, on a right angle base, and the multiplier to it, will give the distance. These may be committed to memory.

Angle 11°, 18′, multiply the base by 5,
" 14, 2 multiply the base by 4,

Angle 18, 26, multiply the base by 3,
" 21, 41, multiply the base by 2.5,
" 26, 34, multiply the base by 2.

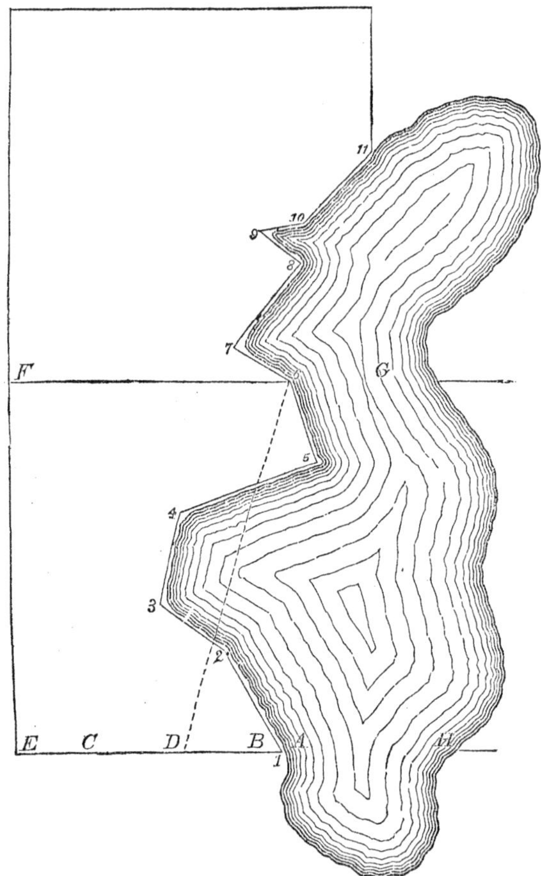

FIGURE 5.

SHORT METHOD OF FINDING THE AREA OF A MULTANGULAR FIELD.

EXAMPLE, SHOWING HOW TO REDUCE THE PLOT OF A MULTANGULAR FIELD TO A FIELD OF EQUAL AREA HAVING ONLY THREE OR FOUR SIDES, BY WHICH ITS CONTENTS MAY BE READILY FOUND.

To reduce such a field the only instruments required, after the meanders are properly laid down, are a good parallel-rule,* and a fine protracting point.

In the preceding figure first extend the base *E H* to an indefinite length; then placing the rule on the angles 1 and 3, move it parallel from the angles 1 and 3 to the angle 2, and mark the exact point of intersection at *A*, on the base *E H*. Now place the rule on *A* and the angle 4, then move it parallel to the angle 3, finding the point *B* on the base *E H*; place the rule on *B* and the angle 5, and move, parallel, to the angle 4, finding the point *C* on the base *E H*. Now place the rule on the point *C*, and the terminating point 6 on the line *F G*, and move the rule, parallel, to the angle 5, finding the point *D* on the base *E H*, from which point draw a line to 6, the process then being complete. The line *D* 6 thus drawn leaves the same area of lake to the left, that there is of land to the right.

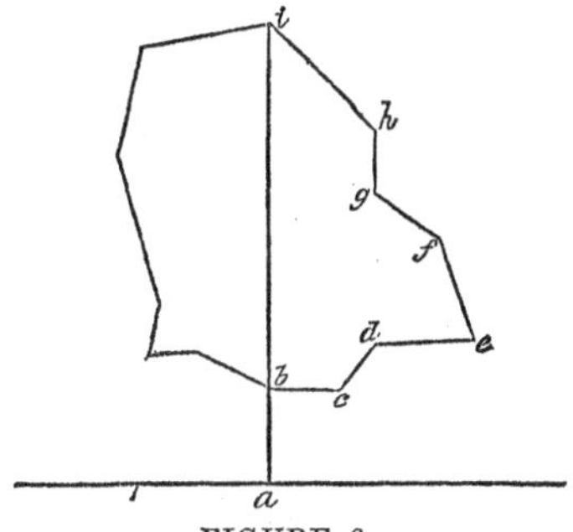

FIGURE 6.

Any figure may be calculated upon the same principle by drawing a base and erecting a perpendicular line from it, passing through

*The triangle and the rule are the best.

the figure. Place the rule at *a* and *c*, then move, parallel, back to *b*, marking the point 1 on the base; then from 1 to *d*, and move forward to *c* and so on to the angle at *i*, leaving a triangle to the right of the perpendicular. Proceed in like manner with that portion of the figure to the left of the perpendicular line, throwing it into two triangles.

CROSSING RIVERS AND LAKES.

In connexion with convenient methods of measuring distances over lakes and rivers, it is proper to take notice of the means employed, by the most experienced surveyors, for the transit of the surveying party over such waters, when fording them, or travelling around their shores, is impracticable, or causes too much delay.

For this purpose floats or rafts made of logs, of the most dry and buoyant timber at hand is used, and when formed into a raft, its length should be about four times its breadth; with this proportion the raft will steer better, and pass through the water with more ease and expedition, than broad and short rafts.

The following is a safe and expeditious method of constructing these floats:—At a convenient place lay two skids, at a suitable distance apart, parallel to the shore, and near to the water, place on these two logs, twenty feet long and one foot diameter, which are to be the outside logs of the raft, and at about two or three feet from the ends of these, make with an axe a *dovetail* notch three or four inches deep, and about as wide on their upper sides; then fit into these notches a cross piece, or tie of a suitable size, and wedge them there firmly, so that these logs will not be separated on the water; then before or after launching this into the water, as convenience may suggest, fill in underneath the cross pieces, between the outside logs, with smaller timber of the same length, and tie them to these pieces, or fasten them by means of a dovetail notch. For crossing deep water, where poles cannot be used, paddles, or oars will be needed; they can be split out of a log and hewn into the proper shape in a few minutes.

With the whole force of the surveying party, it will require from one to two hours to construct a raft of a sufficient size to pass them all over a lake or river at one time.

RUNNING LINES WITH THE SOLAR COMPASS.

In commencing a survey where the latitude, as given by the instrument with which the survey is to be made, is unknown, the surveyor should first determine the latitude of his commencing point. He should remember, that in running any other than an east and west line, he is continually changing his latitude, so that every ninety-two chains and thirty links, of northing, or southing, will change his latitude one minute of a degree, or 5′ 12″ for six miles, and a corresponding change of latitude must be set off on the latitude arc. During the progress of a large survey, the surveyor should determine his latitude daily, if practicable, by the meridian passage of the sun, to test the correctness of the adjustments of the latitude and declination arcs.

It is equally important that the sun's declination be truly set off on the declination arc, for the time and longitude of the station, as it is that the latitude arc be truly adjusted to the latitude of the place of observation.

The following method of preparing the sun's declination, as taken from the Nautical Almanac, for daily use, in any longitude, will be found useful in practice:

EXAMPLE.

To calculate the sun's declination for all hours of the daytime for May 11th, 1854, in latitude 42° N., longitude 120° W., or eight hours before noon, local time, corresponding to Greenwich noon.

12 h. — 8 h. = 4 h. A. M., at the place of observation.

Sun's declination,	17°	52′	11″	at Greenwich noon, as per Nautical Almanac.				
Meridional refraction +			26					
	17	52	37	4 h. A. M.,	17°	57′	41″	at noon.
Hourly difference +			38				38	
	17	53	15	5 h. A. M.,	17	58	19	1 h. P. M.
			38				38	
	17	53	53	6 h. A. M.,	17	58	57	2 h. P. M.
			38				38	
	17	54	31	7 h. A. M.,	17	59	35	3 h. P. M.
			38				38	
	17	55	9	8 h. A. M.,	18	0	13	4 h. P. M.
			38				38	
	17	55	47	9 h. A. M.,	18	0	51	5 h. P. M.
			38				38	
	17	56	25	10 h. A. M.,	18	1	29	6 h. P. M.
			38				38	
	17	57	3	11 h. A. M.,	18	2	7	7 h. P. M.
			38				38	
	17	57	41	12 h. M.,	18	2	45	8 h. P. M.

To calculate the sun's declination for August 25th, 1854, for all hours of the daytime, in latitude 45° N., longitude 90° W., or six hours before noon, local time, corresponding to Greenwich noon.

12 h.—6=6 h. A. M., at the place of observation.

Sun's declination N.,	10°	48′	12″		10°	44′	31″	11 h. A. M.
*Meridional refraction+			39				52	
	10	48	51	6 h. A. M.,	10	43	39	12 h. M.
*Hourly difference—			52				52	
	10	47	59	7 h. A. M.,	10	42	47	1 h. P. M.
			52				52	
	10	47	7	8 h. A. M.,	10	41	55	2 h. P. M.
			52				52	
	10	46	15	9 h. A. M.,	10	41	3	3 h. P. M.
			52				52	
	10	45	23	10 h. A. M.,	10	40	11	4 h. P. M.
			52				52	
	10	44	31	11 h. A. M.,	10	39	19	5 h. P. M.

The calculations for the sun's declination for each hour of the day can be made after the preceding forms, on blank leaves placed in the field book, where they would be required through the day.

In the following table the hourly difference of the sun's declination, as given for the day, in the Nautical Almanac, will be found to the nearest second in the left hand column, and the change of declination for any number of hours to twelve, will be found against it, under the hour at the head of the columns.

This table is useful when the sun's declination is required for any number of hours up to twelve, before or in advance of Greenwich noon.

EXAMPLE.

Suppose the sun's declination is required for September 6, 1854, at 2 h. P. M., in longitude 120° W., or 8 h. in time W. of Greenwich, then 8+2=10 hours. The sun's declination at Greenwich noon is 6° 28′ 52″ N.; hourly difference 56″, against this, in the above table, and under 10 hours, we find 9′ 20″, which subtract from 6° 28′ 52″=6° 19′ 32″ for the sun's declination at the time and place required.

* The hourly difference of the sun's declination must be added when his declination is increasing, and subtracted when it is decreasing; and the meridional refraction must be added to the declination when the latitude is of the same name, and subtracted when of a contrary name. See method of finding Meridional Refraction.

TABLE OF THE INCREASE OR DECREASE OF THE SUN'S DECLINATION FOR HOURLY DIFFERENCES FROM 5″ TO 60″, AND FROM THREE TO TWELVE HOURS OF TIME.

DIFF.	3 H.	4 H.	5 H.	6 H.	7 H.	8 H.	9 H.	10 H.	11 H.	12 H.
″	′ ″	′ ″	′ ″	′ ″	′ ″	′ ″	′ ″	′ ″	′ ″	′ ″
5	15	20	25	30	35	40	45	50	55	1 00
6	18	24	30	36	42	48	54	1 00	1 06	1 12
7	21	28	35	42	49	56	1 03	1 10	1 17	1 24
8	24	32	40	48	56	1 4	1 12	1 20	1 28	1 36
9	27	36	45	54	1 3	1 12	1 21	1 30	1 39	1 48
10	30	40	50	1 00	1 10	1 20	1 30	1 40	1 50	2 00
11	33	44	55	1 6	1 17	1 28	1 39	1 50	2 1	2 12
12	36	48	1 00	1 12	1 24	1 36	1 48	2 00	2 12	2 24
13	39	52	1 5	1 18	1 31	1 44	1 57	2 10	2 23	2 36
14	42	56	1 10	1 24	1 38	1 52	2 6	2 20	2 34	2 48
15	45	1 0	1 15	1 30	1 45	2 00	2 15	2 30	2 45	3 00
16	48	1 4	1 20	1 36	1 52	2 8	2 24	2 40	2 56	3 12
17	51	1 8	1 25	1 42	1 59	2 16	2 33	2 50	3 7	3 24
18	54	1 12	1 30	1 48	2 6	2 24	2 42	3 00	3 18	3 36
19	57	1 16	1 35	1 54	2 13	2 32	2 51	3 10	3 29	3 48
20	1 00	1 20	1 40	2 00	2 20	2 40	3 00	3 20	3 40	4 00
21	1 3	1 24	1 45	2 6	2 27	2 48	3 9	3 30	3 51	4 12
22	1 6	1 28	1 50	2 12	2 34	2 56	3 18	3 40	4 2	4 24
23	1 9	1 32	1 55	2 18	2 41	3 4	3 27	3 50	4 13	4 36
24	1 12	1 36	2 00	2 24	2 48	3 12	3 36	4 00	4 24	4 48
25	1 15	1 40	2 5	2 30	2 55	3 20	3 45	4 10	4 35	5 00
26	1 18	1 44	2 10	2 36	3 2	3 28	3 54	4 20	4 46	5 12
27	1 21	1 48	2 15	2 42	3 9	3 36	4 3	4 30	4 57	5 24
28	1 24	1 52	2 20	2 48	3 16	3 44	4 12	4 40	5 8	5 36
29	1 27	1 56	2 25	2 54	3 23	3 52	4 21	4 50	5 19	5 48
30	1 30	2 00	2 30	3 00	3 30	4 00	4 30	5 00	5 30	6 00
31	1 33	2 4	2 35	3 6	3 37	4 8	4 39	5 10	5 41	6 12
32	1 36	2 8	2 40	3 12	3 44	4 16	4 48	5 20	5 52	6 24
33	1 39	2 12	2 45	3 18	3 51	4 24	4 57	5 30	6 3	6 36
34	1 42	2 16	2 50	3 24	3 58	4 32	5 6	5 40	6 14	6 48
35	1 45	2 20	2 55	3 30	4 5	4 40	5 15	5 50	6 25	7 00
36	1 48	2 24	3 00	3 36	4 12	4 48	5 24	6 00	6 36	7 12
37	1 51	2 28	3 5	3 42	4 19	4 56	5 33	6 10	6 47	7 24
38	1 54	2 32	3 10	3 48	4 26	5 4	5 42	6 20	6 58	7 36
39	1 57	2 36	3 15	3 54	4 33	5 12	5 51	6 30	7 9	7 48
40	2 00	2 40	3 20	4 00	4 40	5 20	6 00	6 40	7 20	8 00
41	2 3	2 44	3 25	4 6	4 47	5 28	6 9	6 50	7 31	8 12
42	2 6	2 48	3 30	4 12	4 54	5 36	6 18	7 00	7 42	8 24
43	2 9	2 52	3 35	4 18	5 1	5 44	6 27	7 10	7 53	8 36
44	2 12	2 56	3 40	4 24	5 8	5 52	6 36	7 20	8 4	8 48
45	2 15	3 0	3 45	4 30	5 15	6 00	6 45	7 30	8 15	9 00
46	2 18	3 4	3 50	4 36	5 22	6 8	6 54	7 40	8 26	9 12
47	2 21	3 8	3 55	4 42	5 29	6 16	7 3	7 50	8 37	9 24
48	2 24	3 12	4 00	4 48	5 36	6 24	7 12	8 00	8 48	9 36
49	2 27	3 16	4 5	4 54	5 43	6 32	7 21	8 10	8 59	9 48
50	2 30	3 20	4 10	5 00	5 50	6 40	7 30	8 20	9 10	10 00
51	2 33	3 24	4 15	5 6	5 57	6 48	7 39	8 30	9 21	10 12
52	2 36	3 28	4 20	5 12	6 4	6 56	7 48	8 40	9 32	10 24
53	2 39	3 32	4 25	5 18	6 11	7 4	7 57	8 50	9 43	10 36
54	2 42	3 36	4 30	5 24	6 18	7 12	8 6	9 00	9 54	10 48
55	2 45	3 40	4 35	5 30	6 25	7 20	8 15	9 10	10 5	11 00
56	2 48	3 44	4 40	5 36	6 32	7 28	8 24	9 20	10 16	11 12
57	2 51	3 48	4 45	5 42	6 39	7 36	8 33	9 30	10 27	11 24
58	2 54	3 52	4 50	5 48	6 46	7 44	8 42	9 40	10 38	11 36
59	2 57	3 56	4 55	5 54	6 53	7 52	8 51	9 50	10 49	11 48
60	3 0	4 0	5 0	6 0	7 0	8 0	9 0	10 0	11 0	12 00

Observations with the solar compass for the purpose of running lines, or to determine the variation of the needle, should not be made when the sun or other celestial object is nearer than 8°, or thirty-two minutes of time from the meridian: nearer than this, the observations may not give the course required sufficiently correct for the ordinary purpose of running lines.

The best part of the year for running lines with the solar compass is the summer season, or when the latitude and the declination of the sun are both of the same name. During this portion of the year there is usually the most fair weather for work of this kind, and the sun's altitude being generally higher through most of the day, affords more frequent opportunities in the forest to adjust the instrument by the sun, to the course of the line. There are, also, more hours of the day in which the solar compass can be used; the advantages of this will be fully realized when running lines in thickly timbered land, or in hilly or mountainous districts, when their summits intervene between the instrument and the sun, until a late hour in the morning and early in the afternoon.

From the principles already given in regard to the use of the solar compass, it will be perceived, that it requires more skill to use it with facility, than it does to use the magnetic compass; therefore, the surveyor should acquire this skill, before entering upon any important survey.

More line can be run with the solar compass in a day, than with the magnetic compass in the same time, if both instruments are properly used; for the reason that it requires less time to adjust the solar compass to the course by the sun, than it does the magnetic compass by the needle.

Much experience has established the fact, that a continual line can be run independently of the needle, through heavy timbered land, without cutting away any timber, except lopping a bush occasionally, between the instrument and the sun. Therefore, lines can be correctly run through any mineral region or other country, however great the local attractions or variations may be on the magnetic needle, with an accuracy not attainable with the magnetic compass. In making the survey of new districts of country, especially where there is considerable local attraction, it is important to determine the variation of the needle frequently, and make a record of the same for future reference.

During the surveys of the mineral region of Lake Superior, it was discovered that all mineral veins in that country had an influence, more or less, on the direction of the magnetic needle, its North end being generally attracted towards the metallic vein. These indications led (and no doubt will to a greater extent in future) to the discovery of mineral veins of various kinds in that and other regions; but the influence of metallic deposits on the magnetic needle, according to their various qualities, courses, distances, depths, &c., from the instrument, are as yet imperfectly understood.

It is to be hoped that this subject *will receive, in future,* that attention which its importance requires.

These aberrations in proximity to metallic deposits, suggest to the mind that they may be caused by galvanic currents, which circulate around the earth, and become deflected out of their general course by the metallic veins being a better conductor than the surrounding medium.

Galvanic currents conducted by any metallic substance always influence the direction of the magnetic needle, and incline it toward a right angle to its course; metallic deposits may also, in connexion with the various rocks and other substances in which they are immediately enclosed, form in themselves, local galvanic batteries, of greater galvanic intensity than is generally circulating in their vicinity, and thus diffuse an influence around them at considerable distances.

If these suggestions are correct, they seem to point to metallic deposits, in connexion with other substances in which they are enclosed, as the producing cause of the galvanic currents which circulate continually around the earth, nearly at right angles to its axis.

CONVENIENT RULES FOR CORRECTING THE COURSE OF RANDOM LINES, WHEN THE CORRECTION DOES NOT EXCEED 200 LINKS TO EACH MILE.

Rule for half a mile, or forty chains.

From the number of links to be corrected in that distance, subtract one-seventh; the difference will be the number of minutes of a degree required for the correction of the course.

EXAMPLE.

Number of links to be corrected, 42—6=36′ answer.

Rule for one mile, or eighty chains.

From half of the number of links to be corrected in that distance, subtract one-seventh; the difference will be the number of minutes of a degree required for the correction of the course.

EXAMPLE.

Number of links to be corrected, 70÷2=35—5=30′ answer.

Rule for three miles.

Divide the whole number of links to be corrected by seven; the quotient will be the number of minutes of a degree required for the correction of the course.

EXAMPLE.

Number of links to be corrected, 297÷7=42$\frac{3}{7}$′ answer.

Rule for six miles.

Divide one half of the number of links to be corrected by seven; the quotient will be the number of minutes required for the correction of the course.

EXAMPLE.

Number of links to be corrected, 370 ÷ 2 = 185 ÷ 7 = 26$\frac{3}{7}$′ ans.*

The distances given for corrections, in the above examples, are those for which corrections are generally made in the surveys of the public lands, and the calculation for the course of the corrected line, can generally be mentally made by the surveyor, while he is occupied in adjusting his instrument.

For other distances, when the correction does not exceed 1° 45′, divide the distance run, by the number of links to be corrected in the length of the line; the quotient will be the natural co-tangent of the correction to be applied to the random course.

In the following table, the angle of correction is given in the first column from 1′ to 1° 40′; and against each angle the departure is given for distances *one, forty, eighty, and two hundred and forty chains,* or *three miles.* These distances may be reckoned as *tens, hundreds, thousands,* if the position of the decimal point in each departure be changed accordingly.

The departure under distance one chain is of course the *natural*

* The above rules are close approximations.

sine of the angle; therefore, if it be multiplied by the distance run on any angle, the product is the departure.

TABLE, SHOWING THE ANGLE OF CORRECTION FOR RANDOM LINES.

Angle.	Number of links in 1 ch.	Links in 40 ch.	Links in 80 chains.	Links in 3 miles.	Angle.	Number of links in 1 ch.	Links in 40 chains.	Links in 80 chains.	Links in 3 miles.
1′	·000291	1·16	2·33	6·90	51′	·014835	59·34	118·68	356·04
2	·000582	2·33	4·66	13·97	52	·015126	60·50	121·01	363·02
3	·000873	3·49	6·98	20·95	53	·015417	61·67	123·34	370·01
4	·001164	4·66	9·31	27·94	54	·015707	62·83	125·66	376·97
5	·001454	5·82	11·63	34·90	55	·015998	63·99	127·98	383·95
6	·001745	6·98	13·96	41·88	56	·016289	65·16	130·31	390·94
7	·002036	8·14	16·29	48·86	57	·016580	66·32	132·64	397·92
8	·002327	9·31	18·62	55·85	58	·016871	67·48	134·97	404·90
9	·002618	10·47	20·94	62·83	59	·017162	68·65	137·30	411·89
10	·002909	11·64	23·27	69·82	1· 0	·017452	69·81	139·62	418·85
11	·003200	12·80	25·60	76·80	1· 1	·017743	70·97	141·94	425·83
12	·003491	13·96	27·93	83·78	1· 2	·018034	72·14	144·27	432·82
13	·003782	15·13	30·26	90·77	1· 3	·018325	73·30	146·60	439·80
14	·004072	16·29	32·58	97·73	1· 4	·018616	74·46	148·93	446·78
15	·004363	17·45	34·90	104·71	1· 5	·018907	75·63	151·26	453·77
16	·004654	18·62	37·23	111·70	1· 6	·019197	76·79	153·58	460·73
17	·004945	19·78	39·56	118·68	1· 7	·019488	77·95	155·90	467·71
18	·005236	20·94	41·89	125·66	1· 8	·019779	79·12	158·23	474·70
19	·005527	22·11	44·22	132·65	1· 9	·020070	80·28	160·56	481·68
20	·005818	23·27	46·54	139·63	1·10	·020361	81·44	162·89	488·66
21	·006109	24·44	48·87	146·62	1·11	·020652	82·61	165·22	495·65
22	·006400	25·60	51·20	153·60	1·12	·020942	83·77	167·54	502·61
23	·006690	26·76	53·52	160·56	1·13	·021233	84·93	169·86	509·59
24	·006981	27·92	55·85	167·54	1·14	·021524	86·10	172·19	516·58
25	·007272	29·09	58·18	174·53	1·15	·021815	87·26	174·52	523·56
26	·007563	30·25	60·50	181·51	1·16	·022106	88·42	176·85	530·54
27	·007854	31·42	62·83	188·50	1·17	·022397	89·59	179·18	537·53
28	·008145	32·58	65·16	195·48	1·18	·022687	90·75	181·50	544·49
29	·008436	33·74	67·49	202·46	1·19	·022978	91·91	183·82	551·47
30	·008726	34·90	69·81	209·42	1·20	·023269	93·08	186·15	558·46
31	·009017	36·07	72·14	216·41	1·21	·023560	94·24	188·48	565·44
32	·009308	37·23	74·46	223·39	1·22	·023851	95·40	190·81	572·24
33	·009599	38·40	76·79	230·38	1·23	·024141	96·56	193·13	579·38
34	·009890	39·56	79·12	237·36	1·24	·024432	97·73	195·46	586·37
35	·010181	40·72	81·45	244·34	1·25	·024723	98·89	197·78	593·35
36	·010472	41·89	83·78	251·33	1·26	·025014	100·06	200·11	600·34
37	·010763	43·05	86·10	258·31	1·27	·025305	101·22	202·44	607·32
38	·011054	44·22	88·43	265·30	1·28	·025595	102·38	204·76	614·28
39	·011344	45·38	90·75	272·26	1·29	·025886	103·54	207·09	621·26
40	·011635	46·54	93·08	279·24	1·30	·026177	104·71	209·42	628·25
41	·011926	47·70	95·41	286·22	1·31	·026468	105·87	211·74	635·23
42	·012217	48·87	97·74	293·21	1·32	·026759	107·04	214·07	642·22
43	·012508	50·03	100·06	300·19	1·33	·027049	108·20	216·39	649·18
44	·012799	51·20	102.39	307·18	1·34	·027340	109·36	218·72	656·16
45	·013090	52·36	104·72	313·16	1·35	·027631	110·52	221·05	663·13
46	·013381	53·52	107·05	321·14	1·36	·027922	111·69	223·38	670·13
47	·013671	54·68	109·37	328·10	1·37	·028212	112·85	225·70	677·09
48	·013962	55·85	111·70	335·09	1·38	·028503	114·01	228·02	684·07
49	·014253	57·01	114·02	342·07	1·39	·028794	115·18	230·35	691·06
50	·014544	58·18	116·35	349·06	1·40	·029085	116·34	232·68	698·04

TABLE OF LATITUDES AND LONGITUDES.

In the use of the Solar Compass, it is necessary to know approximately at least, the Longitude of the place where the instrument is used, for the purpose of taking out of the Nautical Almanac, the Sun's declination, &c., and reducing them to a time, and Longitude of the place of observation.

For this purpose, the following tabular statement of the latitude, and longitude from the meridian of Greenwich, of some of the most important places, in North America, are given.

PLACES.		Latitude North.	Longitude West. In Degrees.	Longitude West. In Time.
		° ′ ″	° ′ ″	H. M. S.
Acapulco, ...	Mex., . .	16 50 19	99 49 9	6 39 16
Albany, (Capitol,)	N. Y., . .	42 39 3	73 44 49	4 54 59
Amherst, (College,)	Mass., . .	42 22 15	72 31 28	4 50 6
Apostle Islands, (Lake Superior,)	. . .	47 00	91 00	6 4
Augusta, (State House,) . . .	Me., . .	44 18 43	69 50	4 39 20
Baltimore, (Monument,) . . .	Md., . .	39 17 48	76 36 39	5 6 26
Bellevue, (Am. Fur. Cos.' Trading Post, Missouri River,) . . .	. . .	38 8 24	95 47 46	6 23 11
Boston, (State House,)	Mass., .	42 21 27	71 3 30	4 44 14
Brazos Santiago, ...	Texas, .	26 6 0	97 12	6 28 48
Brent's Fort, ...	. . .	38 2 38	103 33 15	6 54 13
Burlington, ...	N. J., . .	40 4 51	74 52 37	4 59 30
Burlington, ...	Vt., . .	44 27	73 10	4 52 40
Cape Hancock, (Mouth of Columbia River,)	Oregon, .	46 16 35	124 1 45	8 16 7
Charleston, (St. Mich.'s Ch.,) .	S. C., . .	32 46 33	79 55 38	5 19 42
Chicago, ...	Ill., . .	42 00 00	87 35	5 50 2
Columbus, ...	Ohio, . .	39 57 00	83 3	5 32 12
Concord, (State House,) . . .	N. H., . .	43 12 29	71 29	4 45 56
Dalles of the Columbia Missionary Station, ...	O. T., . .	45 35 55	120 55	8 3 40
Detroit, (St. Paul's Church,) . .	Mich., .	42 19 45	83 2 30	5 32 10
Dover, ...	Del., . .	39 10	75 30	5 2 0
Ewing Harbour, ...	O. T., . .	42 44 22	124 28 52	8 33 55
Falls of St. Anthony, U. S. Cottage,	. . .	44 58 40	93 10 30	6 12 42
False Dungeness Bay, ...	Wash. Ter.,	48 7 52	123 27 21	8 13 49
Fort Boisee, ...	Oregon, .	43 49 22	116 47 3	7 47 8
Fort Gibson, (Old Block House,) .	. . .	35 47 35	95 15 10	6 21
Fort Hall, ...	. . .	43 1 30	112 29 54	7 29 59
Fort Laramie, ...	. . .	42 12 10	104 47 43	6 59 11
Fort Leavenworth, (Landing,) .	. . .	39 21 14	94 44	6 18 56
Fort Nez Perce, ...	O. T., . .	46 3 46		
Frankfort, ...	Ky., . .	38 14	84 40	5 38 40
Frederickton, ...	N. B., . .	46 3	66 45	4 27
Galveston, (Court House,) . .	Texas, .	29 18 14	94 46 34	6 19 6
Granite Island, (Lake Superior,) .	. . .	46 40	87 30	5 50
Great Salt Lake, Island in, . .	. . .	41 10 42	112 21	7 29

PLACES.		Latitude North.	Longitude West. In Degrees.	Longitude West. In Time.
		° ′ ″	° ′ ″	H. M. S.
Halifax, ...	N. S., . .	44 39 20	63 36 40	4 14 26
Harrisburg, ...	Pa., . .	40 16	76 50	5 7 20
Indianapolis, ...	Ind., . .	39 55	86 5	5 44 20
Jackson, ...	Miss., . .	32 23	90 8	6 00 32
Jefferson, ...	Mo., . .	38 36	92 8	6 8 32
Kanzas River, Mouth of, . .	. . .	39 6 3	94 33	6 18 11
Key West Light, ...	Fa., . .	24 33	81 48	5 27 12
Keweenau Point, Lake Superior,)	. . .	47 30	88 30	5 54
Kingston, ...	C. W., . .	44 8	76 40	5 6 40
Little Rock, ...	Ark., . .	34 40	92 12	6 8 48
Mexico, (City of,)	Mex., . .	19 25 45	99 5 6	6 36 20
Milledgeville, ...	Ga., . .	33 7 20	83 19 45	5 33 19
Milwaukie, ...	Wisc., . .	43 3 45	87 57	5 51 48
Mouth of Missouri River, - .	. . .	38 51 36	90 00 40	6 00 3
Mobile, ...	Ala., . .	30 41 26	88 1 29	5 56 2
Monterey, ...	Mex., . .	25 40 13	100 25 36	6 41 42
Montpelier, ...	Vt., . .	44 17	72 36	4 50 24
Montreal, ...	C. E., . .	45 31	73 35	4 54 20
Nebraska, or Platte River, Junction of North and South Forks,	. . . .	41 5 5	101 21 24	6 45 25
New Orleans, (City Hall,) . .	La., . .	29 57 30	90	6
Pittsburg, ...	Pa., . .	40 32	80 2	5 20 8
Point Conception, ...	Cal., . .	34 26 56	120 25 39	8 1 42
Point Hudson, ...	Wash. T., .	48 7 3	122 44 33	8 10 58
Prairie du Chien, Am. Fur Co.'s House, ...	. . . .	43 3 6	91 9 19	6 4 37
Quebec, (Citadel,)	C. E., . .	46 49 12	71 16	4 45 4
Richmond, (Capitol,)	Va., . .	37 32 17	77 27 28	5 9 50
Sacramento City, ...	Cal., . .	38 34 42	120 nearly.	8
Sackett's Harbour, ...	N. Y., . .	43 55	75 57	5 3 48
St. Paul's, ...	Min., . .	44 52 46	93 4 54	6 12 19
St. Vrain's Fort, ...	Indian Ter.,	40 16 52	105 12 23	7 48 1
San Francisco, (Presidio,) . .	Cal., . .	37 47 35	122 26 15	8 9 45
Santa Fe, ...	N. M., . .	35 41 6	106 1 22	7 4 5
Scarboro Harbour, ...	Wash T., .	48 21 49	124 37 12	8 18 29
Snake River, above Amer. Falls, .	. . .	42 47 5	112 40 13	7 30 41
Springfield, ...	Ill., . .	39 48	89 33	5 58 12
Tallahassee, ...	Fa., . .	30 28	84 36	5 38 24
Toronto or York, (Observ.,) . .	C. W., .	43 39 35	79 21 30	5 17 26
Tuscaloosa, ...	Ala., . .	33 12	87 42	5 50 48
Washington, (Capitol,) . . .	D. C., .	38 53 34	77 1 30	5 8 6
York, ...	Me., . .	43 10 0	70 40	4 42 40

The latest and best maps of North America show the longitude of all places within its boundary sufficiently near for the purpose of reducing the sun's declination to their meridians.

LENGTHS IN NAUTICAL MILES AND STATUTE MILES OF DEGREES OF LATITUDE AND LONGITUDE IN DIFFERENT LATITUDES.

DEGREE OF THE PARALLEL.			DEGREE OF THE MERIDIAN.		
Latitude of Parallel.	Nautical miles.	Statute miles.	Latitude of middle point.	Nautical miles.	Statute miles.
20°	56·404	65·018	20°	59·664	68·777
21	56·039	64·598			
22	55·657	64·158			
23	55·258	63·698			
24	54·843	63·219			
25	54·411	62·721	25	59·706	68·825
26	53·962	62·204			
27	53·497	61·668			
28	53·016	61·113			
29	52·518	60·540			
30	52·005	59·948	30	59·749	68·875
31	51·476	59·338			
32	50·931	58·709			
33	50·370	58·063			
34	49·794	57·399			
35	49·203	56·718	35	59·796	68·929
36	48·597	56·019			
37	47·976	55·304			
38	47·341	54·571			
39	46·960	53·822			
40	46·026	53·056	40	59·847	68·987
41	45·348	52·274			
42	44·654	51·476			
43	43·949	50·662			
44	43·230	49·833			
45	42·497	48·988	45	59·899	69·048
46	41·752	48·128			
47	40·993	47·254			
48	40·222	46·365			
49	39·439	45·462			
50	38·643	44·545	50	59·951	69·108

A degree of longitude at the equator = 69·163 statute miles.
A second of time at the equator = 1521·6 feet.

RUNNING PARALLELS OF LATITUDE.

Parallels of latitude are curved lines, and they increase in curvature from the equator to the poles, and cross all meridians at right angles. All lines run at any angle from the meridian, by courses taken at short intervals, partake more or less (according to the angle) of the curvature of parallels of latitude.

When the compass is set to a true east and west course, in any latitude, the line of sight is at right angles to the meridian, and in consequence of the spheroidical figure of the earth, which causes the curvature of the parallels of latitude, this line of sight will converge on the equator. Some correction is therefore due to each course taken between stations, to keep the line on the same parallel of latitude. This correction, however, is too small to make any material error in tracing the parallel, if the stations are not more than 30′′ of longitude apart; but if larger than this, the convergency on the equator should be computed for the distance, and allowed on the side towards the pole. But a more convenient and practical method of running parallels of latitude, or lines at any angle from the meridian, is to back sight on each forward sight, and take half the difference between their courses, when large enough to be perceptible. Thus, the forward and back sights, give double the amount of curvature between the two stations, the one half of which must be set off at the end of the forward sight toward the pole, to keep the line on the same parallel of latitude. Any unusual difference between two equal stations, must be re-examined, and errors corrected if any, as the line advances.

A line run west six miles, or more, with long stations between sights, cannot be retraced by running east in the same manner, for the east line will fall towards the equator; therefore attention should be given to this subject in running the east and west lines of the public lands, when long distances are taken between stations over water, prairies, or open lands.

When running a parallel of latitude, if an object be observed due east or west from any station, the correction of the course to touch the same parallel on the meridian of the object, is equal to one half of the angle of convergency between the two meridians, which pass through the station and the object.

The following table will show the convergency of *six miles apart* on the parallel of each degree of latitude, and *six miles* from them towards the poles of the earth.

TABLE.

Parallel of Latitude.	Links of Convergency.	Angle of Convergency.	Parallel of Latitude.	Links of Convergency.	Angle of Convergency.	Parallel of Latitude.	Links of Convergency.	Angle of Convergency.
°		′ ″	°		′ ″	°		′ ″
10	15·0	1· 4	27	36·9	2·38	44	70·1	5·01
11	15·7	1· 7	28	38·6	2·46	45	72·6	5·12
12	16·5	1·11	29	40·2	2·53	46	75·2	5·23
13	17·3	1·14	30	41·9	3· 0	47	77·8	5·34
14	18·2	1·18	31	43·6	3· 7	48	80·6	5·46
15	19·4	1·23	32	45·4	3·15	49	83·5	5·59
16	20·7	1·29	33	47·2	3·23	50	86·5	6·12
17	22·0	1·34	34	49·1	3·31	51	89·7	6·25
18	23·4	1·40	35	50·9	3·39	52	93·0	6·40
19	24·9	1·47	36	52·7	3·46	53	96·4	6·55
20	26·5	1·54	37	54·7	3·55	54	100·0	7·10
21	27·8	1·59	38	56·8	4· 4	55	103·7	7·26
22	29·3	2· 6	39	58·8	4·13	56	107·6	7·43
23	30·8	2·12	40	60·9	4·22	57	111·8	8·00
24	32·3	2·19	41	63·1	4·31	58	116·2	8·19
25	33·8	2·25	42	65·4	4·41	59	120·9	8·40
26	35·4	2·32	43	67·7	4·51	60	125·7	9·00

EXPLANATION AND USE OF THE ABOVE TABLE.

To find the convergency and angle for the fractional parts of each degree of latitude, increase the convergency and angle, in proportion to the fractional part required. The convergency of equal lengths of meridians with same latitude are in proportion to their distance apart.

The convergency between any two meridians, whose lengths are equal to their mean distance apart, is in proportion to the square of the distance given in the table (six miles) to the square of the length required.

EXAMPLE.

Suppose it is required to find the convergency of two meridians three miles in length and three miles apart, in latitude 42° (6^2) : 65.4 : : 3^2 : 16.35 links.

Suppose a station in latitude 42° N. an object is observed due east eight miles distant; how far north of the object is the same parallel of latitude, of the station from which the observation is made? Proceed as in the above example. 36 : 65. 4 : : 64 : 116. 27. One half of which is 58,14 links nearly, answer. (See rule preceding the above table.) If the angle be required that would touch the same parallel north of the object, it will be given by the following proportion; 6 : 4′ 41″ :: 8 : 6′ 14″. One half of which is 3′ 7″ or N. 89° 56′ 53″ E.*

CONVERGENCY OF MERIDIANS.

Rule.—As the cosine of any given latitude is to a given distance of longitude, in that latitude, so is the cosine of any other latitude, to the distance of a corresponding longitude; the difference of these numbers will be the convergency.

EXAMPLE.

Required the convergency of two range lines that are 6 miles or 480 chains apart, in latitude 42° 30′ north, and extending north ten townships, or to latitude 43° 21′ 48″.

As cosine of lat. 42° 30′	= 9.867631
: Longitude 480 chains,	= 2.681241
:: Cosine of lat. 43° 21′, 48″	= 9.861543
	12.542784
	9.867631
: Log. - -	2.675153 = 473.32 subtract

from 480. chains = 6.68 chains. The convergency.

TO RUN A LINE PARALLEL TO A GIVEN MERIDIAN, AT ANY DISTANCE EAST OR WEST OF IT.

Find the angle of convergency between the meridians for the distance required, then run the line at the angle thus found, east or west of the meridian as the case requires.

AMPLITUDE OF CELESTIAL OBJECTS.

All heavenly bodies will rise and set to the north, or to the south of the east and west points of the horizon, as their declination may be north or south.

In consequence of the horizontal refraction of celestial objects, the proper time of taking their amplitude is when their centers appear about 33′ above the horizon.

TO FIND THE AMPLITUDE.

To the Log, secant of the latitude (rejecting its index,) add the Log-sine of the sun's or star's declination; the sum will be the Log-

* The preceding rules are close approximations to the truth.

sine of the course, the sun or star will rise or set from the east or west point.

EXAMPLE.

Latitude 42° 45′ Log. secant,	.134113
Declination 15° 10,′ Log. sine,	9.417684
Log sine of Amplitude,	9.551797 = 20° 52.

PROBLEMS.

TO FIND THE TIME OF THE SUN RISING OR SETTING.

Rule.—To the tangent of the latitude, add the tangent of the sun's or star's declination, and subtract radius from their sum; the remainder is the cosine of the semi-diurnal arc, when the latitude and declination are of different names; and of the semi-nocturnal arc, when both are of the same name.

EXAMPLE.

Sun's decl. 18° 20′ Tangent =	9.520305
Latitude 41° 50′ Tangent =	9.951896
	19.472201
Subtract radius	10.000000
Cosine	9.472201 = 72° 45′ or 4h. 51min.

Apparent time of sunrise when the latitude and declination are of the same name, or sunset when they are of different names.

TO FIND THE ANGLE THAT THE EQUATORIAL LINES OF THE SOLAR COMPASS, MAKE WITH THE HORIZON IN ANY LATITUDE, WHEN OBSERVING A CELESTIAL OBJECT, AT ANY HOUR ANGLE FROM THE MERIDIAN.

Rule.—As radius is to the cosine of the latitude, so is the sine of the hour angle of the celestial object, to the sine of the angle of the equatorial lines with the horizon.

EXAMPLE.

As radius,	10.000000
: Cosine of lat. 42° 30′	= 9.867631
:: Sine of H'r angle 30° 00′	= 9.698970
: Sine of angle	= 9.566601 = 21° 38′ nearly.

TO FIND THE AZIMUTH OF THE POLE STAR AT THE TIME OF ITS GREATEST ELONGATION.

Rule.—As cosine of the latitude, is to radius, so is the cosine of the declination, to the sign of the azimuth or elongation.

EXAMPLE.

Latitude 40° 20′, declination of the pole star, January 1st, 1854, 88° 32′ 7″.

As cosine of lat. 40° 20′	=	9.882121
: Radius		10.000000
:: Cosine of Decl. 88° 32′ 7″	=	8.407727
		18.407727
		9.882121
: Sine of azimuth		8.525606 = 1°, 55′, 20″.

TO FIND THE MOON'S PARALLAX IN ALTITUDE, AND TO REDUCE IT TO THE QUANTITY TO BE SUBTRACTED FROM HER DECLINATION WHEN HER LATITUDE IS OF THE SAME NAME, AND ADDED TO IT, WHEN OF A CONTRARY NAME.

Rule.—As radius is to the sine of horizontal parallax, so is the cosine of altitude to the sine of parallax in altitude: subtract the refraction in altitude and the meridional refraction; take the proportional part of this difference from table of proportional parts of refraction, and apply it to her declination as above named.

EXAMPLE.

As Radius,		10.000000	
: Sine horizontal parallax 58′	=	8.227134	
: : Cos. altitude 36°	=	9.907958	
: Sine parallax in altitude,	=	8.135092	= 46′, 55″
Refraction in altitude, 1′ 20″			—2′ 8″
Meridional refraction, 48″			
2′ 8″			44.′ 47″ Proportional part in latitude 36° at 3h. from the meridian, 82 = 36′, 43″ to be applied to the Moon's declination.

HOW TO FIND THE MERIDIONAL REFRACTION OF CELESTIAL OBJECTS IN ANY LATITUDE.

EXAMPLE.

In latitude 42° N.,	90°—42=48°,
Sun or star's declination north,	+ 15° 30′
The meridional altitude is	63° 30′.

The refraction of which is 29″, (see table of refraction.)

SECOND EXAMPLE.

In latitude 38° N.,	90°—38°=52° 00′.
Declination south,	— 10° 15′.
The meridional altitude is	41° 45′.

the refraction of which is 1′ 5″

BAROMETER.

In view of the many hilly and mountainous districts yet to be surveyed, and their chorographical and geological characters defined, as well as for other purposes, the following table and theorems as given by Sir George Shuckburgh, will show in what manner the barometer is used for ascertaining the height of Mountains, Hills, &c.*

Thermometer.	Factor.	Thermometer.	Factor.	Thermometer.	Factor.
°		°		°	
30	864·4	47	900·2	64	936·1
31	866·5	48	902·3	65	938·2
32	868·5	49	904·5	66	940·3
33	870·6	50	906·6	67	942·4
34	872·7	51	908·7	68	944·5
35	874·9	52	910·8	69	946·7
36	877·0	53	913·0	70	948·8
37	879·1	54	915·1	71	950·9
38	881·3	55	917·2	72	953·0
39	883·4	56	919·3	73	955·1
40	885·4	57	921·4	74	957·2
41	887·5	58	923·5	75	959·3
42	889·6	59	925·6	76	961·4
43	891·7	60	927·7	77	963·5
44	893·8	61	929·8	78	965·6
45	896·0	62	931·9	79	967·7
46	898·1	63	934·0	80	969·9

DEPRESSION OF MERCURY IN GLASS TUBES, OR CORRECTIONS TO BE ADDED FOR CAPILLARY ATTRACTION.

	INCHES.				
Diameter of Tube,	0· 25	0· 30	0· 40	0· 45	0· 60
Correction, . . .	0·020	0·015	0·007	0·005	0·002

* To perform this operation accurately, two persons should take contemporary observations with two barometers and thermometers, the one at the bottom of the hill, and the other at the top.

RULE.—The difference between the two barometers at the bottom and top of the mountain, multiplied by the height of the barometer at the bottom of the mountain; and that product by the tabular difference corresponding to the mean of the thermometers, and divided by the mean between the readings of the barometers, will equal the amount of elevation in feet.

EXAMPLE.

Suppose the barometer at the bottom of the mountain to stand at 30 inches, thermometer 60°; and the barometer at the top 26.36 inches, thermometer 46°; required the height of the mountain.

As per rule the mean of the two barometers = 28.18 inches, their difference = 3.64 inches; and the mean of the two thermometers = 53°. The number corresponding to 53° in the table is 913.0, hence $(3,64 \times 30 \times 913.0) \div 28.18 = 3537.92 +$. The height of the mountain.

The following are extracts from the remarks of the late eminent Dr. Halley:—

"In calm weather, when the air is inclined to rain, the mercury is commonly low.

"In serene, good weather, the mercury is generally high. Upon very great winds, though they be not accompanied with rain, the mercury sinks lowest of all, with relation to the point of compass the wind blows upon.

"In calm frosty weather, the mercury generally stands high.

"Within the tropics, and near them, there is very little or no variation of the height of mercury in all weathers.

"The greater height of the barometer, is occasioned by two contrary winds blowing towards the place of observation, whereby the air of other places is brought thither and accumulated."

In regard to the course of winds, and their effect on the barometer and weather, they are variable in different countries, and therefore omitted here.

Extracts from a Manual published by J. H. Belville of the Royal Observatory of Greenwich.

"Heat and moisture are the principal causes of the variations in the weight of the atmosphere, and necessarily in the variations in the barometer at the same station."

"The variations of the barometer, are less within the tropics, than in the temperate and polar regions; they vary in different countries in the same latitude, and they are greater in mountainous countries, and islands. In Peru, the range of the mercury is about one-third of an inch—in London two and a half inches, and in St. Petersburg, it exceeds three inches."

"It is not so much the *absolute* height, as the actual rising and falling of the mercury, which determines the kind of weather likely to follow."

"Great depressions at all seasons are followed by change of wind, and by much rain."

"Rain in some quantity may fall with a high pressure, provided the wind be in any of the northerly points."

"No great storm ever sets in with a steady rising barometer."

"The variations of the barometer, are always greater in the winter than in the summer."

"Sudden depressions of the barometer, sometimes occur in weather apparently calm. It is almost an established fact, that storms have a circular motion; and, if when an exhaustion, or sudden diminution of the atmosphere takes place, the mercurial column happens to be in the partial vacuum or centre of motion, the air will be at rest; while the surrounding air at a greater distance from the centre, will be violently agitated with a less fall of the barometer."

N. B.—In all observations for this purpose, the rise and fall of the mercury should be reckoned from its mean height at whatever elevation the station may be above the sea level.

ANEROID BAROMETER.

The Aneroid Barometer is a new instrument for ascertaining the variations of the atmosphere: its action depends on the effect produced by the pressure of the atmosphere on a metallic box, from which the air has been exhausted and then hermetically sealed: the hand of the Aneroid can be set to correspond with the mercurial barometer, by which it should be compared by turning a screw on its back-side. This screw when turned with, or against the sun, alters the position of the hand, *and is not to be touched for any other purpose.*

There is another gilt hand, called the register or index, which moves above the other by a nut or thumb piece which projects through the centre of the glass, to enable the observer to register the barometer hand, by which to refer its movement for another time, or in ascending or descending hills, &c.

The Aneroid Barometer can be carried and used through any country with about the same safety as a watch, and is, therefore, the most suitable barometer of any now in use, for measuring the height of hills and mountains, in new countries.

The corrections for temperature for the Aneroid, are seldom precisely the same as for the mercurial barometer; but the quantity necessary for thermometrical correction can be readily found, by exposing the instrument to the temperature of the external air for twenty or thirty minutes, and set the hands coincident, then place it near the fire until the thermometer is at ninety or a hundred degrees; the variation of the hand, divided by the variation in degrees of the thermometer, will give the quantity for each degree.

MEASUREMENTS OF HEIGHTS WITH THE BAROMETER.

The following table, being an extract from the elaborate table of W. Galbraith, A. M., furnishes another expeditious method for this purpose.

In this table, the third column exhibits numbers in English feet, corresponding to the height of the barometer (shown on its left,) in inches, tenths, and hundredths, the proportional parts to thousandths are given in column headed A.

BAROMETRIC TABLE.

A.	Bar. Inch.	English Feet.	A.	Bar. Inch.	English Feet.	A.	Bar. Inch.	English Feet.
+	28·00	27425·3	+	28·20	27611·3	+	28·40	27795·8
0·9	1	27434·6	0·9	1	27620·6	0·9	1	27805·0
1·9	2	27444·0	1·9	2	27629·8	1·8	2	27814·2
2·8	3	27453·3	2·8	3	27639·1	2·8	3	27823·4
3·7	4	27462·6	3·7	5	27648·3	3·7	4	27832·6
4·7	5	27471·9	4·6	5	27657·6	4·6	5	27841·8
5·6	6	27481·3	5·6	6	27666·8	5·5	6	27851·0
6·5	7	27490·6	6·5	7	27676·1	6·4	7	27860·2
7·5	8	27499·9	7·4	8	27685·3	7·4	8	27869·3
8·4	9	27509·2	8·3	9	27694·6	8·3	9	27878·5
+	28·10	27518·4	+	28·30	27703·7	+	28·50	27887·7
0·9	1	27527·7	0·9	1	27712·9	0·9	1	27896·9
1·9	2	27537·0	1·8	2	27722·2	1·8	2	27906·0
2·8	3	27546·3	2·8	3	27731·4	2·7	3	27915·2
3·7	4	27555·6	3·7	4	27740·6	3·7	4	27924·3
4·6	5	27564·9	4·6	5	27749·8	4·6	5	27933·5
5·6	6	27574·2	5·5	6	27759·1	5·5	6	27942·6
6·5	7	27583·5	6·5	7	27768·3	6·4	7	27951·8
7·4	8	27592·7	7·4	8	27777·5	7·3	8	27960·9
8·4	9	27602·0	8·3	9	27786·7	8·2	9	27970·1

A.	Bar. Inch.	English Feet.	A.	Bar. Inch.	English Feet.	A.	Bar. Inch.	English Feet.
+	28·60	27979·2	+	29·20	28521·7	+	29·80	29053.1
0·9	1	27988·3	0·9	1	28530·6	0·9	1	29061·9
1·8	2	27997·5	1·8	2	28539·6	1·8	2	29070·6
2·7	3	28006·6	2·7	3	28548·5	2·6	3	29079·4
3·7	4	28015·7	3·6	4	28557·5	3 5	4	29088.1
4·6	5	28024·8	4·5	5	28566·4	4·4	5	29096.9
5·5	6	28034·0	5·4	6	28575·4	5·3	6	29105.6
6·4	7	28043.1	6·3	7	28584·3	6·1	7	29114.4
7·3	8	28052.2	7·2	8	28593·2	7·0	8	29123.1
8·2	9	28061.3	8·0	9	28602·2	7·9	9	29131·9
+	28·70	28070·5	+	29·30	28611·1	+	29·90	29140·6
0·9	1	28079·6	0·9	1	28620·0	0·9	1	29149·3
1·8	2	28088·7	1·8	2	28628·9	1·7	2	29158·1
2·7	3	28097·8	2·7	3	28637·8	2·6	3	29166·8
3·6	4	28106·9	3·6	4	28646·7	3·5	4	29175·5
4·5	5	28115·9	4·5	5	28655·6	4·4	5	29184·2
5·5	6	28125·0	5·3	6	28664·5	5·2	6	29193·0
6·4	7	28134·1	6·2	7	28673·4	6·1	7	29201·7
7·3	8	28143·2	7·1	8	28682·3	7·0	8	29210·4
8·2	9	28152·2	8·0	9	28691·2	7·8	9	29219·1
+	28·80	28161·3	+	29·40	28700·0	+	30·00	29227·8
0·9	1	28170·4	0·9	1	28708·9	0.9	1	29236·5
1·8	2	28179·4	1·8	2	28717·8	1·7	2	29245·2
2·7	3	28188·5	2·7	3	28726·6	2·6	3	29253·9
3·6	4	28197·5	3·6	4	28735·5	3·5	4	29262·6
4·5	5	28206·6	4·4	5	28744·4	4·3	5	29271·3
5·4	6	28215·6	5·3	6	28753·3	5·2	6	29280·0
6·3	7	28224.7	6·2	7	28762.1	6·1	7	29288·7
7·2	8	28233·7	7·1	8	28771·0	7·0	8	29297·3
8·1	9	28242·8	8·0	9	28779·9	7·8	9	29306·0
+	28·90	28251·8	+	29·50	28788·7	+	30·10	29314·7
0·9	1	28260·8	0·9	1	28797·5	0·9	1	29323·4
1·8	2	28269·9	1·8	2	28806·4	1·7	2	29332·0
2·7	3	28278·9	2·7	3	28815·2	2·6	3	29340·7
3·6	4	28287·9	3·5	4	28824·1	3·5	4	29349·2
4·5	5	28296.9	4·4	5	28832·9	4·3	5	29358·0
5·4	6	28306·0	5·3	6	28841·8	5·2	6	29366·7
6·3	7	28315·0	6·2	7	28850·6	6·1	7	29375·3
7·2	8	28324·0	7·1	8	28859·4	6·9	8	29384·0
8·1	9	28333·0	8·0	9	28868·2	7·8	9	29392·6
+	29·00	28342·1	+	29·60	28877·1	+	30·20	29401·3
0·9	1	28351·1	0·9	1	28885·9	0·9	1	29409·9
1·8	2	28360·1	1·8	2	28894·7	1·7	2	29418·6
2·7	3	28369·1	2·6	3	28903·6	2·6	3	29427·2
3.6	4	28378·1	3·5	4	28912·4	3·5	4	29435·9
4·5	5	28387·1	4·4	5	28921·2	4·3	5	29444·5
5·4	6	28396·1	5·3	6	28930·0	5·2	6	29453·2
6·3	7	28405·0	6·2	7	28938·8	6·1	7	29461·8
7·2	8	28414·0	7·0	8	28947·6	6·9	8	29470·4
8·1	9	28423·0	7·9	9	28956·4	7·8	9	29479·1
+	29·10	28432·0	+	29·70	28965·2	+	30·30	29487·7
0·9	1	28441·0	0·9	1	28974·0	0·9	1	29496·3
1·8	2	28450·0	1·8	2	28982·8	1·7	2	29504·9
2·7	3	28458·9	2·6	3	28991·6	2·6	3	29513·6
3·6	4	28467·9	3.5	4	29000·4	3·4	4	29522·2
4·5	5	28476·9	4·4	5	29009·1	4·3	5	29530·8
5·4	6	28485·8	5·3	6	29017·9	5·2	6	29539·4
6·3	7	28494·8	6·1	7	29026·7	6·0	7	29548·0
7·2	8	28503·8	7·0	8	29035·5	6·9	8	29556·6
8·1	9	28512·7	7·9	9	29044·2	7·7	9	29565·2

A.	Bar. Inch.	English Feet.	A.	Bar. Inch.	English Feet.	A.	Bar. Inch.	English Feet.
+	30·40	29573·8	+	30·60	29745·0	+	30·80	29915·2
0·9	1	29582·4	0·9	1	29753·5	0·8	1	29923·7
1·7	2	29591·0	1·7	2	29762·1	1·7	2	29932·2
2·6	3	29599·6	2·6	3	29770·6	2·5	3	29940·7
3·4	4	29608·2	3·4	4	29779·1	3·4	4	29949·2
4·3	5	29616·7	4·3	5	29787·6	4·2	5	29957·6
5·2	6	29625·3	5·1	6	29796·2	5·1	6	29966·1
6·0	7	29633·9	6·0	7	29804·7	5·9	7	29974·6
6·9	8	29642·5	6·8	8	29813·2	6·8	8	29983·1
7·7	9	29651·0	7·7	9	29821·7	7·6	9	29991·5
+	30·50	29659·6	+	30·70	29830·2	+	30·90	30000·0
0·9	1	29668·1	0·9	1	29838·7	0·8	1	30008·5
1·7	2	29676·7	1·7	2	29847·2	1·7	2	30016·9
2·6	3	29685·2	2·5	3	29855·7	2·5	3	30025·4
3·4	4	29693·8	3·4	4	29864·2	3·4	4	30033·8
4·3	5	29702·3	4·3	5	29872·7	4·2	5	30042·3
5·1	6	29710·9	5·1	6	29881·2	5·1	6	30050·7
6·0	7	29719·4	6·0	7	29889·7	5·9	7	30059·2
6·8	8	29727·9	6·8	8	29898·2	6·8	8	30067·6
7·7	9	29736·5	7·7	9	29906·7	7·6	9	30076·1

EXAMPLE.

At the foot of a hill the barometer indicates 29.54 inches, then carried immediately to the top of the hill reads 28.70 inches. In the table at 29.54 we find 28824.1 feet,
at 28.70 we find —28070.5 feet,

Height of hill 753.6 feet.

To perform this operation accurately, when the interval of time exceeds ten minutes between the two *observations*, two persons should take contemporary observations, with two Aneroid Barometers, one at the foot, and the other at the top of the hill, and correct each for temperature.

SYSTEM OF SURVEYS OF THE U. S. LANDS.

The public lands of the United States are surveyed in a uniform mode established by law, by lines run by the cardinal points of the compass; the north and south lines coinciding with the true meridian, and the east and west lines intersecting them at right angles, giving to the tracts thus surveyed the rectangular form.

The public lands are laid off and surveyed, primarily, into tracts of six miles square as near as practicable, called *townships*, containing 23040 acres each. The townships are subdivided into thirty-six tracts, called *sections*, each of which is one mile square, as near as may be, and contains 640 acres. Any number, or series, of contiguous townships, situated north or south of each other, constitute a *Range*.

To obtain and preserve a convenient and uniform mode of numbering the ranges and townships, it is usual in commencing the survey of an insulated body of public lands to run, or assume two *standard lines*, as the basis of the survey to be made therein. One of these standard lines is run due north and south, and is called the *principal meridian*, to which the ranges are parallel, and from which they are numbered eastward and westward. The other standard line is run due east and west, and is called the *base line*, from which the townships are numbered, northward and southward.

To distinguish from each other, the systems, or series of surveys thus formed, the several principal meridians are designated by progressive numbers. Thus the meridian running north from the mouth of the Great Miami river, is called the *first* principal meridian: the meridian running north through the centre of the State of Indiana, is called the *second* principal meridian: that running north from the mouth of the Ohio river, through the State of Illinois, is called the *third* principal meridian: that running north from the mouth of the Illinois river, through the States of Illinois and Wisconsin, is called the *fourth* principal meridian: and that running north from the mouth of the Arkansas river, through the states of Missouri and Iowa, is called the *fifth* principal meridian.

Correction Lines correct the error that would otherwise arise from the convergency of meridians, and arrest that proceeding from the inaccuracies of measurement. They are run due east and west at stated distances, generally at the end of every tenth township, and each forms a base for the townships north of it. Each range of townships should be made as much over six miles in width on each base and correction line as it will fall short of the same width where it closes, on the next correction line north, the excess or deficiency of width being always thrown into the last half mile, on all of the lines closing out to the west boundary of each township.

This mode of executing the public surveys, conduces more, perhaps, than any other which could be devised, to the simplicity, regularity, and symmetry of the work, and to the ease and certainty with which any tract may be identified.

The public lands are surveyed by Deputy Surveyors, appointed by the Surveyor General of the State or Territory, in which the district assigned to each deputy may be situated; their duties are prescribed by general and special instructions.

OF SUBDIVIDING TOWNSHIPS.

Each township is laid off and surveyed into thirty-six sections as near as may be of one mile square, by lines running north and south, crossed by others running east and west. The sections are known and designated by progressive numbers, beginning at the north-east corner of the township, and numbering westward and eastward alternately, as shown in the following diagram.

6	5	4	3	2	1
7	8	9	10	11	12
18	17	16	15	14	13
19	20	21	22	23	24
30	29	28	27	26	25
31	32	33	34	35	36

Quarter section corners are established equidistant between the section corners, except on the section lines closing on the north and west boundaries of townships, on which they are set at forty chains from the last section corner, and the excess or deficiency of measure (if any) is carried out into the last half mile, and cast upon the north and west sides of the township, as required by law. Various instructions have been given by Surveyors General, to Deputy Surveyors, for the purpose of accomplishing an equitable and lawful subdivision of townships into sections, none of which, it is believed, will effect this object better than the system adopted in 1850, by the surveyor general of Ohio, Indiana, and Michigan; by which the true course and measurement of every line is given, and the inequalities of measurement proportionally carried to every sectional line. This, together with the closing of the section lines at post on the north and west boundaries of the townships, (which were formerly closed at the intersection of the lines run to them, whether at post or not,) has much improved the symmetry and equality of the subdivision of townships.

An act of Congress of the 24th May, 1824, authorizes a departure from the ordinary mode of surveying the public land, on any river, lake, or bayou, whenever, in the opinion of the President of the United States, the public interest would be promoted thereby, so as to survey such lands, in tracts of two acres in width fronting on such river, lake, or bayou, and running back to the depth of forty acres.

ON SUBDIVIDING SECTIONS, AND RE-ESTABLISHING OF EXTINCT LINES AND CORNERS: DEDUCED FROM THE ACTS OF CONGRESS, IN REGARD TO THE SURVEYS OF THE PUBLIC LANDS; AND THE CONSTRUCTIONS AND USAGES THEREON.

The general principles on which the public lands are surveyed, have already been given; but the county surveyors and purchasers of these lands, are more immediately interested in the proper method of subdividing sections into such tracts as are sold to purchasers from the United States land offices; and the re-establishing of extinct lines and corners, when from any cause they are lost or cannot be found.

In the regular surveys of the public lands, no other lines are actually run and marked by the Deputy Surveyors of the United States, than township lines, and sections, or subdivisional lines of townships, into sections; on all of these lines, no other than section and quarter section corners are established; except meander corners at the end of all fractional section lines which close on rivers, lakes, &c.

All sections in a full township, except those which are bounded by its north and west sides, are treated as full sections in their sales and subdivisions; and also, the south half of sections on the north boundary, and the east half of sections on the west boundary of each full township, are sold and subdivided as full half sections. Section *sixteen* in each township is reserved for *school purposes*, and is not, therefore, subject to private entry.

From various causes (elsewhere treated of in this work) section lines do not always correctly coincide with the cardinal points; nor will their measurement in all cases be found exactly eighty chains or one mile in length. (See article on measurement with the chain.) Quarter section corners, especially in the older surveys, may not always be found equidistant between the section corners. This defect arises in most cases, it is believed, from difficult or careless measurement with the chain.

Notwithstanding such errors, all corners that can be identified by the original field notes, or other unquestionable testimony, must be regarded as the original corners, and for that purpose should be perpetuated with new posts and bearings when the old ones decay.

EXTINCT LINES AND CORNERS.

When a *Section corner* cannot be identified by the original field notes, or by clear and unquestionable testimony, run a right line between the nearest noted station trees, north and south, and east and west of the lost section corner, if there be any such trees within the distance of the nearest quarter section, or section corners; but if no station trees be found, then between the nearest quarter section or section corners, and at the point of intersection of these two lines re-establish the section corner, with new bearings from it to the nearest and most durable objects; which of course should be recorded with the survey.

Extinct Quarter Section corners, except on *fractional section lines*, if not identified as above stated *for section corners*, must be re-established equidistant between the section corners, in a right line between the nearest noted station trees each side of it, if there be any; but if none are found, then in right line between the section corners.

Extinct Quarter Section corners, on section line, which close on the *north* and *west boundaries* of townships, must be re-established according to the original measurement thereof, at forty chains from the last interior section corner towards the township line. For an example, suppose the line between sections 3 and 4, or 18 and 19, to be 81.30 chains, according to the original survey, and by the measurement of the county survey, 80.90 chains. Then say as 81.30: 80.90:: 40.00 to = 39.81½. Thus 39 chains and 81½ links is the distance the quarter section corner must be established from the last interior section corner, according to the measure of the county surveyor.

Lost or extinct Township corners, except on correction lines, should be restored in the same manner as already given for section corners; and extinct quarter section corners on township lines, should be restored in the same manner as those on interior section lines.

In subdividing townships into sections, the section lines which close on the north and west boundaries of townships, have not always been closed at the section corners which were established on the survey of the township lines; but at such points on their boundaries, as the first lines run to them may have intersected.

Wherever this has been done on the north and west boundaries of townships, a new quarter section corner must be established, equidistant between the corners of all such irregular closing lines; for, the section and quarter section corners established on the survey of these boundaries, belong exclusively to the adjoining township. Consequently, to restore lost or extinct section corners, that were established on the north and west boundaries of townships, during their subdivisions into sections as above mentioned, the section lines closing at these corners, must be retraced to them. But to restore lost section, or quarter section corners, that were established on the original survey of the township lines, these boundaries should be carefully retraced and measured, and the lost section and quarter section corners should be re-established at their proportional distance from each other, between known corners. The only exception to this rule is, when it is clear that the section lines have been regularly run according to instructions, and can be correctly retraced to the township line. The section corners should then be re-established at such intersections.

Extinct or obliterated lines may be restored by running right lines between re-established and other known corners; except noted station trees be found between them, when the lines between corners must conform to the noted station trees.

It may be remarked here that no surveyor can legally alter or correct the original surveys. It is his duty to restore them as far as practicable to their original condition.—In making resurveys of the public lands, such directions and absolute length must be given to each line as were given to them by the original surveyor, whether the retracing, courses and measurements, agree with the original survey or not; except otherwise directed by the Surveyor General, or the Commissioner of the general Land Office.

BEARING TREES, &c.

Bearing trees, to corners, have a blaze with a notch in them near the ground and facing the corner; sometimes the letters B T are found in the blaze above or below the notch, which are the initials of Bearing Tree. Their size, kind of timber, course and distance from the corner post, is given in the field notes of the survey.

Section, and quarter section trees are "faced off" on the side towards the corner, four or five feet from the ground. The quarter section trees are marked thus ¼ S. At section corners these trees are marked with the number of the Range, Township and Section, thus, R. 24 W. T. 45 N. S. 15.* There is no note made of these trees in the field books, unless they are bearing trees also; they are marked for the purpose of giving information at the corner, of the number of the *sections* which corner there, and also, the number of the *township* and *range*. Station trees on the lines, are notched with two notches on each side in the direction of the line, and their size, kind of timber, and distance from the last section corner are given in the field notes.

SUBDIVISION LINES OF SECTIONS.

The subdivisions of whole sections into such tracts as are sold by the land officers of the United States, to purchasers of public lands, are made by running right lines between the quarter section corners, on the north and south, and east and west sides of the section; and at the intersection of these lines is established the common corner for

* In prairies, the posts set in mounds for corners are marked in like manner.

its four quarters, without regard to the quantity of land contained in each of them. These quarter sections are sold as containing 160 acres each, and are designated as the N. E., N. W., S. E. and S. W. quarters.

Quarter sections are divided into halves, by a north and south line, equidistant by measurement between its east and west corners. These tracts of land are supposed to contain eighty acres each, and are designated as the east and west half of the quarter section.

Furthermore a quarter section is, also, divided into quarters by lines run north and south, and east and west, equidistant between its four corners, and at the intersection of these lines at the centre of the quarter section, is established the common corner to its four quarters. These quarters of a quarter section are supposed to contain forty acres each, and are described as the N. E., N. W., S. E. and S. W. quarters of the quarter section.

The following diagram of the subdivision of a whole section, will more clearly show the method of subdividing such section.

79·68
39·84 19·91 19·91
20·00 20.01 20·02
39.96 40·00
19·96½ 19·96½
20·00 20·01 20·02
79·92 40·04 20·02 20·02 80·08
20·02 20·02
19·98 19·99
20·07 40·00 40·02 40·04
19·98 19·99
20·12 20·12 20·12 20·12
80·48

Quarter sections adjoining the north and west boundaries of townships, are deemed to be fractional, and therefore, may contain more or less land, than is given to other quarter sections within the townships; they are sold or surveyed according to their plats in the land offices.

ON SUBDIVIDING FRACTIONAL SECTIONS.

Fractional section lines which close on meandered rivers and lakes, or on reservations, &c., are required by law to be run north and south, or east and west, as the case requires. These lines like those before mentioned in the subdivisions of townships into sections, may not precisely agree with the cardinal points of the compass.—Therefore, in subdividing fractional sections embraced by fractional section lines, which close on meandered streams, lakes, &c.; the quarter section line should be run with an intermediate course between the section lines; and the fractional quarter sections thereof, should be divided in like manner.

The subdivisions of fractional sections, are indicated on the maps of surveys, in the land offices.

AN ACT CONCERNING THE MODE OF SURVEYING THE PUBLIC LANDS OF THE UNITED STATES.

§ I. *Be it enacted, &c.* That the Surveyor General shall cause all those lands north of the river Ohio, which, by virtue of the act entitled, "An act providing for the sale of the lands of the United States in the territory northwest of the river Ohio, and above the mouth of the Kentucky river," were subdivided, by running through the townships parallel lines, each way, at the end of every two miles, and by marking a corner on each of the said lines, at the end of every mile, to be subdivided into sections, by running straight lines, from the mile corners thus marked to the opposite corresponding corners, and by marking, on each of the said lines, intermediate corners, as near as possible equidistant from the corners of the sections of the same. And the said Surveyor General shall also cause the boundaries of all the half section, which had been purchased previous to the first day of July last, and on which the surveying fees had been paid according to law by the purchaser, to be surveyed and marked, by running straight lines from the half mile corners heretofore marked, to the opposite corresponding corners; and intermediate corners shall at the same time, be marked on each of the said dividing lines, as nearly as possible equidistant from the corners of the half section on the same line: *Provided*, That the whole expense of surveying and marking the lines, shall not exceed three dollars for every mile which has not yet been surveyed, and which shall be actually run, surveyed and marked, by virtue of this section. And the expense of making the subdivisions directed by this section, shall be defrayed out of the moneys appropriated, or which may be hereafter appropriated, for completing the surveys of the public lands of the United States.

§ II. That the boundaries and contents of the several sections, half sections, and quarter sections, of the public lands of the United States, shall be ascertained in conformity with the following principles, any act or acts to the contrary notwithstanding:—1st. All the corners marked in the surveys returned by the Surveyor General, or by the surveyor of the land south of the state of Tennessee respectively, shall be established as the proper corners of sections, or subdivisions of sections, which they were intended to designate; and the corners of half and quarter sections, not marked on said surveys, shall be placed as nearly as possible equidistant from those two corners which stand on the same line. 2d. The boundary lines, actually run and marked in the surveys returned by the Surveyor General, or by the surveyor of the land south of the state of Tennessee, respectively, shall be established as the proper boundary lines of the sections, or subdivisions for which they were intended; and the length of such lines, as returned by either of the surveyors aforesaid, shall be held and considered as the true length thereof. And the boundary lines which shall not have been actually run and marked as aforesaid, shall be ascertained by running straight lines from the established corners to the opposite corresponding corners; but in those portions of the fractional townships, when no such opposite corresponding corners have been or can be fixed, the said boundary lines shall be ascertained by running from the established corners, due north and south or east and west lines, as the case may be, to the water course, Indian boundary line, or other external boundary of such fractional township. 3d. Each section, or subdivision of section, the contents whereof shall have been, or by virtue of the first section of this act, shall be returned by the Surveyor General, or by the surveyor of the public lands south of the state of Tennessee, respectively, shall be held and considered as containing the exact quantity expressed in such return or returns; and the half sections and quarter sections, the contents whereof shall not have been thus returned, shall be held and considered as containing the one half, or the one fourth part, respectively, of the returned contents of the section of which they make part.

§ III. That so much of the act, entitled "An act making provision for the disposal of the lands of Indiana territory, and for other purposes," as provides the mode of ascertaining the true contents of sections, or subdivisions of sections, and prevents the issue of final certificates, unless the said contents shall have been ascertained, and a plot certified by the District Surveyor, lodged with the register, be, and the same is hereby repealed. [Approved, February 11, 1805.]

GEOLOGICAL AND TOPOGRAPHICAL IN CONNEXION WITH LINEAR SURVEYS.

GEOLOGICAL SURVEYS.

In connexion with the linear surveys of new districts of country, the surveyors have good opportunities to make geological examinations, and to collect specimens of minerals that may be discovered in the course of their work. Such specimens, when submitted to a scientific and practical geologist, will enable him to determine the true character of such new districts, and what kinds of products may be expected to be derived from them.

It is, therefore, of much importance, that surveyors of the public land should possess or acquire, at least a sufficient knowledge of geology, to enable them to make a proper collection of geological specimens; and also, to observe the character, stratifications, dip, &c., of any rocks in place, or other mineral deposits.

Such services afford pleasure and profit to the surveyors, while they contribute to the public interest, and to science. A system of surveys for this purpose has been partly tested; but while in successful progress, it was interrupted by the death of the geologist, the lamented Dr. Douglass Houghton, while he was engaged in prosecuting a geological, in connexion with the linear survey, of the south coast of Lake Superior.

This system possesses many advantages over any other that has been adopted, for obtaining a general geological knowledge of new and unsettled countries, the expense of which is trifling compared with an independent geological survey; also, such surveys are of great value when known, in directing emigrants to the country suitable for their occupation or enterprise, and thus effect an early and judicious development of its resources.

A system of linear and geological surveys may be satisfactorily prosecuted, by the appointment of a competent geologist to a clerkship in each Surveyor General's office: the Deputy Surveyors being made assistant geologists to execute the field work, under a well digested system for that purpose, who should make their report, and return their specimens to the Surveyor General, when the geologist under him can investigate such reports, and embody the whole in one connected geological report, so far as such surveys extend.

By this system, it will be seen that the position of all mineral deposits from which specimens are taken, may be precisely located by measure on the survey, and be as easily found as the various section, quarter section, and other subdivisions themselves, a consideration of much importance, which any independent geological or other system yet adopted fails to do.

TOPOGRAPHICAL SURVEYS.

The general topographical features of new districts of country are of much interest to the public, and especially to emigrants. Such surveys can be made with but little expenditure of time while the linear surveys are in progress, by a proper use of the Aneroid Barometer, for the purpose of determining on the lines, the height of hills, ledges, &c., above the valleys; (see article on the use of the Aneroid Barometer) and by observing also, the course and angle of elevation or depression of distant noticeable objects, on the summits of hills, mountains, ledges, &c., and in the valleys below them; which can be seen from two or more stations on the lines at the time they are being run. And further, when running the meanderings of the shores of rivers or lakes, bearings and angles of elevation may be also taken to conspicuous objects on islands, rocks, sand-bars, &c., which can be seen from their shores. To these observations should be added, sketches of landscapes, ledges, and whatever else may interest the inquiring mind.

Such bearings, and angles of elevation and depression, form triangles with a given base to each, which are good data for mapping, or trigonometrical calculations, to establish the course, distance, elevation or depression, from a fixed point within the survey of every object with which they are connected.*

* In making calculations for the heights of distant objects, the table for corrections for curvature and refraction, will give the number of feet to be added to their height; on the account of the difference of the apparent and true level from the point of observation.

TABLE OF CORRECTIONS FOR CURVATURE AND REFRACTION, SHOWING THE DIFFERENCE OF THE APPARENT AND TRUE LEVEL, IN FEET AND DECIMALS OF A FOOT, FOR DISTANCES IN FEET AND MILES.

Distances in Feet.	Correction in Feet.			Distances in Miles.	Correction in Feet.		
	For Curvature.	For Refraction.	For Curvature and Refraction.		For Curvature.	For Refraction.	For Curvature and Refraction.
100	·00024	·00004	·00020	¼	·0417	·0060	·0357
150	·00054	·00008	·00046	½	·1668	·0238	·1430
200	·00094	·00013	·00083	¾	·3752	·0536	·3216
250	·00149	·00021	·00128	1	·6670	·0953	·5717
300	·00215	·00031	·00184	1½	1·5008	·2144	1·2864
350	·00293	·00042	·00251	2	2·6680	·3811	2·2869
400	·00383	·00055	·00328	2½	4·1688	·5955	3·5733
450	·00484	·00069	·00415	3	6·0030	·8561	5·1469
500	·00598	·00085	·00513	3½	8·1708	1·1673	7·0035
550	·00724	·00103	·00621	4	10·6720	1·5246	9·1474
600	·00861	·00123	·00738	4½	13·5468	1·9295	11·5773
650	·01010	·00144	·00866	5	16·6750	2·3821	14·2929
700	·01172	·00167	·01005	5½	20·1769	2·8824	17·2945
750	·01345	·00192	·01153	6	24·0120	3·4303	20·5817
800	·01531	·00219	·01312	6½	28·1809	4·0258	24·1551
850	·01728	·00247	·01481	7	32·6830	4·6690	28·0143
900	·01938	·00277	·01661	7½	37·5190	5·3599	32·1591
950	·02159	·00308	·01851	8	42·6880	6·0997	36·5883
1000	·02392	·00333	·02059	8½	48·1910	6·8844	41·3066
1050	·02638	·00377	·02261	9	54·0270	7·7181	46·3089
1100	·02895	·00414	·02481	9½	60·1971	8·5996	51·5975
1150	·03164	·00452	·02712	10	66·7000	9·5286	57·1714
1200	·03445	·00492	·02953	11	80·7070	11·5296	69·1774
1250	·03738	·00534	·03204	12	96·0480	13·7211	82·3269
1300	·04043	·00578	·03465	13	112·7230	16·1033	96·6197
1350	·04361	·00623	·03738	14	130·7320	18·6760	112·0560
1400	·04689	·00670	·04019	15	150·0750	21·4393	128·6357
1450	·05030	·00719	·04311	16	170·7520	24·3931	146·3589
1500	·05383	·00769	·04614	17	192·7630	27·5376	165·2254
1550	·05748	·00821	·04927	18	216·1086	30·8727	185·2359
1600	·06125	·00875	·05250	19	240·7870	34·3981	206·3889
1650	·06514	·00931	·05583	20	266·8000	38·1143	228·6857
1700	·06914	·00988	·05926				
1750	·07327	·01047	·06280				
1800	·07792	·01107	·06645				
1850	·08188	·01170	·07018				
1900	·08637	·01234	·07403				
1950	·09098	·01300	·07798				
2000	·09570	·01367	·08203				

For a very close approximation.

Correction for Curvature in feet $= \frac{2 D^2}{3}$

D being distance in miles.

A useful application of a series of triangles can be made across lakes, bays, harbours, &c., commencing from a correctly measured base, on or near their coasts, so connected with every point or object on their shores or within their waters, that the meanderings of their shores, and position of islands, sand-bars, soundings or other objects can be correctly delineated on a map, by course and distance from any known point of survey. A full description of the above principles with proper examples, would occupy too much space to be admitted here, but it is believed that the well qualified practical surveyor, will find but little, if any difficulty in applying these principles to any survey that may require their use.

OUTFIT FOR A SURVEYING COMPANY OF SIX MEN FOR FOUR MONTHS IN THE PUBLIC SURVEYS.

SUPPLIES OF PROVISION.

The following quantity and kinds, or a substitute for them, is generally required.

8 barrels of flour.
2½ do. of clear pork.
3 bushels of beans.
2 do. of dried apples.
120 lbs. of good dry sugar.
70 lbs. of ground coffee, or a substitute for it.
10 lbs. of saleratus, or its substitute.
1 lb. of ground pepper.
1 small bag of table salt.
25 lbs. of rice.
4 lbs. of Castile soap.

CAMP FURNITURE.

1 large tent for the surveying company.
1 small tent for the packmen.
6 Mackinaw blankets.
3 common blankets to spread underneath them.
2 dozen boxes of matches. (best kind.)
1 good chopping axe.
4 tin pails, made to fit into each other.
14 tin basins.
1 set of knives and forks. (Small size.)
1 butcher, or meat knife.

7 spoons.
3 light frying pans.
2 half round cans, made to fit inside of the pails,—for lard and saleratus.
2 tin pepper boxes, with covers to fit closely over the sieve.
6 "soldiers' drinking cups," also needles, awls, thread, twine, small cord, &c.
2 mixing cloths, made of heavy cotton drilling, one yard square each.
4 papers of 3 oz. tacks for nailing boots.

FOR PACKING, ETC.*

1 or 2 good horses, or mules, as circumstances require; one pack saddle; a bell and spancil for each.
20 stout bags, that hold one and a half bushels each.
4 linen bags, for pork.
6 small bags, for beans, dried apples, knives and forks, &c.
3 India Rubber bags for sugar and coffee. (Should be lined.)
2 strong drilling cloths, two or two and one half, yards square, to do up the camp equipage into packs; also, strap and cords, to secure the packs to the horse and saddle.

SURVEYING INSTRUMENTS, ETC.

1 solar compass.
1 case of drawing instruments.
1 measuring chain.
1 standard chain.
11 tally pins.
1 tape measure.
1 Telescope 16 or 18 inches in length.
2 marking tools.
2 pocket compasses.
2 marking axes, weighing three and a half pounds each.
1 hatchet, and two whetstones.
2 three-cornered files, for sharpening axes, &c.
2 small round files for sharpening marking tools. Also, field books, mapping and writing paper, ink, pens, pencils, India rubber, mouth glue, and a small valise (or box) to carry them in.

* "Packing." This word is used by surveyors of the public lands, both for making up and conveying packs.

Remarks.

Camp pails, or kettles, should be made of heavy tin, and the covers and ears riveted, where they would be likely to separate when exposed to the fire.

The most approved form of a camp pail is an elliptical, or oblong bottom, with upright sides. The largest pail should be made about nine inches in depth, and to hold twelve quarts, or more; the other three of a less size, so as to fit inside of the largest one.

The basins are made six or seven inches in diameter, and one and a half inch in depth; they serve in the place of plates, cups, soup and meat dishes, &c. The knives, forks, and spoons, should be of a small size, except one large spoon for mixing bread, &c.

Flour is mixed for bread on a cloth of cotton drilling, of about one yard square. It is done as follows:—

Spread the cloth on a blanket, folded and laid on the ground; pour enough flour upon it for a mixing, and make a hollow in it; then pour in some lard from the can, and add saleratus and salt dissolved in warm water, stirring the flour with a spoon to a proper consistency for kneading with the hand, taking care not to reach the bottom of the flour so as to wet the cloth.—Bake the loaves in the frying pans before the fire, and when done, fold the cloth, and lay it aside for future use.

TENTS.

The soldiers' tent made of good firm cotton drilling, will answer the purpose very well, in any country. The *Marquée*, however, is better in a prairie country. Another tent, much approved by some surveyors, for a timbered country, is made of good cotton drilling: when pitched, nearly resembles a little more than one half of a steep roofed building, with its share of the ends. It can be quickly pitched with poles, and crotches, by having suitable eyelets, and strings at the bottom, and at the ridge, and front. It has four or five breadths of cloth, about four and one third yards in length; the end may be made of cotton sheeting, of the form above indicated. This tent possesses the advantage of being less in weight and bulk, than any other in use among surveyors: therefore, very suitable to be used when the carrying is done by men.

CONVEYING PACKS WITH HORSES OR MULES.

The man who manages the pack horse, should be an experienced woodsman, capable of finding his way with the help of a pocket compass, to any point within the district to be surveyed, that may be designated by the surveyor.

The "sack Indian saddle" is the best in use for the purpose of packing, but pack saddles may be made in the form of those used by most of the Indian tribes. They should have attached to them a stout girt, breast strap, and breeching, and be well padded, or have a folded blanket under it, when in use.

Suitable straps with buckles should be provided, to tightly buckle around near the ends of each bag, or articles done up with a wrapper and cord, which are intended for side packs. Before buckling these straps, a loop made of strong cord about ten inches long, should be slipped on to each; after buckling the straps, these loops will be hitched over the horns of the saddle, and wound around them if too long: thus each side pack lies lengthwise of the horse, suspended by the loops from the horns of the saddle.

Between the side packs other loose articles may be placed, such as tin pails, frying pans, &c. These bags and other loading, should be well balanced, and bound to the horse and saddle, with a cord of suitable length. That the horse may be easily found when not at work, a small bell should be fastened to his neck, with a strap and buckle. A "spancil" should also be provided, made of leather with two buckles, for fastening the forefeet of the animal nearly together, that he may not go astray.

CONVEYING PACKS WITH MEN.

When packs cannot be carried on horses, or mules, men are employed for that purpose, and should be provided with suitable pack straps. The "portage strap" is sometimes used: it is made of leather, and is ten or twelve feet in length; the middle part is two feet long, and three inches broad in the middle, and tapers each way; at each end of this broad part is secured a thong of leather, sufficiently strong to support the pack. Each end of the portage strap is tied around the pack to be carried: the broad part passes over the forehead, or chest of the person who conveys it.

Another "pack strap" in common use, is made of five straps;

two of them are buckled around the pack near each end, and two are slipped under them and sewed together, in such a manner, as when buckled, to form shoulder straps; the fifth strap is about three inches broad at the middle, and tapers each way, and is buckled at each end to the other straps, in such a manner as to pass over the forehead when in use; the pack is put on in a similar manner to that of a peddler's pack.

Packs which are carried by men, to supply a surveying company in the field, usually weigh from seventy-five to a hundred and twenty pounds each.

SURVEYORS' WEARING APPAREL.

The common wool hat is best for any season of the year, especially in timbered land.

Trowsers should be made large, and of strong cloth.

A light coat, or frock, should be provided, well supplied with water-proof pockets, to keep books and papers dry in wet weather, and a light India rubber, or water proof cape should also be provided to keep the compass dry, when travelling in wet weather.

Flannel for under clothes, is preferable to cotton, for all seasons and kinds of weather.

Boots may be made of good kip skin, and rather larger than for ordinary use; the fronts of the legs should be cut narrower, and the backs wider, than is usual to cut them. A thick single sole projecting about one quarter of an inch from under the upper leather, and well nailed over the bottoms with sparables, or tacks, are the most durable. The nails keep the feet from slipping, and the broad sole protects the upper leather from wearing against bushes, grass, &c. A large silk handkerchief, of any colour but red, to tie over the ears and neck, is a good protection from flies and musquitoes.

DEPOTS IN ADVANCE OF A SURVEY.

Much difficulty has sometimes been experienced by surveyors in new and unsettled countries, in providing an ample supply of provisions for their parties while engaged in large surveys of exterior township lines. This difficulty can be overcome in a great measure by the use of the solar compass. The latitude of the township corner, which is to be the commencing point of the survey, must be determined with the instrument to be used in executing the work;

then convey the supplies by the most feasible route, to the desired position within the district to be surveyed, and deposit it securely from storms and wild animals, on or near some stream, lake, Indian trail, or other conspicuous object that can be recognised, in the latitude of any east and west township line; which may be determined by allowing 5′ 12″ of latitude for each township of six miles north or south of the commencing corner of the survey. If the township line, when run, should pass a few chains to the right or left of the depot thus made, it can be found in a few minutes.

This method of depositing supplies of provisions in advance of the surveyed lines, has been successfully practised by the author.

TRAVERSE TABLE,

SHOWING THE DIFFERENCE OF

LATITUDE AND DEPARTURE

FOR

DISTANCES BETWEEN 1 AND 100;

AND FOR

ANGLES TO QUARTER DEGREES BETWEEN 1° AND 90°,

AND

NATURAL SIGNS AND TANGENTS

TO EVERY DEGREE AND MINUTE OF THE QUADRANT.

Distance.	¼ Deg.		½ Deg.		¾ Deg.		Distance.
	Lat.	Dep.	Lat.	Dep.	Lat.	Dep.	
1	1·00	0·00	1·00	0·01	1·00	0·01	1
2	2·00	0·01	2·00	0·02	2·00	0·03	2
3	3·00	0·01	3·00	0·03	3·00	0·04	3
4	4·00	0·02	4·00	0·03	4·00	0·05	4
5	5·00	0·02	5·00	0·04	5·00	0·07	5
6	6·00	0·03	6·00	0·05	6·00	0·08	6
7	7·00	0·03	7·00	0·06	7·00	0·09	7
8	8·00	0·03	8·00	0·07	8·00	0·10	8
9	9·00	0·04	9·00	0·08	9·00	0·12	9
10	10·00	0·04	10.00	0·09	10·00	0.13	10
11	11·00	0·05	11·00	0·10	11·00	0·14	11
12	12·00	0·05	12·00	0·10	12·00	0·16	12
13	13·00	0·06	13·00	0.11	13·00	0·17	13
14	14·00	0·06	14·00	0·12	14·00	0.18	14
15	15·00	0·07	15·00	0·13	15·00	0·20	15
16	16·00	0·07	16·00	0·14	16.00	0·21	16
17	17·00	0·07	17·00	0·15	17·00	0·22	17
18	18·00	0·08	18·00	0·16	18·00	0·24	18
19	19·00	0·08	19·00	0·17	19·00	0·25	19
20	20·00	0·09	20·00	0·17	20·00	0·26	20
21	21·00	0·09	21·00	0·18	21·00	0·27	21
22	22·00	0·10	22·00	0·19	22·00	0·29	22
23	23·00	0·10	23·00	0·20	23·00	0·30	23
24	24·00	0·10	24·00	0·21	24·00	0·31	24
25	25·00	0·11	25·00	0·22	25·00	0·33	25
26	26·00	0·11	26·00	0·23	26.00	0·34	26
27	27·00	0·12	27·00	0·24	27·00	0·35	27
28	28·00	0·12	28·00	0·24	28·00	0·37	28
29	29·00	0.13	29·00	0·25	29·00	0·38	29
30	30·00	0·13	30·00	0·26	30·00	0·39	30
31	31·00	0·14	31·00	0.27	31·00	0·41	31
32	32·00	0·14	32·00	0·28	32·00	0·42	32
33	33·00	0·14	33·00	0·29	33·00	0·43	33
34	34·00	0·15	34·00	0·30	34·00	0·45	34
35	35·00	0·15	35·00	0·31	35·00	0·46	35
36	36·00	0·16	36·00	0·31	36·00	0·47	36
37	37·00	0·16	37·00	0·32	37·00	0·48	37
38	38·00	0·17	38·00	0·33	38·00	0·50	38
39	39·00	0·17	39·00	0·34	39·00	0·51	39
40	40·00	0·17	40·00	0·35	40·00	0·52	40
41	41·00	0·18	41·00	0.36	41·00	0·54	41
42	42·00	0·18	42·00	0·37	42·00	0·55	42
43	43·00	0·19	43·00	0·38	43·00	0·56	43
44	44·00	0·19	44·00	0.38	44·00	0·58	44
45	45·00	0·20	45·00	0·39	45·00	0·59	45
46	46·00	0·20	46·00	0·40	46·00	0·60	46
47	47·00	0·21	47·00	0·41	47·00	0·62	47
48	48·00	0·21	48·00	0·42	48·00	0·63	48
49	49·00	0·21	49·00	0·43	49·00	0·64	49
50	50·00	0·22	50·00	0·44	50·00	0·65	50
Distance.	Dep.	Lat.	Dep.	Lat.	Dep.	Lat.	Distance.
	89¾ Deg.		89½ Deg.		89¼ Deg.		

Distance.	¼ Deg.		½ Deg.		¾ Deg.		Distance.
	Lat.	Dep.	Lat.	Dep.	Lat.	Dep.	
51	51·00	0·22	51·00	0·45	51·00	0·67	51
52	52·00	0·23	52·00	0·45	52·00	0·68	52
53	53·00	0·23	53·00	0·46	53·00	0·69	53
54	54·00	0·24	54·00	0·47	54·00	0·71	54
55	55·00	0·24	55·00	0·48	55·00	0·72	55
56	56·00	0·24	56·00	0·49	56·00	0·73	56
57	57·00	0·25	57·00	0·50	57·00	0·75	57
58	58·00	0·25	58·00	0·51	57·99	0·76	58
59	59·00	0·26	59·00	0·51	58·99	0·77	59
60	60·00	0·26	60·00	0·52	59·99	0·79	60
61	61·00	0·27	61·00	0·53	60·99	0·80	61
62	62·00	0·27	62·00	0·54	61·99	0·81	62
63	63·00	0·27	63·00	0·55	62·99	0·82	63
64	64·00	0·28	64·00	0·56	63·99	0·84	64
65	65·00	0·28	65·00	0·57	64·99	0·85	65
66	66·00	0·29	66·00	0·58	65·99	0·86	66
67	67·00	0·29	67·00	0·58	66·99	0·88	67
68	68·00	0·30	68·00	0·59	67·99	0·89	68
69	69·00	0·30	69·00	0·60	68·99	0·90	69
70	70·00	0·31	70·00	0·61	69·99	0·92	70
71	71·00	0·31	71·00	0·62	70·99	0·93	71
72	72·00	0·31	72·00	0·63	71·99	0·94	72
73	73·00	0·32	73·00	0·64	72·99	0·96	73
74	74·00	0·32	74·00	0·65	73·99	0·97	74
75	75·00	0·33	75·00	0·65	74·99	0·98	75
76	76·00	0·33	76·00	0·66	75·99	0·99	76
77	77·00	0·34	77·00	0·67	76·99	1·01	77
78	78·00	0·34	78·00	0·68	77·99	1·02	78
79	79·00	0·34	79·00	0·69	78·99	1·03	79
80	80·00	0·35	80·00	0·70	79·99	1·05	80
81	81·00	0·35	81·00	0·71	80·99	1·06	81
82	82·00	0·36	82·00	0·72	81·99	1·07	82
83	83·00	0·36	83·00	0·72	82·99	1·09	83
84	84·00	0·37	84·00	0·73	83·99	1·10	84
85	85·00	0·37	85·00	0·74	84·99	1·11	85
86	86·00	0·38	86·00	0·75	85·99	1·13	86
87	87·00	0·38	87·00	0·76	86·99	1·14	87
88	88·00	0·38	88·00	0·77	87·99	1·15	88
89	89·00	0·39	89·00	0·78	88·99	1·16	89
90	90·00	0·39	90·00	0·79	89·99	1·18	90
91	91·00	0·40	91·00	0·79	90·99	1·19	91
92	92·00	0·40	92·00	0·80	91·99	1·20	92
93	93·00	0·41	93·00	0·81	92·99	1·22	93
94	94·00	0·41	94·00	0·82	93·99	1·23	94
95	95·00	0·41	95·00	0·83	94·99	1·24	95
96	96·00	0·42	96·00	0·84	95·99	1·26	96
97	97·00	0·42	97·00	0·85	96·99	1·27	97
98	98·00	0·43	98·00	0·86	97·99	1·28	98
99	99·00	0·43	99·00	0·86	98·99	1·30	99
100	100·00	0·44	100·00	0·87	99·99	1·31	100
	Dep.	Lat.	Dep.	Lat.	Dep.	Lat.	
Distance.	89¾ Deg.		89½ Deg.		89¼ Deg.		Distance.

Distance.	1 Deg.		1¼ Deg.		1½ Deg.		1¾ Deg.		Distance.
	Lat.	Dep.	Lat.	Dep.	Lat.	Dep.	Lat.	Dep.	
1	1·00	0·02	1·00	0·02	1·00	0·03	1·00	0·03	1
2	2·00	0·03	2·00	0 04	2·00	0·05	2·00	0·06	2
3	3·00	0·05	3·00	0·07	3·00	0·08	3·00	0·09	3
4	4·00	0·07	4·00	0·09	4·00	0·10	4·00	0·12	4
5	5·00	0·09	5·00	0.11	5·00	0·13	5·00	0·15	5
6	6·00	0·10	6·00	0·13	6·00	0·16	6·00	0·18	6
7	7·00	0·12	7·00	0·15	7·00	0·18	7·00	0·21	7
8	8·00	0·14	8·00	0·17	8·00	0·21	8·00	0·25	8
9	9·00	0·16	9·00	0·20	9·00	0·24	9·00	0·28	9
10	10·00	0·17	10·00	0·22	10·00	0·26	10·00	0·31	10
11	11·00	0·19	11·00	0·24	11·00	0·28	10·99	0·34	11
12	12 00	0·21	12·00	0·26	12·00	0·31	11·99	0·37	12
13	13·00	0·23	13·00	0·28	13·00	0·34	12·99	0·40	13
14	14·00	0·24	14·00	0·31	14·00	0·37	13·99	0·43	14
15	15·00	0·26	15·00	0·33	14·99	0·39	14·99	0·46	15
16	16·00	0·28	16·00	0·35	15·99	0·42	15·99	0·49	16
17	17·00	0·30	17·00	0·37	16·99	0·45	16·99	0·52	17
18	18·00	0·31	18·00	0·39	17·99	0·47	17·99	0·55	18
19	19·00	0 33	19·00	0·41	18·99	0·50	18·99	0·58	19
20	20·00	0·35	20·00	0·44	19·99	0·52	19·99	0·61	20
21	21·00	0·37	21·00	0·46	20·99	0·55	20·99	0·64	21
22	22 00	0·38	21 99	0·48	21·99	0·58	21·99	0·67	22
23	23·00	0·40	22·99	0·50	22·99	0·60	22·99	0·70	23
24	24·00	0·42	23·99	0·52	23·99	0·63	23·99	0·73	24
25	25·00	0·44	24·99	0·55	24·99	0·65	24·99	0·76	25
26	26·00	0·45	25·99	0·57	25·99	0·68	25·99	0·79	26
27	27·00	0·47	26·99	0·59	26·99	0·71	26·99	0·83	27
28	28·00	0·49	27·99	0·61	27·99	0·73	27·99	0·86	28
29	29·00	0·51	28·99	0·63	28·99	0·76	28·99	0·89	29
30	30·00	0·52	29·99	0·65	29·99	0·79	29.99	0·92	30
31	31·00	0·54	30·99	0·68	30·99	0·81	30.99	0·95	31
32	32·00	0·56	31·99	0·70	31·99	0·84	31·99	0·98	32
33	32·99	0·58	32·99	0·72	32·99	0·86	32·98	1·01	33
34	33·99	0·59	33·99	0·74	33·99	0·89	33·98	1 04	34
35	34·99	0·61	34·99	0·76	34·99	0·92	34·98	1·07	35
36	35·99	0·63	35·99	0·79	35·99	0·94	35·98	1·10	36
37	36·99	0·65	36·99	0·81	36·99	0·97	36·98	1·13	37
38	37·99	0·66	37·99	0·83	37·99	0·99	37·98	1·16	38
39	38·99	0·68	38·99	0·85	38·99	1·02	38·98	1·19	39
40	39·99	0·70	39·99	0·87	39·99	1·05	39·98	1·22	40
41	40.99	0·72	40·99	0.89	40·99	1·07	40·98	1·25	41
42	41·99	0·73	41·99	0·92	41·99	1·10	41·98	1·28	42
43	42·99	0·75	42·99	0·94	42·99	1·13	42·98	1·31	43
44	43·99	0·77	43·99	0·96	43·99	1·15	43·98	1 34	44
45	44·99	0·79	44·99	0·98	44·99	1·18	44·98	1·37	45
46	45·99	0·80	45·99	1·00	45·99	1·20	45·98	1·40	46
47	46·99	0·82	46·99	1·03	46·99	1·23	46·98	1·44	47
48	47·99	0·84	47·99	1·05.	47·99	1·26	47·98	1·47	48
49	48·99	0·86	48·99	1·07	48·99	1·28	48·98	1·50	49
50	49·99	0·87	49·99	1·09	49·99	1·31	49·98	1·53	50
Distance.	Dep.	Lat.	Dep.	Lat.	Dep.	Lat.	Dep.	Lat.	Distance.
	89 Deg.		88¾ Deg.		88½ Deg.		88¼ Deg.		

Distance.	1 Deg.		1¼ Deg.		1½ Deg.		1¾ Deg.		Distance.
	Lat.	Dep.	Lat.	Dep.	Lat.	Dep.	Lat.	Dep.	
51	50·99	0·89	50·99	1·11	50·98	1·34	50·98	1·56	51
52	51·99	0·91	51·99	1·13	51·98	1·36	51·98	1·59	52
53	52·99	0·92	52·99	1·16	52·98	1·39	52·98	1·62	53
54	53·99	0·94	53·99	1·18	53·98	1·41	53·97	1·65	54
55	54·99	0·96	54·99	1·20	54·98	1·44	54·97	1·68	55
56	55·99	0·98	55·99	1·22	55·98	1·47	55·97	1·71	56
57	56·99	0·99	56·99	1·24	56·98	1·49	56·97	1·74	57
58	57·99	1·01	57·99	1·27	57·98	1·52	57·97	1·77	58
59	58·99	1·03	58·99	1·29	58·98	1·54	58·97	1·80	59
60	59·99	1·05	59·99	1·31	59·98	1·57	59·97	1·83	60
61	60·99	1·06	60·99	1·33	60·98	1·60	60·97	1·86	61
62	61·99	1·08	61·99	1·35	61·98	1·62	61·97	1·89	62
63	62·99	1·10	62·99	1·37	62·98	1·65	62·97	1·92	63
64	63·99	1·12	63·98	1·40	63·98	1·68	63·97	1·95	64
65	64·99	1·13	64·98	1·42	64·98	1·70	64·97	1·99	65
66	65·99	1·15	65·98	1·44	65·98	1·73	65·97	2·02	66
67	66·99	1·17	66·98	1·46	66·98	1·75	66·97	2·05	67
68	67·99	1·19	67·98	1·48	67·98	1·78	67·97	2·08	68
69	68·99	1·20	68·98	1·51	68·98	1·81	68·97	2·11	69
70	69·99	1·22	69·98	1·53	69·98	1·83	69·97	2·14	70
71	70·99	1·24	70·98	1·55	70·98	1·86	70·97	2·17	71
72	71·99	1·26	71·98	1·57	71·98	1·88	71·97	2·20	72
73	72·99	1·27	72·98	1·59	72·97	1.91	72·97	2·23	73
74	73·99	1·29	73·98	1·61	73·97	1·94	73·97	2·26	74
75	74·99	1·31	74·98	1·64	74·97	1·96	74·97	2·29	75
76	75·99	1·33	75·98	1·66	75·97	1·99	75·96	2·32	76
77	76·99	1·34	76·98	1·68	76·97	2·02	76·96	2·35	77
78	77·99	1·36	77·98	1·70	77·97	2·04	77·96	2·38	78
79	78·99	1·38	78·98	1·72	78·97	2·07	78·96	2·41	79
80	79·99	1·40	79·98	1·75	79·97	2·09	79·96	2·44	80
81	80·99	1·41	80·98	1·77	80·97	2·12	80·96	2·47	81
82	81·99	1·43	81·98	1·79	81·97	2·15	81·96	2·50	82
83	82·99	1·45	82·98	1·81	82·97	2·17	82·96	2·53	83
84	83·99	1·47	83·98	1·83	83·97	2·20	83·96	2·57	84
85	84·99	1·48	84·98	1·85	84·97	2·23	84·96	2·60	85
86	85·99	1·50	85·98	1·88	85·97	2·25	85·96	2·63	86
87	86·99	1·52	86·98	1·90	86·97	2·28	86·96	2·66	87
88	87·99	1·54	87·98	1·92	87·97	2·30	87·96	2·69	88
89	88·99	1·55	88·98	1·94	88·97	2·33	88·96	2·72	89
90	89·99	1·57	89·98	1·96	89·97	2·36	89·96	2·75	90
91	90·99	1·59	90·98	1·99	90·97	2·38	90·96	2·78	91
92	91·99	1·61	91·98	2·01	91·97	2·41	91·96	2·81	92
93	92·99	1·62	92·98	2·03	92·97	2·43	92·96	2·84	93
94	93·99	1·64	93·98	2·05	93·97	2·46	93·96	2·87	94
95	94·99	1·66	94·98	2·07	94·97	2·49	94·96	2·90	95
96	95·99	1·68	95·98	2·09	95·97	2·51	95·96	2·94	96
97	96·99	1·69	96·98	2·12	96·97	2.54	96·95	2·96	97
98	97·99	1·71	97·98	2·14	97·97	2·57	97·95	2·99	98
99	98·98	1·73	98·98	2·16	98·97	2·59	98·95	3·02	99
100	99·98	1·75	99·98	2·18	99·97	2·62	99·95	3·05	100
Distance.	Dep.	Lat.	Dep.	Lat.	Dep.	Lat.	Dep.	Lat.	Distance.
	89 Deg.		88¾ Deg.		88½ Deg.		88¼ Deg.		

Distance.	2 Deg.		2¼ Deg.		2½ Deg.		2¾ Deg.		Distance.
	Lat.	Dep.	Lat.	Dep.	Lat.	Dep.	Lat.	Dep.	
1	1·00	0·03	1·00	0·04	1·00	0·04	1·00	0·05	1
2	2·00	0·07	2·00	0·08	2·00	0·09	2·00	0·10	2
3	3·00	0·10	3·00	0·12	3·00	0·13	3·00	0·14	3
4	4·00	0·14	4·00	0·16	4·00	0·17	4·00	0·19	4
5	5·00	0·17	5·00	0·20	5·00	0·22	4·99	0·24	5
6	6·00	0·21	6·00	0·24	5·99	0·26	5·99	0·29	6
7	7·00	0·24	6·99	0·27	6·99	0·31	6·99	0·34	7
8	7·99	0·28	7·99	0·31	7·99	0·35	7·99	0·38	8
9	8·99	0·31	8·99	0·35	8·99	0·39	8·99	0·43	9
10	9·99	0·35	9·99	0·39	9·99	0·44	9·99	0·48	10
11	10·99	0·38	10·99	0·43	10·99	0·48	10·99	0·53	11
12	11·99	0·42	11·99	0·47	11·99	0·52	11·99	0·58	12
13	12·99	0·45	12·99	0·51	12·99	0·57	12·99	0·62	13
14	13·99	0·49	13·99	0·55	13·99	0·61	13·98	0·67	14
15	14·99	0·52	14·99	0·59	14·99	0·65	14·98	0·72	15
16	15·99	0·56	15·99	0·63	15·99	0·70	15·98	0·77	16
17	16·99	0·59	16·99	0·67	16·98	0·74	16·98	0·82	17
18	17·99	0·63	17·99	0·71	17·98	0·79	17·98	0·86	18
19	18·99	0·66	18·99	0·75	18·98	0·83	18·98	0·91	19
20	19·99	0·70	19·98	0·79	19·98	0·87	19·98	0·96	20
21	20·99	0·73	20·93	0·82	20·98	0·92	20·98	1·01	21
22	21·99	0·77	21·93	0·86	21·98	0·96	21·97	1·06	22
23	22·99	0·80	22·98	0·90	22·98	1·00	22·97	1·10	23
24	23·99	0·84	23·98	0·94	23·98	1·05	23·97	1·15	24
25	24·98	0·87	24·98	0·98	24·98	1·09	24·97	1·20	25
26	25·98	0·91	25·98	1·02	25·98	1·13	25·97	1·25	26
27	26·98	0·94	26·98	1·06	26·97	1·18	26·97	1·30	27
28	27·98	0·98	27·98	1·10	27·97	1·22	27·97	1·34	28
29	28·98	1·01	28·98	1·14	28·97	1·26	28·97	1·39	29
30	29·98	1·05	29·98	1·18	29·97	1·31	29·97	1·44	30
31	30·98	1·08	30·98	1·22	30·97	1·35	30·96	1·49	31
32	31·98	1·12	31·98	1·26	31·97	1·40	31·96	1·54	32
33	32·98	1·15	32·97	1·30	32·97	1·44	32·96	1·58	33
34	33·98	1·19	33·97	1·33	33·97	1·48	33·96	1·63	34
35	34·98	1·22	34·97	1·37	34·97	1·53	34·96	1·68	35
36	35·98	1.26	35·97	1·41	35·97	1·57	35·96	1·73	36
37	36·98	1·29	36·97	1·45	36·96	1·61	36·96	1·78	37
38	37·98	1·33	37·97	1·49	37·96	1·66	37·96	1·82	38
39	38·98	1·36	38·97	1·53	38·96	1·70	38·96	1·87	39
40	39·98	1·40	39·97	1·57	39·96	1·75	39·95	1·92	40
41	40·98	1·43	40·97	1·61	40·96	1·77	40·95	1·97	41
42	41·97	1·47	41·97	1·65	41·96	1·83	41·95	2·02	42
43	42·97	1·50	42·97	1·69	42·96	1·88	42·95	2·06	43
44	43·97	1·54	43·97	1·73	43·96	1·92	43.95	2·11	44
45	44·97	1·57	44·97	1·77	44·96	1·96	44·95	2·16	45
46	45·97	1·61	45·96	1·81	45·96	2·01	45·95	2·21	46
47	46·97	1·64	46·96	1·85	46·96	2·05	46·95	2·25	47
48	47·97	1·68	47·96	1·88	47·95	2·09	47·95	2·30	48
49	48·97	1·71	48·96	1·92	48·95	2·14	48·94	2·35	49
50	49·97	1·74	49·96	1·96	49·95	2·18	49·94	2·40	50
Distance.	Dep.	Lat.	Dep.	Lat.	Dep.	Lat.	Dep.	Lat.	Distance.
	88 Deg.		87¾ Deg.		87½ Deg.		87¼ Deg.		

Distance.	2 Deg.		2¼ Deg.		2½ Deg.		2¾ Deg.		Distance.
	Lat.	Dep.	Lat.	Dep.	Lat.	Dep.	Lat.	Dep.	
51	50·97	1·78	50·96	2·00	50·95	2·22	50·94	2·45	51
52	51·97	1·81	51·96	2·04	51·95	2·27	51·94	2·50	52
53	52·97	1·85	52·96	2·08	52·95	2·31	52·94	2·54	53
54	53·97	1·88	53·96	2·12	53·95	2·36	53·94	2·59	54
55	54·97	1·92	54·96	2·16	54·95	2·40	54·94	2·64	55
56	55·97	1·95	55·96	2·20	55·95	2·44	55·94	2·69	56
57	56·97	1·99	56·96	2·24	56·95	2·49	56·93	2·73	57
58	57·96	2·02	57·96	2·28	57·94	2·53	57·93	2·78	58
59	58·96	2·06	58·95	2·32	58·94	2·57	58·93	2·83	59
60	59·96	2·09	59·95	2·36	59·94	2·62	59·93	2·88	60
61	60·96	2·13	60·95	2·39	60·94	2·66	60·93	2·93	61
62	61·96	2·16	61·95	2·43	61·94	2·70	61·93	2·97	62
63	62·96	2·20	62·95	2·47	62·94	2·75	62·93	3·02	63
64	63·96	2·23	63·95	2·51	63·94	2·79	63·93	3·07	64
65	64·96	2·27	64·95	2·55	64·94	2·84	64·93	3·12	65
66	65·96	2·30	65·95	2·59	65·94	2·88	65·92	3·17	66
67	66·96	2·34	66·95	2·63	66·94	2·92	66·92	3·21	67
68	67·96	2·37	67·95	2·67	67·94	2·97	67·92	3·26	68
69	68·96	2·41	68·95	2·71	68·93	3·01	68·92	3·31	69
70	69·96	2·44	69·95	2·75	69·93	3·05	69·92	3·36	70
71	70·96	2·48	70·95	2·79	70·93	3·10	70·92	3·41	71
72	71·96	2·51	71·94	2·83	71·93	3·14	71·92	3·45	72
73	72·96	2·55	72·94	2·87	72·93	3·18	72·92	3·50	73
74	73·95	2·58	73·94	2·91	73·93	3·23	73·91	3·55	74
75	74·95	2·62	74·94	2·94	74·93	3·27	74·91	3·60	75
76	75·95	2·65	75·94	2·98	75·93	3·31	75·91	3·65	76
77	76·95	2·69	76·94	3·02	76·93	3·36	76·91	3·70	77
78	77·95	2·72	77·94	3·06	77·93	3·40	77·91	3·74	78
79	78·95	2·76	78·94	3·10	78·92	3·45	78·91	3·79	79
80	79·95	2·79	79·94	3·14	79·92	3·49	79·91	3·84	80
81	80·95	2·83	80·94	3·18	80·92	3·53	80·91	3·89	81
82	81·95	2·86	81·94	3·22	81·92	3·58	81·91	3·93	82
83	82·95	2·90	82·94	3·26	82·92	3·62	82·90	3·98	83
84	83·95	2·93	83·94	3·30	83·92	3·66	83·90	4·03	84
85	84·95	2·97	84·93	3·34	84·92	3·71	84·90	4·08	85
86	85·95	3·00	85·93	3·38	85·92	3·75	85·90	4·13	86
87	86·95	3·04	86·93	3·42	86·92	3·79	86·90	4·17	87
88	87·95	3·07	87·93	3·45	87·92	3·84	87·90	4·22	88
89	88·95	3·11	88·93	3·49	88·92	3·88	88·90	4·27	89
90	89·95	3·14	89·93	3·53	89·91	3·93	89·90	4·32	90
91	90·95	3·18	90·93	3·57	90·91	3·97	90·90	4·37	91
92	91·94	3·21	91·93	3·61	91·91	4·01	91·89	4·41	92
93	92·94	3·25	92·93	3·65	92·91	4·06	92·89	4·46	93
94	93·94	3·28	93·93	3·69	93·91	4·10	93·89	4·51	94
95	94·94	3·32	94·93	3·73	94·91	4·14	94·89	4·56	95
96	95·94	3·35	95·93	3·77	95·91	4·19	95·89	4·61	96
97	96·94	3·39	96·93	3·81	96·91	4·23	96·89	4·65	97
98	97·94	3·42	97·92	3·85	97·91	4·27	97·89	4·70	98
99	98·94	3·46	98·92	3·89	98·91	4·32	98·89	4·75	99
100	99·94	3·49	99·92	3·93	99·91	4·36	99·88	4·80	100
Distance.	Dep.	Lat.	Dep.	Lat.	Dep.	Lat.	Dep.	Lat.	Distance.
	88 Deg.		87¾ Deg.		87½ Deg.		87¼ Deg.		

Distance.	3 Deg.		3¼ Deg.		3½ Deg.		3¾ Deg.		Distance.
	Lat.	Dep.	Lat.	Dep.	Lat.	Dep.	Lat.	Dep.	
1	1·00	0·05	1·00	0·06	1·00	0·06	1·00	0·06	1
2	2·00	0·10	2·00	0·11	2·00	0·12	2·00	0·13	2
3	3·00	0·16	3·00	0·17	2·99	0·18	2·99	0·20	3
4	3·99	0·21	3·99	0·23	3·99	0·24	3·99	0·26	4
5	4·99	0·26	4·99	0·28	4·99	0·31	4·99	0·33	5
6	5·99	0·31	5·99	0·34	5·99	0·37	5·99	0·39	6
7	6·99	0·37	6·99	0·40	6·99	0·43	6·99	0·46	7
8	7·99	0·42	7·99	0·45	7·99	0·49	7·98	0·52	8
9	8·99	0·47	8·99	0·51	8·98	0·55	8·98	0·59	9
10	9·99	0·52	9·98	0·57	9·98	0·61	9·98	0·65	10
11	10·98	0·58	10·98	0·62	10·98	0·67	10·98	0·72	11
12	11·98	0·63	11·98	0·68	11·98	0·73	11·97	0·78	12
13	12·98	0·68	12·98	0·73	12·98	0·79	12·97	0·85	13
14	13·98	0·73	13·98	0·79	13·97	0·85	13·97	0·92	14
15	14·98	0·79	14·98	0·85	14·97	0·92	14·97	0·98	15
16	15·98	0·84	15.97	0·91	15·97	0·98	15·97	1·05	16
17	16·98	0·89	16·97	0·96	16·97	1·04	16·96	1·11	17
18	17·98	0·94	17·97	1·02	17·97	1·10	17·96	1·18	18
19	18·98	0·99	18·97	1·08	18·96	1·16	18·96	1·24	19
20	19·97	1·05	19·97	1·13	19·96	1·22	19·96	1·31	20
21	20·97	1·10	20·97	1·19	20·96	1·28	20·96	1·37	21
22	21·97	1·15	21·96	1·25	21·96	1·34	21·95	1·44	22
23	22·97	1·20	22·96	1·30	22·96	1·40	22·95	1·50	23
24	23·97	1·26	23·96	1·36	23·96	1·47	23·95	1·57	24
25	24·97	1·31	24·96	1·42	24·95	1·53	24·95	1·64	25
26	25·96	1·36	25·96	1·47	25·95	1·59	25·94	1·70	26
27	26·96	1·41	26·96	1·53	26·95	1·65	26·94	1·77	27
28	27·96	1·47	27·95	1·59	27·95	1·71	27·94	1·83	28
29	28·96	1·52	28·95	1·64	28·95	1·77	28·94	1·90	29
30	29·96	1·57	29·95	1·70	29·94	1·83	29·94	1·96	30
31	30·96	1·62	30·95	1·76	30·94	1·89	30·93	2·03	31
32	31·96	1·67	31·95	1·81	31·94	1·95	31·93	2·09	32
33	32·95	1·73	32·95	1·87	32·94	2·01	32·93	2·16	33
34	33·95	1·78	33·95	1·93	33·94	2·08	33·93	2·22	34
35	34·95	1·83	34·94	1·98	34·93	2·14	34·92	2·29	35
36	35·95	1·88	35·94	2·04	35·93	2·20	35·92	2·35	36
37	36·95	1·94	36·94	2·10	36·93	2·26	36·92	2·42	37
38	37·95	1·99	37·94	2·15	37·93	2·32	37·92	2·49	38
39	38·95	2·04	38·94	2·21	38·93	2·38	38·92	2·55	39
40	39·95	2·09	39·94	2·27	39·93	2·44	39·91	2·62	40
41	40·94	2·15	40·93	2·32	40·92	2·50	40·91	2·68	41
42	41·94	2·20	41·93	2·38	41·92	2·56	41·91	2·75	42
43	42·94	2·25	42·93	2·44	42·92	2·63	42·91	2·81	43
44	43·94	2·30	43·93	2·49	43·92	2·69	43·91	2·88	44
45	44·94	2·36	44·93	2·55	44·92	2·75	44·90	2·94	45
46	45·94	2·41	45·93	2·61	45·91	2·81	45·90	3·01	46
47	46·94	2·46	46·92	2·66	46·91	2·87	46·90	3·07	47
48	47·93	2·51	47·92	2·72	47·91	2·93	47·90	3·14	48
49	48·93	2.56	48·92	2·78	48·91	2·99	48·90	3·20	49
50	49·93	2·62	49·92	2·83	49·91	3·05	49·89	3·27	50
Distance.	Dep.	Lat.	Dep.	Lat.	Dep.	Lat.	Dep.	Lat.	Distance.
	87 Deg.		86¾ Deg.		86½ Deg.		86¼ Deg.		

Distance.	3 Deg.		3¼ Deg.		3½ Deg.		3¾ Deg.		Distance.
	Lat.	Dep.	Lat.	Dep.	Lat.	Dep.	Lat.	Dep.	
51	50·93	2·67	50·92	2·89	50·90	3·11	50·89	3·34	51
52	51·93	2·72	51·92	2·95	51·90	3·17	51·89	3·40	52
53	52·93	2·77	52·91	3·00	52·90	3·24	52·89	3·47	53
54	53·93	2·83	53·91	3·06	53·90	3·30	53·88	3·53	54
55	54·92	2·88	54·91	3·12	54·90	3·36	54·88	3·60	55
56	55·92	2·93	55·91	3·17	55·90	3·42	55·88	3·66	56
57	56·92	2·98	56·91	3·23	56·89	3·48	56·88	3·73	57
58	57·92	3·04	57·91	3·29	57·89	3·54	57·88	3·79	58
59	58·92	3·09	58·91	3·34	58·89	3·60	58·87	3·86	59
60	59·92	3·14	59·90	3·40	59·89	3·66	59·87	3·92	60
61	60·92	3·19	60·90	3·46	60·89	3·72	60·87	3·99	61
62	61·92	3·24	61·90	3·51	61·88	3·79	61·87	4·05	62
63	62·91	3·30	62·90	3·57	62·88	3·85	62·87	4·12	63
64	63·91	3·35	63·90	3·63	63·88	3·91	63·86	4·19	64
65	64·91	3·40	64·90	3·69	64·88	3·97	64·86	4·25	65
66	65·91	3·45	65·89	3·74	65·88	4·03	65·86	4·32	66
67	66·91	3·51	66·89	3·80	66·88	4·09	66·86	4·38	67
68	67·91	3·56	67·89	3·86	67·87	4·15	67·85	4·45	68
69	68·91	3·61	68·89	3·91	68·87	4·21	68·85	4·51	69
70	69·90	3·66	69·89	3·97	69·87	4·27	69·85	4·58	70
71	70·90	3·72	70·89	4·03	70·87	4·33	70·85	4·64	71
72	71·90	3·77	71·88	4·08	71·87	4·40	71·85	4·71	72
73	72·90	3·82	72·88	4·14	72·86	4·46	72·84	4·77	73
74	73·90	3·87	73·88	4·20	73·86	4·52	73·84	4·84	74
75	74·90	3·93	74·88	4·25	74·86	4·58	74·84	4·91	75
76	75·90	3·98	75·88	4·31	75·86	4·64	75·84	4·97	76
77	76·89	4·03	76·88	4·37	76·86	4·70	76·84	5·04	77
78	77·89	4·08	77·87	4·42	77·85	4·76	77·83	5·10	78
79	78·89	4·13	78·87	4·48	78·85	4·82	78·83	5·17	79
80	79·89	4·19	79·87	4·54	79·85	4·88	79·83	5·23	80
81	80·89	4·24	80·87	4·59	80·85	4·94	80·83	5·30	81
82	81·89	4·29	81·87	4·65	81·85	5·01	81·82	5·36	82
83	82·89	4·34	82·87	4·71	82·85	5·07	82·82	5·43	83
84	83·88	4·40	83·86	4·76	83·84	5·13	83·82	5·49	84
85	84·88	4·45	84·86	4·82	84·84	5·19	84·82	5·56	85
86	85·88	4·50	85·86	4·88	85·84	5·25	85·82	5·62	86
87	86·88	4·55	86·86	4·93	86·84	5·31	86·81	5·69	87
88	87·88	4·61	87·86	4·99	87·84	5·37	87·81	5·76	88
89	88·88	4·66	88·86	5·05	88·83	5·43	88·81	5·82	89
90	89·88	4·71	89·86	5·10	89·83	5·49	89·81	5·89	90
91	90·88	4·76	90·85	5·16	90·83	5·56	90·81	5·95	91
92	91·87	4·81	91·85	5·22	91·83	5·62	91·80	6·02	92
93	92·87	4·87	92·85	5·27	92·83	5·68	92·80	6·08	93
94	93·87	4·92	93·85	5·33	93·82	5·74	93·80	6·15	94
95	94·87	4·97	94·85	5·39	94·82	5·80	94·80	6·21	95
96	95·87	5·02	95·85	5·44	95·82	5·86	95·79	6·28	96
97	96·87	5·08	96·84	5·50	96·82	5·92	96·79	6·34	97
98	97·87	5·13	97·84	5·56	97·82	5·98	97·79	6·41	98
99	98·86	5·18	98·84	5·61	98·82	6·04	98·79	6·47	99
100	99·86	5·23	99·84	5·67	99·81	6·10	99·79	6·54	100
	Dep.	Lat.	Dep.	Lat.	Dep.	Lat.	Dep.	Lat.	
Distance.	87 Deg.		86¾ Deg.		86½ Deg.		86¼ Deg.		Distance.

Distance.	4 Deg.		4¼ Deg.		4½ Deg.		4¾ Deg.		Distance.
	Lat.	Dep.	Lat.	Dep.	Lat.	Dep.	Lat.	Dep.	
1	1·00	0·07	1·00	0·07	1·00	0·08	1·00	0·08	1
2	2·00	0·14	1·99	0·15	1·99	0·16	1·99	0·17	2
3	2·99	0·21	2·99	0·22	2·99	0·24	2·99	0·25	3
4	3·99	0·28	3·99	0·30	3·98	0·31	3·98	0·33	4
5	4·99	0·35	4·99	0·37	4·98	0·39	4·98	0·41	5
6	5·99	0·42	5·98	0·44	5·98	0·47	5·98	0·50	6
7	6·98	0·49	6·98	0·52	6·98	0·55	6·97	0·58	7
8	7·98	0·56	7·98	0·59	7·98	0·63	7·97	0·66	8
9	8·98	0·63	8·98	0·67	8·97	0·71	8·97	0·75	9
10	9·98	0·70	9·97	0·74	9·97	0·78	9·97	0·83	10
11	10·97	0·77	10·97	0·82	10·97	0·86	10·96	0·91	11
12	11·97	0·84	11·97	0·89	11·96	0·94	11·96	0·99	12
13	12·97	0·91	12·96	0·96	12·96	1·02	12·96	1·08	13
14	13·97	0·98	13·96	1·04	13·96	1·10	13·95	1·16	14
15	14·96	1·05	14·96	1·11	14·95	1·18	14·95	1·24	15
16	15·96	1·12	15·96	1·19	15·95	1·26	15·95	1·32	16
17	16·96	1·19	16·95	1·26	16·95	1·33	16·94	1·41	17
18	17·96	1·26	17·95	1·33	17·94	1·41	17·94	1·49	18
19	18·95	1·33	18·95	1·40	18·94	1·49	18·93	1·57	19
20	19·95	1·40	19·95	1·48	19·94	1·57	19·93	1·66	20
21	20·95	1·46	20·94	1·56	20·94	1·65	20·93	1·74	21
22	21·95	1·53	21·94	1·63	21·93	1·73	21·92	1·82	22
23	22·94	1·60	22·94	1·70	22·93	1·80	22·92	1·90	23
24	23·94	1·67	23·93	1·78	23·93	1·88	23·92	1·99	24
25	24·94	1·74	24·93	1·85	24·92	1·96	24·91	2·07	25
26	25·94	1·81	25·93	1·93	25·92	2·04	25·91	2·15	26
27	26·93	1·88	26·93	2·00	26·92	2·12	26·91	2·24	27
28	27·93	1·95	27·92	2·08	27·91	2·20	27·90	2·32	28
29	28·93	2·02	28·92	2·15	28·91	2·28	28·90	2·40	29
30	29·93	2·09	29·92	2·22	29·91	2·35	29·90	2·48	30
31	30·92	2·16	30·91	2·30	30·90	2·43	30·89	2·57	31
32	31·92	2·23	31·91	2·37	31·90	2·51	31·89	2·65	32
33	32·92	2·30	32·91	2·45	32·90	2·59	32·89	2·73	33
34	33·92	2·37	33·91	2·52	33·90	2·67	33·88	2·82	34
35	34·91	2·44	34·90	2·59	34·89	2·75	34·88	2·90	35
36	35·91	2·51	35·90	2·67	35·89	2·82	35·88	2·98	36
37	36·91	2·58	36·90	2·74	36·89	2·90	36·87	3·06	37
38	37·91	2·65	37·90	2·82	37·88	2·98	37·87	3·15	38
39	38·90	2·72	38·89	2·89	38·88	3·06	38·87	3·23	39
40	39·90	2·79	39·89	2·96	39·88	3·14	39·86	3·31	40
41	40·90	2·86	40·89	3·04	40·87	3·22	40·86	3·40	41
42	41·90	2·93	41·88	3 11	41·87	3·30	41·86	3·48	42
43	42·90	3·00	42·88	3·19	42·87	3·37	42·85	3·56	43
44	43·89	3·07	43·88	3·26	43·86	3·45	43·85	3·64	44
45	44·89	3·14	44·88	3·33	44·86	3·53	44·85	3·73	45
46	45·89	3·21	45·87	3·41	45·86	3·61	45·84	3·81	46
47	46·89	3·28	46·87	3·48	46·86	3·69	46·84	3·89	47
48	47·88	3·35	47·87	3·56	47·85	3·77	47·84	3·97	48
49	48·88	3·42	48·87	3·63	48·85	3·84	48·83	4·06	49
50	49·88	3·49	49·86	3·71	49·85	3·92	49·83	4·14	50
Distance.	Dep.	Lat.	Dep.	Lat.	Dep.	Lat.	Dep.	Lat.	Distance.
	86 Deg.		85¾ Deg.		85½ Deg.		85¼ Deg.		

Distance.	4 Deg.		4¼ Deg.		4½ Deg.		4¾ Deg.		Distance.
	Lat.	Dep.	Lat.	Dep.	Lat.	Dep.	Lat.	Dep.	
51	50·88	3·56	50·86	3·78	50·84	4·00	50·82	4·22	51
52	51·87	3·63	51·86	3·85	51·84	4·08	51·82	4·31	52
53	52·87	3·70	52·85	3·93	52·84	4·16	52·82	4·39	53
54	53·87	3·77	53·85	4·00	53·83	4·24	53·81	4·47	54
55	54·87	3·84	54·85	4·08	54·83	4·32	54·81	4·55	55
56	55·86	3·91	55·85	4·15	55·83	4·39	55·81	4·64	56
57	56·86	3·98	56·84	4·22	56·82	4·47	56·80	4·72	57
58	57·86	4·05	57·84	4·30	57·82	4·55	57·80	4·80	58
59	58·86	4·12	58·84	4·37	58·82	4·63	58·80	4·89	59
60	59·85	4·19	59·84	4·45	59·82	4·71	59·79	4·97	60
61	60·85	4·26	60·83	4·52	60·81	4·79	60·79	5·05	61
62	61·85	4·32	61·83	4·59	61·81	4·86	61·79	5·13	62
63	62·85	4·39	62·83	4·67	62·81	4·94	62·78	5·22	63
64	63·84	4·46	63·82	4·74	63·80	5·02	63·78	5·30	64
65	64·84	4·53	64·82	4·82	64·80	5·10	64·78	5·38	65
66	65·84	4·60	65·82	4·89	65·80	5·18	65·77	5·47	66
67	66·84	4·67	66·82	4·97	66·79	5·26	66·77	5·55	67
68	67·83	4·74	67·81	5·04	67·79	5·34	67·77	5·63	68
69	68·83	4·81	68·81	5·11	68·79	5·41	68·76	5·71	69
70	69·83	4·88	69·81	5·19	69·78	5·49	69·76	5·80	70
71	70·83	4·95	70·80	5·26	70·78	5·57	70·76	5·88	71
72	71·82	5·02	71·80	5·34	71·78	5·65	71·75	5·96	72
73	72·82	5·09	72·80	5·41	72·77	5·73	72·75	6·04	73
74	73·82	5·16	73·80	5·48	73·77	5·81	73·75	6·13	74
75	74·82	5·23	74·79	5·56	74·77	5·88	74·74	6·21	75
76	75·81	5·30	75·79	5·63	75·77	5·96	75·74	6·29	76
77	76·81	5·37	76·79	5·71	76·76	6·04	76·74	6·38	77
78	77·81	5·44	77·79	5·78	77·76	6·12	77·73	6·46	78
79	78·81	5·51	78·78	5·85	78·76	6·20	78·73	6·54	79
80	79·81	5·58	79·78	5·93	79·75	6·28	79·73	6·62	80
81	80·80	5·65	80·78	6·00	80·75	6·36	80·72	6·71	81
82	81·80	5·72	81·78	6·08	81·75	6·43	81·72	6·79	82
83	82·80	5·79	82·77	6·15	82·74	6·51	82·71	6·87	83
84	83·80	5·86	83·77	6·23	83·74	6·59	83·71	6·96	84
85	84·79	5·93	84·77	6·30	84·74	6·67	84·71	7·04	85
86	85·79	6·00	85·76	6·37	85·73	6·75	85·70	7·12	86
87	86·79	6·07	86·76	6·45	86·73	6·83	86·70	7·20	87
88	87·79	6·14	87·76	6·52	87·73	6·90	87·70	7·29	88
89	88·78	6·21	88·76	6·60	88·73	6·98	88·70	7·37	89
90	89·78	6·28	89·75	6·67	89·72	7·06	89·69	7·45	90
91	90·78	6·35	90·75	6·74	90·72	7·14	90·69	7·54	91
92	91·78	6·42	91·75	6·82	91·72	7·22	91·68	7·62	92
93	92·77	6·49	92·74	6·89	92·71	7·30	92·68	7·70	93
94	93·77	6·56	93·74	6·97	93·71	7·38	93·68	7·78	94
95	94·77	6·63	94·74	7·04	94·71	7·45	94·67	7·87	95
96	95·77	6·70	95·74	7·11	95·70	7·53	95·67	7·95	96
97	96·76	6·77	96·73	7·19	96·70	7·61	96·67	8·03	97
98	97·76	6·84	97·73	7·26	97·70	7·69	97·66	8·12	98
99	98·76	6·91	98·73	7·34	98·69	7·77	98·66	8·20	99
100	99·76	6·98	99·73	7·41	99·69	7·85	99·66	8·28	100
Distance.	Dep.	Lat.	Dep.	Lat.	Dep.	Lat.	Dep.	Lat.	Distance.
	86 Deg.		85¾ Deg.		85½ Deg.		85¼ Deg.		

Distance.	5 Deg.		5¼ Deg.		5½ Deg.		5¾ Deg.		Distance.
	Lat.	Dep.	Lat.	Dep.	Lat.	Dep.	Lat.	Dep.	
1	1·00	0·09	1·00	0·09	1·00	0·10	0·99	0·10	1
2	1·99	0·17	1·99	0·18	1·99	0·19	1·99	0·20	2
3	2·99	0·26	2·99	0·27	2·99	0·29	2·98	0·30	3
4	3·98	0·35	3·98	0·37	3·98	0·38	3·98	0·40	4
5	4·98	0·44	4·98	0·46	4·98	0·48	4·97	0·50	5
6	5·98	0·52	5·97	0·55	5·97	0·58	5·97	0·60	6
7	6·97	0·61	6·97	0·64	6·97	0·67	6·96	0·70	7
8	7·97	0·70	7·97	0·73	7·96	0·76	7·96	0·80	8
9	8·97	0·78	8·96	0·82	8·96	0·86	8·95	0·90	9
10	9·96	0·87	9·96	0·92	9·95	0·96	9·95	1·00	10
11	10·96	0·96	10·95	1·01	10·95	1·05	10·94	1·10	11
12	11·95	1·05	11·95	1·10	11·94	1·15	11·94	1·20	12
13	12·95	1·13	12·95	1·19	12·94	1·25	12·93	1·30	13
14	13·95	1·22	13·94	1·28	13·94	1·34	13·93	1·40	14
15	14·94	1·31	14·94	1·37	14·93	1·44	14·92	1·50	15
16	15·94	1·39	15.93	1·46	15·93	1·53	15·92	1·60	16
17	16·94	1·48	16·93	1·56	16·92	1·63	16·91	1·70	17
18	17·93	1·57	17·92	1·65	17·92	1·73	17·91	1·80	18
19	18·93	1·66	18·92	1·74	18·91	1·82	18·90	1·90	19
20	19·92	1·74	19·92	1·83	19·91	1·92	19·90	2·00	20
21	20·92	1·83	20·91	1·92	20·90	2·01	20·89	2·10	21
22	21·92	1·92	21·91	2·01	21·90	2·11	21·89	2·20	22
23	22·91	2·00	22·90	2·10	22·89	2·20	22·88	2·30	23
24	23·91	2·09	23·90	2·20	23·89	2·30	23·88	2·40	24
25	24·90	2·18	24·90	2·29	24·88	2·40	24·87	2·50	25
26	25·90	2·27	25·89	2·38	25·88	2·49	25·87	2·60	26
27	26·90	2·35	26·89	2·47	26·88	2·59	26·86	2·71	27
28	27·89	2·44	27·88	2·56	27·87	2·68	27·86	2·81	28
29	28·89	2·53	28·88	2·65	28·87	2·78	28·85	2·91	29
30	29·89	2·61	29·87	2·75	29·86	2·88	29·85	3·01	30
31	30·88	2·70	30·87	2·84	30·86	2·97	30·84	3·11	31
32	31·88	2·79	31·87	2·93	31·85	3·07	31·84	3·21	32
33	32·87	2·88	32·86	3·02	32·85	3·16	32·83	3·31	33
34	33·87	2·96	33·86	3·11	33·84	3·26	33·83	3·41	34
35	34·87	3·05	34·85	3·20	34·84	3·35	34·82	3·51	35
36	35·86	3·14	35·85	3·29	35·83	3·45	35·82	3·61	36
37	36·86	3·22	36·84	3·39	36·83	3·55	36·81	3·71	37
38	37·86	3·31	37·84	3·48	37·83	3·64	37·81	3·81	38
39	38·85	3·40	38·84	3·57	38·82	3·74	38·80	3·91	39
40	39·85	3·49	39·83	3·66	39·82	3·83	39·80	4·01	40
41	40·84	3·57	40·82	3·75	40·81	3·93	40·79	4·11	41
42	41·84	3·66	41·82	3·84	41·81	4·03	41·79	4·21	42
43	42·84	3·75	42·82	3·93	42·80	4·12	42·78	4·31	43
44	43·83	3·83	43·82	4·03	43·80	4·22	43·78	4·41	44
45	44·83	3·92	44·81	4·12	44·79	4·31	44·77	4·51	45
46	45·82	4·01	45·81	4·21	45·79	4·41	45·77	4·61	46
47	46·82	4·10	46·80	4·30	46·78	4·50	46·76	4·71	47
48	47·82	4·18	47·80	4·39	47·78	4·60	47·76	4·81	48
49	48·81	4.27	48·79	4·48	48·77	4·70	48·75	4·91	49
50	49·81	4·36	49·79	4·58	49·77	4·79	49·75	5·01	50
	Dep.	Lat.	Dep.	Lat.	Dep.	Lat.	Dep.	Lat.	
Distance.	85 Deg.		84¾ Deg.		84½ Deg.		84¼ Deg.		Distance.

Distance.	5 Deg.		5¼ Deg.		5½ Deg.		5¾ Deg.		Distance.
	Lat.	Dep.	Lat.	Dep.	Lat.	Dep.	Lat.	Dep.	
51	50·81	4·44	50·79	4·67	50·77	4·89	50·74	5·11	51
52	51·80	4·53	51·78	4·76	51·76	4·98	51·74	5·21	52
53	52·80	4·62	52·78	4·85	52·76	5·08	52·73	5·31	53
54	53·79	4·71	53·77	4·94	53·75	5·18	53·73	5·41	54
55	54·79	4·79	54·77	5·03	54·75	5·27	54·72	5·51	55
56	55·79	4·88	55·77	5·12	55·74	5·37	55·72	5·61	56
57	56·78	4·97	56·76	5·22	56·74	5·46	56·71	5·71	57
58	57·78	5·06	57·76	5·31	57·73	5·56	57·71	5·81	58
59	58·78	5·14	58·75	5·40	58·73	5·65	58·70	5·91	59
60	59·77	5·23	59·75	5·49	59·72	5·75	59·70	6·01	60
61	60·77	5·32	60·74	5·58	60·72	5·85	60·69	6·11	61
62	61·76	5·40	61·74	5·67	61·71	5·94	61·69	6·21	62
63	62·76	5·49	62·74	5·76	62·71	6·04	62·68	6·31	63
64	63·76	5·58	63·73	5·86	63·71	6·13	63·68	6·41	64
65	64·75	5·67	64·73	5·95	64·70	6·23	64·67	6·51	65
66	65·75	5·75	65·72	6·04	65·70	6·33	65·67	6·61	66
67	66·75	5·84	66·72	6·13	66·69	6·42	66·66	6·71	67
68	67·74	5·93	67·71	6·22	67·69	6·52	67·66	6·81	68
69	68·74	6·01	68·71	6·31	68·68	6·61	68·65	6·91	69
70	69·73	6·10	69·71	6·41	69·68	6·71	69·65	7·01	70
71	70·73	6·19	70·70	6·50	70·67	6·81	70·64	7·11	71
72	71·73	6·28	71·70	6·59	71·67	6·90	71·64	7·21	72
73	72·72	6·36	72·69	6·68	72·66	7·00	72·63	7·31	73
74	73·72	6·45	73·69	6·77	73·66	7·09	73·63	7·41	74
75	74·71	6·54	74·69	6·86	74·65	7·19	74·62	7·51	75
76	75·71	6·62	75·68	6·95	75·65	7·28	75·62	7·61	76
77	76·71	6·71	76·68	7·05	76·65	7·38	76·61	7·71	77
78	77·70	6·80	77·67	7·14	77·64	7·48	77·61	7·81	78
79	78·70	6·89	78·67	7·23	78·64	7·57	78·60	7·91	79
80	79·70	6·97	79·66	7·32	79·63	7·67	79·60	8·02	80
81	80·69	7·06	80·66	7·41	80·63	7·76	80·59	8·12	81
82	81·69	7·15	81·66	7·50	81·62	7·86	81·59	8·22	82
83	82·68	7·23	82·65	7·59	82·62	7·96	82·58	8·32	83
84	83·68	7·32	83·65	7·69	83·61	8·05	83·58	8·42	84
85	84·68	7·41	84·64	7·78	84·61	8·15	84·57	8·52	85
86	85·67	7·50	85·64	7·87	85·60	8·24	85·57	8·62	86
87	86·67	7·58	86·64	7·96	86·60	8·34	86·56	8·72	87
88	87·67	7·67	87·63	8·05	87·59	8·43	87·56	8·82	88
89	88·66	7·76	88·63	8·14	88·59	8·53	88·55	8·92	89
90	89·66	7·84	89·62	8·24	89·59	8·63	89·55	9·02	90
91	90·65	7·93	90·62	8·33	90·58	8·72	90·54	9·12	91
92	91·65	8·02	91·61	8·42	91·58	8·82	91·54	9·22	92
93	92·65	8·11	92·61	8·51	92·57	8·91	92·53	9·32	93
94	93·64	8·19	93·61	8·60	93·57	9·01	93·53	9·42	94
95	94·64	8·28	94·60	8·69	94·56	9·11	94·52	9·52	95
96	95·63	8·37	95·60	8·78	95·56	9·20	95·52	9·62	96
97	96·63	8·45	96·59	8·88	96·55	9·30	96·51	9·72	97
98	97·63	8·54	97·59	8·97	97·55	9·39	97·51	9·82	98
99	98·62	8·63	98·59	9·06	98·54	9·49	98·50	9·92	99
100	99·62	8·72	99·58	9·15	99·54	9·58	99·50	10·02	100
Distance.	Dep.	Lat.	Dep.	Lat.	Dep.	Lat.	Dep.	Lat.	Distance.
	85 Deg.		84¾ Deg.		84½ Deg.		84¼ Deg.		

Distance.	6 Deg.		6¼ Deg.		6½ Deg.		6¾ Deg.		Distance.
	Lat.	Dep.	Lat.	Dep.	Lat.	Dep.	Lat.	Dep.	
1	0·99	0·10	0·99	0·11	0·99	0·11	0·99	0·12	1
2	1·99	0·21	1·99	0·22	1·99	0·23	1·99	0·24	2
3	2·98	0·31	2·98	0·33	2·98	0·34	2·98	0·35	3
4	3·98	0·41	3·98	0·44	3·97	0·45	3·97	0·47	4
5	4·97	0·52	4·97	0·54	4·97	0·57	4·97	0·59	5
6	5·97	0·63	5·96	0·65	5·96	0·68	5·96	0·71	6
7	6·96	0·73	6·96	0·76	6·96	0·79	6·95	0·82	7
8	7·96	0·84	7·95	0·87	7·95	0·91	7·94	0·94	8
9	8·95	0·94	8·95	0·98	8·94	1·02	8·94	1·06	9
10	9·95	1·05	9·94	1·09	9·94	1·13	9·93	1·18	10
11	10·94	1·15	10·93	1·20	10·93	1·25	10·92	1·29	11
12	11·93	1·25	11·93	1·31	11·92	1·36	11·92	1·41	12
13	12·93	1·36	12·92	1·42	12·92	1·47	12·91	1·53	13
14	13·92	1·46	13·92	1·52	13·91	1·59	13·90	1·65	14
15	14·92	1·57	14·91	1·63	14·90	1·70	14·90	1·76	15
16	15·91	1·67	15·90	1·74	15·90	1·81	15·89	1·88	16
17	16·91	1·78	16·90	1·85	16·89	1·92	16·88	2·00	17
18	17·90	1·88	17·89	1·96	17·88	2·04	17·88	2·12	18
19	18·90	1·99	18·89	2·07	18·88	2·15	18·87	2·23	19
20	19·89	2·09	19·88	2·18	19·87	2·26	19·86	2·35	20
21	20·88	2·20	20·88	2·29	20·87	2·38	20·85	2·47	21
22	21·88	2·30	21·87	2·40	21·86	2·49	21·85	2·59	22
23	22·87	2·40	22·86	2·50	22·85	2·60	22·84	2·70	23
24	23·87	2·51	23·86	2·61	23·85	2·72	23·83	2·82	24
25	24·86	2·61	24·85	2·72	24·84	2·83	24·83	2·94	25
26	25·86	2·72	25·85	2·83	25·83	2·94	25·82	3·06	26
27	26·85	2·82	26·84	2·94	26·83	3·06	26·81	3·17	27
28	27·85	2·93	27·83	3·05	27·82	3·17	27·81	3·29	28
29	28·84	3·03	28·83	3·16	28·81	3·28	28·80	3·41	29
30	29·84	3·14	29·82	3·27	29·81	3·40	29·79	3·53	30
31	30·83	3·24	30·82	3·37	30·80	3·51	30·79	3·64	31
32	31·82	3·34	31·81	3·48	31·79	3·62	31·78	3·76	32
33	32·82	3·45	32·80	3·59	32·79	3·74	32·77	3·88	33
34	33·81	3·55	33·80	3·70	33·78	3·85	33·76	4·00	34
35	34·81	3·66	34·79	3·81	34·78	3·96	34·76	4·11	35
36	35·80	3·76	35·79	3·92	35·77	4·08	35·75	4·23	36
37	36·80	3·87	36·78	4·03	36·76	4·19	36·75	4·35	37
38	37·79	3·97	37·77	4·14	37·76	4·30	37·74	4·47	38
39	38·79	4·08	38·77	4·25	38·75	4·41	38·73	4·58	39
40	39·78	4·18	39·76	4·35	39·74	4·53	39·72	4·70	40
41	40·78	4·29	40·76	4·46	40·74	4·64	40·72	4·82	41
42	41·77	4·39	41·75	4·57	41·73	4·76	41·71	4·94	42
43	42·76	4·49	42·74	4·68	42·72	4·87	42·70	5·05	43
44	43·76	4·60	43·74	4·79	43·72	4·98	43.70	5·17	44
45	44·75	4·70	44·73	4·90	44·71	5·09	44·69	5·29	45
46	45·75	4·81	45·73	5·01	45·70	5·21	45·68	5·41	46
47	46·74	4·91	46·72	5·12	46·70	5·32	46·67	5·52	47
48	47·74	5·02	47·71	5·23	47·69	5·43	47·67	5·64	48
49	48·73	5·12	48·71	5·34	48·69	5·55	48·66	5·76	49
50	49·73	5·23	49·70	5·44	49·68	5·66	49·65	5·88	50
	Dep.	Lat.	Dep.	Lat.	Dep.	Lat.	Dep.	Lat.	
Distance.	84 Deg.		83¾ Deg.		83½ Deg.		Deg. 83¼		Distance.

Distance.	6 Deg.		6¼ Deg.		6½ Deg.		6¾ Deg.		Distance.
	Lat.	Dep.	Lat.	Dep.	Lat.	Dep.	Lat.	Dep.	
51	50·72	5·33	50·70	5·55	50·67	5·77	50·65	5·99	51
52	51·72	5·44	51·69	5·66	51·67	5·89	51·64	6·11	52
53	52·71	5·54	52·68	5·77	52·66	6·00	52·63	6·23	53
54	53·70	5·64	53·68	5·88	53·65	6·11	53·63	6·35	54
55	54·70	5·75	54·67	5·99	54·65	6·23	54·62	6·46	55
56	55·69	5·85	55·67	6·10	55·64	6·34	55·61	6·58	56
57	56·69	5·96	56·66	6·21	56·63	6·45	56·60	6·70	57
58	57·68	6·06	57·66	6·31	57·63	6·57	57·60	6·82	58
59	58·68	6·17	58·65	6·42	58·62	6·68	58·59	6·93	59
60	59·67	6·27	59·64	6·53	59·61	6·79	59·58	7·05	60
61	60·67	6·38	60·64	6·64	60·61	6·91	60·58	7·17	61
62	61·66	6·48	61·63	6·75	61·60	7·02	61·57	7·29	62
63	62·65	6·59	62·63	6·86	62·60	7·13	62·56	7·40	63
64	63·65	6·69	63·62	6·97	63·59	7·25	63·56	7·52	64
65	64·64	6·79	64·61	7·08	64·58	7·36	64·55	7·64	65
66	65·64	6·90	65·61	7·19	65·58	7·47	65·54	7·76	66
67	66·63	7·00	66·60	7·29	66·57	7·58	66·54	7·88	67
68	67·63	7·11	67·60	7·40	67·56	7·70	67·53	7·99	68
69	68·62	7·21	68·59	7·51	68·56	7·81	68·52	8·11	69
70	69·62	7·32	69·58	7·62	69·55	7·92	69·51	8·23	70
71	70·61	7·42	70·58	7·73	70·54	8·04	70·51	8·35	71
72	71·61	7·53	71·57	7·84	71·54	8·15	71·50	8·46	72
73	72·60	7·63	72·57	7·95	72·53	8·26	72·49	8·58	73
74	73·59	7·74	73·56	8·06	73·52	8·38	73·49	8·70	74
75	74·59	7·84	74·55	8·17	74·52	8·49	74·48	8·82	75
76	75·58	7·94	75·55	8·27	75·51	8·60	75·47	8·93	76
77	76·58	8·05	76·54	8·38	76·51	8·72	76·47	9·05	77
78	77·57	8·15	77·54	8·49	77·50	8·83	77·46	9·17	78
79	78·57	8·26	78·53	8·60	78·49	8·94	78·45	9·29	79
80	79·56	8·36	79·53	8·71	79·49	9·06	79·45	9·40	80
81	80·56	8·47	80·52	8·82	80·48	9·17	80·44	9·52	81
82	81·55	8·57	81·51	8·93	81·47	9·28	81·43	9·64	82
83	82·55	8·68	82·51	9·04	82·47	9·40	82·42	9·76	83
84	83·54	8·78	83·50	9·14	83·46	9·51	83·42	9·87	84
85	84·53	8·88	84·50	9·25	84·45	9·62	84·41	9·99	85
86	85·53	8·99	85·49	9·36	85·45	9·74	85·40	10·11	86
87	86·52	9·09	86·48	9·47	86·44	9·85	86·40	10·23	87
88	87·52	9·20	87·48	9·58	87·43	9·96	87·39	10·34	88
89	88·51	9·30	88·47	9·69	88·43	10·08	88·38	10·46	89
90	89·51	9·41	89·47	9·80	89·42	10·19	89·38	10·58	90
91	90·50	9·51	90·46	9·91	90·42	10·30	90·37	10·70	91
92	91·50	9·62	91·45	10·02	91·41	10·41	91·36	10·81	92
93	92·49	9·72	92·45	10·12	92·40	10·53	92·36	10·93	93
94	93·49	9·83	93·44	10·23	93·40	10·64	93·35	11·05	94
95	94·48	9·93	94·44	10·34	94·39	10·75	94·34	11·17	95
96	95·47	10·03	95·43	10·45	95·38	10·87	95·33	11·28	96
97	96·47	10·14	96·42	10·56	96·38	10·98	96·33	11·40	97
98	97·46	10·24	97·42	10·67	97·37	11·09	97·32	11·52	98
99	98·46	10·35	98·41	10·78	98·36	11·21	98·31	11·64	99
100	99·45	10·45	99·41	10·89	99·36	11·32	99·31	11·75	100
	Dep.	Lat.	Dep.	Lat.	Dep.	Lat.	Dep.	Lat.	
Distance.	84 Deg.		83¾ Deg.		83½ Deg.		83¼ Deg.		Distance.

Distance.	7 Deg.		7¼ Deg		7½ Deg.		7¾ Deg.		Distance.
	Lat.	Dep.	Lat.	Dep.	Lat.	Dep.	Lat.	Dep.	
1	0·99	0·12	0·99	0·13	0·99	0·13	0·99	0·13	1
2	1·99	0·24	1·98	0·25	1·98	0·26	1·98	0·27	2
3	2·98	0·37	2·98	0·38	2·97	0·39	2·97	0·40	3
4	3·97	0·49	3·97	0·50	3·97	0·52	3·96	0·54	4
5	4·96	0·61	4·96	0·63	4·96	0·65	4·95	0·67	5
6	5·96	0·73	5·95	0·76	5·95	0·78	5·95	0·81	6
7	6·95	0·85	6·94	0·88	6·94	0·91	6·94	0·94	7
8	7·94	0·97	7·94	1·01	7·93	1·04	7·93	1·08	8
9	8·93	1·10	8·93	1·14	8·92	1·17	8·92	1·21	9
10	9·93	1·22	9·92	1·26	9·91	1·31	9·91	1·35	10
11	10·92	1·34	10·91	1·39	10·91	1·44	10·90	1·48	11
12	11·91	1·46	11·90	1·51	11·90	1·57	11·89	1·62	12
13	12·90	1·58	12·90	1·64	12·89	1·70	12·88	1·75	13
14	13·90	1·71	13·89	1·77	13·88	1·83	13·87	1·89	14
15	14·89	1·83	14·88	1·89	14·87	1·96	14·86	2·02	15
16	15·88	1·95	15·87	2·02	15·86	2·09	15·85	2·16	16
17	16·87	2·07	16·86	2·15	16·85	2·22	16·84	2·29	17
18	17·87	2·19	17·86	2·27	17·85	2·35	17·84	2·43	18
19	18·86	2·32	18·85	2·40	18·84	2·48	18·83	2·56	19
20	19·85	2·44	19·84	2·52	19·83	2·61	19·82	2·70	20
21	20·84	2·56	20·83	2·65	20·82	2·74	20·81	2·83	21
22	21·84	2·68	21·82	2·78	21·81	2·87	21·80	2·97	22
23	22·83	2·80	22·82	2·90	22·80	3·00	22·79	3·10	23
24	23·82	2·92	23·81	3·03	23·79	3·13	23·78	3·24	24
25	24·81	3·05	24·80	3·15	24·79	3·26	24·77	3·37	25
26	25·81	3·17	25·79	3·28	25·78	3·39	25·76	3·51	26
27	26·80	3·29	26·78	3·41	26·77	3·52	26·75	3·64	27
28	27·79	3·41	27·78	3·53	27·76	3·65	27·74	3·78	28
29	28·78	3·53	28·77	3·66	28·75	3·79	28·74	3·91	29
30	29·78	3·66	29·76	3·79	29·74	3·92	29·73	4·05	30
31	30·77	3·78	30·75	3·91	30·73	4·05	30·72	4·18	31
32	31·76	3·90	31·74	4·04	31·73	4·18	31·71	4·32	32
33	32·75	4·02	32·74	4·16	32·72	4·31	32·70	4·45	33
34	33·75	4·14	33·73	4·29	33·71	4·44	33·69	4·58	34
35	34·74	4·27	34·72	4·42	34·70	4·57	34·68	4·72	35
36	35·73	4·39	35·71	4·54	35·69	4·70	35·67	4·85	36
37	36·72	4·51	36·70	4·67	36·68	4·83	36·66	4·99	37
38	37·72	4·63	37·70	4·80	37·67	4·96	37·65	5·12	38
39	38·71	4·75	38·69	4·92	38·67	5·09	38·64	5·26	39
40	39·70	4·87	39·68	5·05	39·66	5·22	39·63	5·39	40
41	40·70	5·00	40·67	5·17	40·65	5·35	40·63	5·53	41
42	41·69	5·12	41·66	5·30	41·64	5·48	41·62	5·66	42
43	42·68	5·24	42·66	5·43	42·63	5·61	42·61	5·80	43
44	43·67	5·36	43·65	5·55	43·62	5·74	43·60	5·93	44
45	44·67	5·48	44·64	5·68	44·62	5·87	44·59	6·07	45
46	45·66	5·61	45·63	5·81	45·61	6·00	45·58	6·20	46
47	46·65	5·73	46·62	5·93	46·60	6·13	46·57	6·34	47
48	47·64	5·85	47·62	6·06	47·59	6·27	47·56	6·47	48
49	48·63	5·97	48·61	6·18	48·58	6·40	48·55	6·61	49
50	49·63	6·09	49·60	6·31	49·57	6·53	49·54	6·74	50
Distance.	Dep.	Lat.	Dep.	Lat.	Dep.	Lat.	Dep.	Lat.	Distance.
	83 Deg.		82¾ Deg.		82½ Deg.		82¼ Deg.		

Distance.	7 Deg.		7¼ Deg.		7½ Deg.		7¾ Deg.		Distance.
	Lat.	Dep.	Lat.	Dep.	Lat.	Dep.	Lat.	Dep.	
51	50·62	6·22	50·59	6·44	50·56	6·66	50·53	6·88	51
52	51·61	6·34	51·58	6·56	51·56	6·79	51·53	7·01	52
53	52·60	6·46	52·58	6·69	52·55	6·92	52·52	7·15	53
54	53·60	6·58	53·57	6·81	53·54	7·05	53·51	7·28	54
55	54·59	6·70	54·56	6·94	54·53	7·18	54·50	7·42	55
56	55·58	6·82	55·55	7·07	55·52	7·31	55·49	7·55	56
57	56·58	6·95	56·54	7·19	56·51	7·44	56·48	7·69	57
58	57·57	7·07	57·54	7·32	57·50	7·57	57·47	7·82	58
59	58·56	7·19	58·53	7·45	58·50	7·70	58·46	7·96	59
60	59·55	7·31	59·52	7·57	59·49	7·83	59·45	8·09	60
61	60·55	7·43	60·51	7·70	60·48	7·96	60·44	8·23	61
62	61·54	7·56	61·50	7·82	61·47	8·09	61·43	8·36	62
63	62·53	7·68	62·50	7·95	62·46	8·22	62·42	8·50	63
64	63·52	7·80	63·49	8·08	63·45	8·35	63·42	8·63	64
65	64·52	7·92	64·48	8·20	64·44	8·48	64·41	8·77	65
66	65·51	8·04	65·47	8·33	65·44	8·61	65·40	8·90	66
67	66·50	8·17	66·46	8·46	66·43	8·75	66·39	9·04	67
68	67·49	8·29	67·46	8·58	67·42	8·88	67·38	9·17	68
69	68·49	8·41	68·45	8·71	68·41	9·01	68·37	9·30	69
70	69·48	8·53	69·44	8·83	69·40	9·14	69·36	9·44	70
71	70·47	8·65	70·43	8·96	70·39	9·27	70·35	9·57	71
72	71·46	8·77	71·42	9·09	71·38	9·40	71·34	9·71	72
73	72·46	8·90	72·42	9·21	72·38	9·53	72·33	9·84	73
74	73·45	9·02	73·41	9·34	73·37	9·66	73·32	9·98	74
75	74·44	9·14	74·40	9·46	74·36	9·79	74·31	10·11	75
76	75·43	9·26	75·39	9·59	75·35	9·92	75·31	10·25	76
77	76·43	9·38	76·38	9·72	76·34	10·05	76·30	10·38	77
78	77·42	9·51	77·38	9·84	77·33	10·18	77·29	10·52	78
79	78·41	9·63	78·37	9·97	78·32	10·31	78·28	10·65	79
80	79·40	9·75	79·36	10·10	79·32	10·44	79·27	10·79	80
81	80·40	9·87	80·35	10·22	80·31	10·57	80·26	10·92	81
82	81·39	9·99	81·34	10·35	81·30	10·70	81·25	11·06	82
83	82·38	10·12	82·34	10·47	82·29	10·83	82·24	11·19	83
84	83·37	10·24	83·33	10·60	83·28	10·96	83·23	11·33	84
85	84·37	10·36	84·32	10·73	84·27	11·09	84·22	11·46	85
86	85·36	10·48	85·31	10·85	85·26	11·23	85·21	11·60	86
87	86·35	10·60	86·30	10·98	86·26	11·36	86·21	11·73	87
88	87·34	10·72	87·30	11·11	87·25	11·49	87·20	11·87	88
89	88·34	10·85	88·29	11·23	88·24	11·62	88·19	12.00	89
90	89·33	10·97	89·28	11·36	89·23	11·75	89·18	12·14	90
91	90·32	11·09	90·27	11·48	90·22	11·88	90·17	12·27	91
92	91·31	11·21	91·26	11·61	91·21	12·01	91·16	12·41	92
93	92·31	11·33	92·26	11·74	92·20	12·14	92·15	12·54	93
94	93·30	11·46	93·25	11·86	93·20	12·27	93·14	12·68	94
95	94·29	11·58	94·24	11·99	94·19	12·40	94·13	12·81	95
96	95·28	11·70	95·23	12·12	95·18	12·53	95·12	12·95	96
97	96·28	11·82	96·22	12·24	96·17	12·66	96·11	13·08	97
98	97·27	11·94	97·22	12·37	97·16	12·79	97·10	13·22	98
99	98·26	12·07	98·21	12·49	98·15	12·92	98·10	13·35	99
100	99·25	12·19	99·20	12·62	99·14	13·05	99.09	13·49	100
Distance.	Dep.	Lat.	Dep.	Lat.	Dep.	Lat.	Dep.	Lat.	Distance.
	83 Deg.		82¾ Deg.		82½ Deg.		82¼ Deg.		

Distance.	8 Deg.		8¼ Deg.		8½ Deg.		8¾ Deg.		Distance.
	Lat.	Dep.	Lat.	Dep.	Lat.	Dep.	Lat.	Dep.	
1	0·99	0·14	0·99	0·14	0·99	0·15	0·99	0·15	1
2	1·98	0·28	1·98	0·29	1·98	0·30	1·98	0·30	2
3	2·97	0·42	2·97	0·43	2·97	0·44	2·97	0·46	3
4	3·96	0·56	3·96	0·57	3·96	0·59	3·95	0·61	4
5	4·95	0·70	4·95	0·72	4·95	0·74	4·94	0·76	5
6	5·94	0·84	5·94	0·86	5·93	0·89	5·93	0·91	6
7	6·93	0·97	6·93	1·00	6·92	1·03	6·92	1·06	7
8	7·92	1·11	7·92	1·15	7·91	1·18	7·91	1·22	8
9	8·91	1·25	8·91	1·29	8·90	1·33	8·90	1·37	9
10	9.90	1·39	9·90	1·43	9·89	1·48	9·88	1·52	10
11	10·89	1·53	10·89	1·58	10·88	1·63	10·87	1·67	11
12	11·88	1·67	11·88	1·72	11·87	1·77	11·86	1·83	12
13	12·87	1·81	12·87	1·87	12·86	1·92	12·85	1·98	13
14	13·86	1·95	13·86	2·01	13·85	2·07	13·84	2·13	14
15	14·85	2·09	14·85	2·15	14·84	2·22	14·83	2·28	15
16	15·84	2·23	15·84	2·30	15·82	2·36	15·81	2·43	16
17	16·83	2·37	16·83	2·44	16·81	2·51	16·80	2·59	17
18	17·82	2·51	17·81	2·58	17·80	2·66	17·79	2·74	18
19	18·82	2·64	18·80	2·73	18·79	2·81	18·78	2·89	19
20	19·81	2·78	19·79	2·87	19·78	2·96	19·77	3·04	20
21	20·80	2·92	20·78	3·01	20·77	3·10	20·76	3·19	21
22	21·79	3·06	21·77	3·16	21·76	3·25	21·74	3·35	22
23	22·78	3·20	22·76	3·30	22·75	3·40	22·73	3·50	23
24	23·77	3·34	23·75	3·44	23·74	3·55	23·72	3·65	24
25	24·76	3·48	24·74	3·59	24·73	3·70	24·71	3·80	25
26	25·75	3·62	25·73	3·73	25·71	3·84	25·70	3·96	26
27	26·74	3·76	26·72	3·87	26·70	3·99	26·69	4·11	27
28	27·73	3·90	27·71	4·02	27·69	4·14	27·67	4·26	28
29	28·72	4·04	28·70	4·16	28·68	4·29	28·66	4·41	29
30	29·71	4·18	29·69	4·30	29·67	4·43	29·65	4·56	30
31	30·70	4·31	30·68	4·45	30·66	4·58	30·64	4·72	31
32	31·69	4·45	31·67	4·59	31·65	4·73	31·63	4·87	32
33	32·68	4·59	32·66	4·74	32·64	4·88	32·62	5·02	33
34	33·67	4·73	33·65	4·88	33·63	5·03	33·60	5·17	34
35	34·66	4·87	34·64	5·02	34·62	5·17	34·59	5·32	35
36	35·65	5·01	35·63	5·17	35·60	5·32	35·58	5·48	36
37	36·64	5·15	36·62	5·31	36·59	5·47	36·57	5·63	37
38	37·63	5·29	37·61	5·45	37·58	5·62	37·56	5·78	38
39	38·62	5·43	38·60	5·60	38·57	5·76	38·55	5·93	39
40	39·61	5·57	39·59	5·74	39·56	5·91	39·53	6·08	40
41	40·60	5·71	40·58	5·88	40·55	6·06	40·52	6·24	41
42	41·59	5·85	41·57	6·03	41·54	6·21	41·51	6·39	42
43	42·58	5·98	42·56	6·17	42·53	6·36	42·50	6·54	43
44	43·57	6·12	43·54	6·31	43·52	6·50	43·49	6·69	44
45	44·56	6·26	44·53	6·46	44·51	6·65	44·48	6·85	45
46	45·55	6·40	45·52	6·60	45·49	6·80	45·46	7·00	46
47	46·54	6·54	46·51	6·74	46·48	6·95	46·45	7·15	47
48	47·53	6·68	47·50	6·89	47·47	7·09	47·44	7·30	48
49	48·52	6·82	48·49	7·03	48·46	7·24	48·43	7·45	49
50	49·51	6·96	49·48	7·17	49·45	7·39	49·42	7·61	50
Distance.	Dep.	Lat.	Dep.	Lat.	Dep.	Lat.	Dep.	Lat.	Distance.
	82 Deg.		81¾ Deg.		81½ Deg.		81¼ Deg.		

Distance.	8 Deg.		8¼ Deg.		8½ Deg.		8¾ Deg.		Distance.
	Lat.	Dep.	Lat.	Dep.	Lat.	Dep.	Lat.	Dep.	
51	50·50	7·10	50·47	7·32	50·44	7·54	50·41	7·76	51
52	51·49	7·24	51·46	7·46	51·43	7·69	51·39	7·91	52
53	52·48	7·38	52·45	7·61	52·42	7·83	52·38	8·06	53
54	53·47	7·52	53·44	7·75	53·41	7·98	53·37	8·21	54
55	54·46	7·65	54·43	7·89	54·40	8·13	54·36	8·37	55
56	55·46	7·79	55·42	8·04	55·38	8·28	55·35	8·52	56
57	56·45	7·93	56·41	8·18	56·37	8·43	56·34	8·67	57
58	57·44	8·07	57·40	8·32	57·36	8·57	57·32	8·82	58
59	58·43	8·21	58·39	8·47	58·35	8·72	58·31	8·98	59
60	59·42	8·35	59·38	8·61	59·34	8·87	59·30	9·13	60
61	60·41	8·49	60·37	8·75	60·33	9·02	60·29	9·28	61
62	61·40	8·63	61·36	8·90	61·32	9·16	61·28	9·43	62
63	62·39	8·77	62·35	9·04	62·31	9·31	62·27	9·58	63
64	63·38	8·91	63·34	9·18	63·30	9·46	63·26	9·74	64
65	64·37	9·05	64·33	9·33	64·29	9·61	64·24	9·89	65
66	65·36	9·19	65·32	9·47	65·28	9·76	65·23	10·04	66
67	66·35	9·32	66·31	9·61	66·26	9·90	66·22	10·19	67
68	67·34	9·46	67·30	9·76	67·25	10·05	67·21	10·34	68
69	68·33	9·60	68·29	9·90	68·24	10·20	68·20	10·50	69
70	69·32	9·74	69·28	10·04	69·23	10·35	69·19	10·65	70
71	70·31	9·88	70·27	10·19	70·22	10·49	70·17	10·80	71
72	71·30	10·02	71·25	10·33	71·21	10·64	71·16	10·95	72
73	72·29	10·16	72·24	10·47	72·20	10·79	72·15	11·10	73
74	73·28	10·30	73·23	10·62	73·19	10·94	73·14	11·26	74
75	74·27	10·44	74·22	10·76	74·18	11·09	74·13	11·41	75
76	75·26	10·58	75·21	10·91	75·17	11·23	75·12	11·56	76
77	76·25	10·72	76·20	11·05	76·15	11·38	76·10	11·71	77
78	77·24	10·86	77·19	11·19	77·14	11·53	77·09	11·87	78
79	78·23	10·99	78·18	11·34	78·13	11·68	78·08	12·02	79
80	79·22	11·13	79·17	11·48	79·12	11·82	79·07	12·17	80
81	80·21	11·27	80·16	11·62	80·11	11·97	80·06	12·32	81
82	81·20	11·41	81·15	11·77	81·10	12·12	81·05	12·47	82
83	82·19	11·55	82·14	11·91	82·09	12·27	82·03	12·63	83
84	83·18	11·69	83·13	12·05	83·08	12·42	83·02	12·78	84
85	84·17	11·83	84·12	12·20	84·07	12·56	84·01	12·93	85
86	85·16	11·97	85·11	12·34	85·06	12·71	85·00	13·08	86
87	86·15	12·11	86·10	12·48	86·04	12·86	85·99	13·23	87
88	87·14	12·25	87·09	12·63	87·03	13·01	86·98	13·39	88
89	88·13	12·39	88·08	12·77	88·02	13·16	87·96	13·54	89
90	89·12	12·53	89·07	12·91	89·01	13·30	88·95	13·69	90
91	90·11	12·66	90·06	13·06	90·00	13·45	89·94	13·84	91
92	91·10	12·80	91·05	13·20	90·99	13·60	90·93	14·00	92
93	92·09	12·94	92·04	13·34	91·98	13·75	91·92	14·15	93
94	93·09	13·08	93·03	13·49	92·97	13·89	92·91	14·30	94
95	94·08	13·22	94·02	13·63	93·96	14·04	93·89	14·45	95
96	95·07	13·36	95·01	13·78	94·95	14·19	94·88	14·60	96
97	96·06	13·50	96·00	13·92	95·93	14·34	95·87	14·76	97
98	97·05	13·64	96·99	14·06	96·92	14·49	96·86	14·91	98
99	98·04	13·78	97·98	14·21	97·91	14·63	97·85	15·06	99
100	99·03	13·92	98·97	14·35	98·90	14·78	98·84	15·21	100
Distance.	Dep.	Lat.	Dep.	Lat.	Dep.	Lat.	Dep.	Lat.	Distance.
	82 Deg.		81¾ Deg.		81½ Deg.		81¼ Deg.		

Distance.	9 Deg.		9¼ Deg.		9½ Deg.		9¾ Deg.		Distance.
	Lat.	Dep.	Lat.	Dep.	Lat.	Dep.	Lat.	Dep.	
1	0·99	0·16	0·99	0·16	0·99	0·17	0·99	0·17	1
2	1·98	0·31	1·97	0·32	1·97	0·33	1·97	0·34	2
3	2·96	0·47	2·96	0·48	2·96	0·50	2·96	0·51	3
4	3·95	0·63	3·95	0·64	3·95	0·66	3·94	0·68	4
5	4·94	0·78	4·93	0·80	4·93	0·83	4·93	0·85	5
6	5·93	0·94	5·92	0·96	5·92	0·99	5·91	1·02	6
7	6·91	1·10	6·91	1·13	6·90	1·16	6·90	1·19	7
8	7·90	1·25	7·90	1·29	7·89	1·32	7·88	1·35	8
9	8·89	1·41	8·88	1·45	8·88	1·49	8·87	1·52	9
10	9·88	1·56	9·87	1·61	9·86	1·65	9·86	1·69	10
11	10·86	1·72	10·86	1·77	10·85	1·82	10·84	1·86	11
12	11·85	1·88	11·84	1·93	11·84	1·98	11·83	2·03	12
13	12·84	2·03	12·83	2·09	12·82	2·15	12·81	2·20	13
14	13·83	2·19	13·82	2·25	13·81	2·31	13·80	2·37	14
15	14·82	2·35	14·80	2·41	14·79	2·48	14·78	2·54	15
16	15·80	2·50	15·79	2·57	15·78	2·64	15·77	2·71	16
17	16·79	2·66	16·78	2·73	16·77	2·81	16·75	2·88	17
18	17·78	2·82	17·77	2·89	17·75	2·97	17·74	3·05	18
19	18·77	2·97	18·75	3·05	18·74	3·14	18·73	3·22	19
20	19·75	3·13	19·74	3·21	19·73	3·30	19·71	3·39	20
21	20·74	3·29	20·73	3·38	20·71	3·47	20·70	3·56	21
22	21·73	3·44	21·71	3·54	21·70	3·63	21·68	3·73	22
23	22·72	3·60	22·70	3·70	22·68	3·80	22·67	3·90	23
24	23·70	3·75	23·69	3·86	23·67	3·96	23·65	4·06	24
25	24·69	3·91	24·67	4·02	24·66	4·13	24·64	4·23	25
26	25·68	4·07	25·66	4·18	25·64	4·29	25·62	4·40	26
27	26·67	4·22	26·65	4·34	26·63	4·46	26·61	4·57	27
28	27·66	4·38	27·64	4·50	27·62	4·62	27·60	4·74	28
29	28·64	4·54	28·62	4·66	28·60	4·79	28·58	4·91	29
30	29·63	4·69	29·61	4·82	29·59	4·95	29·57	5·08	30
31	30·62	4·85	30·60	4·98	30·57	5·12	30·55	5·25	31
32	31·61	5·01	31·58	5·14	31·56	5·28	31·54	5·42	32
33	32·59	5·16	32·57	5·30	32·55	5·45	32·52	5·59	33
34	33·58	5·32	33·56	5·47	33·53	5·61	33·51	5·76	34
35	34·57	5·48	34·54	5·63	34·52	5·78	34·49	5·93	35
36	35·56	5·63	35·53	5·79	35·51	5·94	35·48	6·10	36
37	36·54	5·79	36·52	5·95	36·49	6·11	36·47	6·27	37
38	37·53	5·94	37·51	6·11	37·48	6·27	37·45	6·44	38
39	38·52	6·10	38·49	6·27	38·47	6·44	38·44	6·60	39
40	39·51	6·26	39·48	6·43	39·45	6·60	39·42	6·77	40
41	40·50	6·41	40·47	6·59	40·44	6·77	40·41	6·94	41
42	41·48	6·57	41·45	6·75	41·42	6·92	41·39	7·11	42
43	42·47	6·73	42·44	6·91	42·41	7·10	42·38	7·28	43
44	43·46	6·88	43·43	7·07	43·40	7·26	43·36	7·45	44
45	44·45	7·04	44·41	7·23	44·38	7·43	44·35	7·62	45
46	45·43	7·20	45·40	7·39	45·37	7·59	45·34	7·79	46
47	46·42	7·35	46·39	7·55	46·36	7·76	46·32	7·96	47
48	47·41	7·51	47·38	7·72	47·34	7·92	47·31	8·13	48
49	48·40	7·67	48·36	7·88	48·33	8·09	48·29	8·30	49
50	49·38	7·82	49·35	8·04	49·32	8·25	49·28	8·47	50
	Dep.	Lat.	Dep.	Lat.	Dep.	Lat.	Dep.	Lat.	
Distance.	81 Deg.		80¾ Deg.		80½ Deg.		80¼ Deg.		Distance.

Distance.	9 Deg.		9¼ Deg.		9½ Deg.		9¾ Deg.		Distance.
	Lat.	Dep.	Lat.	Dep.	Lat.	Dep.	Lat.	Dep.	
51	50·37	7·98	50·34	8·20	50·30	8·42	50·26	8·64	51
52	51·36	8·13	51·32	8·36	51·29	8·58	51·25	8·81	52
53	52·35	8·29	52·31	8·52	52·27	8·75	52·23	8·98	53
54	53·34	8·45	53·30	8·68	53·26	8·91	53·22	9·14	54
55	54·32	8·60	54·28	8·84	54·25	9·08	54·21	9·31	55
56	55·31	8·76	55·27	9·00	55·23	9·24	55·19	9·48	56
57	56·30	8·92	56·26	9·16	56·22	9·41	56·18	9·65	57
58	57·29	9·07	57·25	9·32	57·20	9·57	57·16	9·82	58
59	58·27	9·23	58·23	9·48	58·19	9·74	58·15	9·99	59
60	59·26	9·39	59·22	9·64	59·18	9·90	59·13	10·16	60
61	60·25	9·54	60·21	9·81	60·16	10·07	60·12	10·33	61
62	61·24	9·70	61·19	9·97	61·15	10·23	61·10	10·50	62
63	62·22	9·86	62·18	10·13	62·14	10·40	62·09	10·67	63
64	63·21	10·01	63·17	10·29	63·12	10·56	63·08	10·84	64
65	64·20	10·17	64·15	10·45	64·11	10·73	64·06	11·01	65
66	65·19	10·32	65·14	10·61	65·09	10·89	65·05	11·18	66
67	66·18	10·48	66·13	10·77	66·08	11·06	66·03	11·35	67
68	67·16	10·64	67·12	10·93	67·07	11·22	67·02	11·52	68
69	68·15	10·79	68·10	11·09	68·05	11·39	68·00	11·69	69
70	69·14	10·95	69·09	11·25	69·04	11·55	68·99	11·85	70
71	70·13	11·11	70·08	11·41	70·03	11·72	69·97	12·02	71
72	71·11	11·26	71·06	11·57	71·01	11·88	70·96	12·19	72
73	72·10	11·42	72·05	11·73	72·00	12·05	71·95	12·36	73
74	73·09	11·58	73·04	11·89	72·99	12·21	72·93	12·53	74
75	74·08	11·73	74·02	12·06	73·97	12·38	73·92	12·70	75
76	75·06	11·89	75·01	12·22	74·96	12·54	74·90	12·87	76
77	76·05	12·05	76·00	12·38	75·94	12·71	75·89	13·04	77
78	77·04	12·20	76·99	12·54	76·93	12·87	76·87	13·21	78
79	78·03	12·36	77·97	12·70	77·92	13·04	77·86	13·38	79
80	79·02	12·51	78·96	12·86	78·90	13·20	78·84	13·55	80
81	80·00	12·67	79·95	13·02	79·89	13·37	79·83	13·72	81
82	80·99	12·83	80·93	13·18	80·88	13·53	80·82	13·89	82
83	81·98	12·98	81·92	13·34	81·86	13·70	81·80	14·06	83
84	82·97	13·14	82·91	13·50	82·85	13·86	82·79	14·23	84
85	83·95	13·30	83·89	13·66	83·83	14·03	83·77	14·39	85
86	84·94	13·45	84·88	13·82	84·82	14·19	84·76	14·56	86
87	85·93	13·61	85·87	13·98	85·81	14·36	85·74	14·73	87
88	86·92	13·77	86·86	14·15	86·79	14·52	86·73	14·90	88
89	87·90	13·92	87·84	14·31	87·78	14·69	87·71	15·07	89
90	88·89	14·08	88·83	14·47	88·77	14·85	88·70	15·24	90
91	89·88	14·24	89·82	14·63	89·75	15·02	89·69	15·41	91
92	90·87	14·39	90·80	14·79	90·74	15·18	90·67	15·58	92
93	91·86	14·55	91·79	14·95	91·72	15·35	91·66	15·75	93
94	92·84	14·70	92·78	15·11	92·71	15·51	92·64	15·92	94
95	93·83	14·86	93·76	15·27	93·70	15·68	93·63	16·09	95
96	94·82	15·02	94·75	15·43	94·68	15·84	94·61	16·26	96
97	95·81	15·17	95·74	15·59	95·67	16·01	95·60	16·43	97
98	96·79	15·33	96·73	15·75	96·66	16·17	93·58	16·60	98
99	97·78	15·49	97·71	15·91	97·64	16·34	97·57	16·77	99
100	98·77	15·64	98·70	16·07	98·63	16·50	98·56	16·93	100
	Dep.	Lat.	Dep.	Lat.	Dep.	Lat.	Dep.	Lat.	
Distance.	81 Deg.		80¾ Deg.		80½ Deg.		80¼ Deg.		Distance.

Distance.	10 Deg.		10¼ Deg.		10½ Deg.		10¾ Deg.		Distance.
	Lat.	Dep.	Lat.	Dep.	Lat.	Dep.	Lat.	Dep.	
1	0·98	0·17	0·98	0·18	0·98	0·18	0·98	0·19	1
2	1·97	0·35	1·97	0·36	1·97	0·36	1·96	0·37	2
3	2·95	0·52	2·95	0·53	2·95	0·55	2·95	0·56	3
4	3·94	0·69	3·94	0·71	3·93	0·73	3·93	0·75	4
5	4·92	0·87	4·92	0·89	4·92	0·91	4·91	0·93	5
6	5·91	1·04	5·90	1·07	5·90	1·09	5·89	1·12	6
7	6·89	1·22	6·89	1·25	6·88	1·28	6·88	1·31	7
8	7·88	1·39	7·87	1·42	7·87	1·46	7·86	1·49	8
9	8·86	1·56	8·86	1·60	8·85	1·64	8·84	1·68	9
10	9·85	1·74	9·84	1·78	9·83	1·82	9·82	1·87	10
11	10·83	1·91	10·82	1·96	10·82	2·00	10·81	2·05	11
12	11·82	2·08	11·81	2·14	11·80	2·19	11·79	2·24	12
13	12·80	2·26	12·79	2·31	12·78	2·37	12·77	2·42	13
14	13·79	2·43	13·78	2·49	13·77	2·55	13·75	2·61	14
15	14·77	2·60	14·76	2·67	14·75	2·73	14·74	2·80	15
16	15·76	2·78	15·74	2·85	15·73	2·92	15·72	2·98	16
17	16·74	2·95	16·73	3·03	16·72	3·10	16·70	3·17	17
18	17·73	3·13	17·71	3·20	17·70	3·28	17·68	3·36	18
19	18·71	3·30	18·70	3·38	18·68	3·46	18·67	3·54	19
20	19·70	3·47	19·68	3·56	19·67	3·64	19·65	3·73	20
21	20·68	3·65	20·66	3·74	20·65	3·83	20·63	3·92	21
22	21·67	3·82	21·65	3·91	21·63	4·01	21·61	4·10	22
23	22·65	3·99	22·63	4·09	22·61	4·19	22·60	4·29	23
24	23·64	4·17	23·62	4·27	23·60	4·37	23·58	4·48	24
25	24·62	4·34	24·60	4·45	24·58	4·56	24·56	4·66	25
26	25·61	4·51	25·59	4·63	25·56	4·74	25·54	4·85	26
27	26·59	4·69	26·57	4·80	26·55	4·92	26·53	5·04	27
28	27·57	4·86	27·55	4·98	27·53	5·10	27·51	5·22	28
29	28·56	5·04	28·54	5·16	28·51	5·28	28·49	5·41	29
30	29·54	5·21	29·52	5·34	29·50	5·47	29·47	5·60	30
31	30·53	5·38	30·51	5·52	30·48	5·65	30·46	5·78	31
32	31·51	5·56	31·49	5·69	31·46	5·83	31·44	5·97	32
33	32·50	5·73	32·47	5·87	32·45	6·01	32·42	6·16	33
34	33·48	5·90	33·46	6·05	33·43	6·20	33·40	6·34	34
35	34·47	6·08	34·44	6·23	34·41	6·38	34·39	6·53	35
36	35·45	6·25	35·43	6·41	35·40	6·56	35·37	6·71	36
37	36·44	6·42	36·41	6·58	36·38	6·74	36·35	6·90	37
38	37·42	6·60	37·39	6·76	37·36	6·92	37·33	7·09	38
39	38·41	6·77	38·38	6·94	38·35	7·11	38·32	7·27	39
40	39·39	6·95	39·36	7·12	39·33	7·29	39·30	7·46	40
41	40·38	7·12	40·35	7·30	40·31	7·47	40·28	7·65	41
42	41·36	7·29	41·33	7·47	41·30	7·65	41·26	7·83	42
43	42·35	7·47	42·31	7·65	42·28	7·84	42·25	8·02	43
44	43·33	7·64	43·30	7·83	43·26	8·02	43·23	8·21	44
45	44·32	7·81	44·28	8·01	44·25	8·20	44·21	8·39	45
46	45·30	7·99	45·27	8·19	45·23	8·38	45·19	8·58	46
47	46·29	8·16	46·25	8·36	46·21	8·57	46·18	8·77	47
48	47·27	8·34	47·23	8·54	47·20	8·75	47·16	8·95	48
49	48·26	8·51	48·22	8·72	48·18	8·93	48·14	9·14	49
50	49·24	8·68	49·20	8·90	49·16	9·11	49·12	9·33	50
	Dep.	Lat.	Dep.	Lat.	Dep.	Lat.	Dep.	Lat.	
Distance.	80 Deg.		79¾ Deg.		79½ Deg.		79¼ Deg.		Distance.

Distance.	10 Deg.		10¼ Deg.		10½ Deg.		10¾ Deg.		Distance.
	Lat.	Dep.	Lat.	Dep.	Lat.	Dep.	Lat.	Dep.	
51	50·23	8·86	50·19	9·08	50·15	9·29	50·10	9·51	51
52	51·21	9·03	51·17	9·25	51·13	9·48	51·09	9·70	52
53	52·19	9.20	52·15	9·43	52·11	9·66	52·07	9·89	53
54	53·18	9.38	53·14	9·61	53·10	9·84	53·05	10·07	54
55	54·16	9·55	54·12	9·79	54·08	10·02	54·03	10·26	55
56	55·15	9·72	55·11	9·96	55·06	10·21	55·02	10·45	56
57	56·13	9·90	56·09	10·14	56·05	10·39	56·00	10·63	57
58	57·12	10·07	57·07	10·32	57·03	10.57	56·98	10·82	58
59	58·10	10·25	58·06	10·50	58·01	10·75	57·96	11·00	59
60	59·09	10·42	59·04	10·68	59·00	10·93	58·95	11·19	60
61	60·07	10·59	60·03	10·85	59·98	11·12	59·93	11·38	61
62	61·06	10·77	61·01	11·03	60·96	11·30	60·91	11·56	62
63	62·04	10·94	61·99	11·21	61·95	11·48	61·89	11·75	63
64	63·03	11·11	62·98	11·39	62·93	11·66	62·88	11·94	64
65	64·01	11·29	63·96	11·57	63·91	11·85	63·86	12·12	65
66	65·00	11·46	64·95	11·74	64·89	12·03	64·84	12·31	66
67	65·98	11·63	65·93	11·92	65·88	12·21	65·82	12·50	67
68	66·97	11·81	66·91	12·10	66·86	12·39	66·81	12·68	68
69	67·95	11 98	67·90	12·28	67·84	12·57	67·79	12·87	69
70	68·94	12·16	68·88	12·46	68·83	12·76	68·77	13·06	70
71	69·92	12·33	69·87	12·63	69·81	12·94	69·75	13·24	71
72	70·91	12·50	70·85	12·81	70·79	13·12	70·74	13·43	72
73	71·89	12·68	71·83	12·99	71·78	13·30	71·72	13·62	73
74	72·88	12·85	72·82	13·17	72·76	13·49	72·70	13·80	74
75	73·86	13·02	73·80	13·35	73·74	13·67	73·68	13·99	75
76	74·85	13·20	74·79	13·52	74·73	13·85	74·67	14·18	76
77	75·83	13·37	75·77	13·70	75·71	14·03	75·65	14·36	77
78	76·82	13·54	76·76	13·88	76·69	14·21	76·63	14·55	78
79	77·80	13·72	77·74	14·06	77·68	14·40	77·61	14·74	79
80	78·78	13·89	78·72	14·24	78·66	14·58	78·60	14·92	80
81	79·77	14·07	79·71	14·41	79·64	14·76	79·58	15·11	81
82	80·75	14·24	80·69	14·59	80·63	14·94	80·56	15·29	82
83	81·74	14·41	81·68	14·77	81·61	15·13	81·54	15·48	83
84	82·72	14·59	82·66	14·95	82·59	15·31	82·53	15·67	84
85	83·71	14·76	83·64	15·13	83·58	15·49	83 51	15·85	85
86	84·69	14·93	84·63	15·30	84·56	15·67	84·49	16·04	86
87	85·68	15·11	85·61	15·48	85·54	15·85	85·47	16·23	87
88	86·66	15·28	86·60	15·66	86·53	16·04	86·46	16·41	88
89	87·65	15·45	87·58	15·84	87·51	16·22	87·44	16·60	89
90	88·63	15·63	88·56	16·01	88·49	16·40	88·42	16·79	90
91	89·62	15·80	89·55	16·19	89·48	16·58	89·40	16·97	91
92	90·60	15·98	90·53	16·37	90·46	16·77	90·39	17·16	92
93	91·59	16·15	91·52	16·55	91·44	16·95	91·37	17·35	93
94	92·57	16·32	92·50	16·73	92·43	17·13	92·35	17·53	94
95	93·56	16·50	93·48	16·90	93·41	17·31	93·33	17·72	95
96	94·54	16·67	94·47	17·08	94·39	17·49	94·32	17·91	96
97	95 53	16·84	95·45	17·26	95·38	17·68	95·30	18·09	97
98	96·51	17·02	96·44	17·44	96·36	17·86	96·28	18·28	98
99	97·50	17·19	97·42	17·62	97·34	18·04	97·26	18·47	99
100	98·48	17·36	98·40	17·79	98·33	18·22	98·25	18·65	100
	Dep.	Lat.	Dep.	Lat.	Dep.	Lat.	Dep.	Lat.	
Distance.	80 Deg.		79¾ Deg.		79½ Deg.		79¼ Deg.		Distance.

Distance.	11 Deg.		11¼ Deg.		11½ Deg.		11¾ Deg.		Distance.
	Lat.	Dep.	Lat.	Dep.	Lat.	Dep.	Lat.	Dep.	
1	0·98	0·19	0·98	0·20	0·98	0·20	0·98	0·20	1
2	1·96	0·38	1·96	0·39	1·96	0·40	1·96	0·41	2
3	2·94	0·57	2·94	0·59	2·94	0·60	2·94	0·61	3
4	3·93	0·76	3·92	0·78	3·92	0·80	3·92	0·82	4
5	4·91	0·95	4·90	0·98	4·90	1·00	4·90	1·02	5
6	5·89	1·14	5·88	1·17	5·88	1·20	5·87	1·22	6
7	6·87	1·34	6·87	1·37	6·86	1·40	6·85	1·43	7
8	7·85	1·53	7·85	1·56	7·84	1·59	7·83	1·63	8
9	8·83	1·72	8·83	1·76	8·82	1·79	8·81	1·83	9
10	9·82	1·91	9·81	1·95	9·80	1·99	9·79	2·04	10
11	10·80	2·10	10·79	2·15	10·78	2·19	10·77	2·24	11
12	11·78	2·29	11·77	2·34	11·76	2·39	11·75	2·44	12
13	12·76	2·48	12·75	2·54	12·74	2·59	12·73	2·65	13
14	13·74	2·67	13·73	2·73	13·72	2·79	13·71	2·85	14
15	14·72	2·86	14·71	2·93	14·70	2·99	14·69	3·06	15
16	15·71	3·05	15·69	3·12	15·68	3·19	15·66	3·26	16
17	16·69	3·24	16·67	3·32	16·66	3·39	16·64	3·46	17
18	17·67	3·43	17·65	3·51	17·64	3·59	17·62	3·66	18
19	18·65	3·63	18·63	3·71	18·62	3·79	18·60	3·87	19
20	19·63	3·82	19·62	3·90	19·60	3·99	19·58	4·07	20
21	20·61	4·01	20·60	4·10	20·58	4·19	20·56	4·28	21
22	21·60	4·20	21·58	4·29	21·56	4·39	21·54	4·48	22
23	22·58	4·39	22·56	4·49	22·54	4·59	22·52	4·68	23
24	23·56	4·58	23·54	4·68	23·52	4·78	23·50	4·89	24
25	24·54	4·77	24·52	4·88	24·50	4·98	24·48	5·09	25
26	25·52	4·96	25·50	5·07	25·48	5·18	25·46	5·30	26
27	26·50	5·15	26·48	5·27	26·46	5·38	26·43	5·50	27
28	27·49	5·34	27·46	5·46	27·44	5·58	27·41	5·70	28
29	28·47	5·53	28·44	5·66	28·42	5·78	28·39	5·91	29
30	29·45	5·72	29·42	5·85	29·40	5·98	29·37	6·11	30
31	30·43	5·92	30·40	6·05	30·38	6·18	30·35	6·31	31
32	31·41	6·11	31·39	6·24	31·36	6·38	31·33	6·52	32
33	32·39	6·30	32·37	6·44	32·34	6·58	32·31	6·72	33
34	33·38	6·49	33·35	6·63	33·32	6·78	33·29	6·92	34
35	34·36	6·68	34·33	6·83	34·30	6·98	34·27	7·13	35
36	35·34	6·87	35·31	7·02	35·28	7·18	35·25	7·33	36
37	36·32	7·06	36·29	7·22	36·26	7·38	36·22	7·53	37
38	37·30	7·25	37·27	7·41	37·24	7·58	37·20	7·74	38
39	38·28	7·44	38·25	7·61	38·22	7·78	38·18	7·94	39
40	39·27	7·63	39·23	7·80	39·20	7·97	39·16	8·15	40
41	40·25	7·82	40·21	8·00	40·18	8·17	40·14	8·35	41
42	41·23	8·01	41·19	8·19	41·16	8·37	41·12	8·55	42
43	42·21	8·20	42·17	8·39	42·14	8·57	42·10	8·76	43
44	43·19	8·40	43·15	8·58	43·12	8·77	43·08	8·96	44
45	44·17	8·59	44·14	8·78	44·10	8·97	44·06	9·16	45
46	45·15	8·78	45·12	8·97	45·08	9·17	45·04	9·37	46
47	46·14	8·97	46·10	9·17	46·06	9·37	46·02	9·57	47
48	47·12	9·16	47·08	9·36	47·04	9·57	46·99	9·78	48
49	48·10	9·35	48·06	9·56	48·02	9·77	47·97	9·98	49
50	49·08	9·54	49·04	9·75	49·00	9·97	48·95	10·18	50
Distance.	Dep.	Lat.	Dep.	Lat.	Dep.	Lat.	Dep.	Lat.	Distance.
	79 Deg.		78¾ Deg.		78½ Deg.		78¼ Deg.		

Distance.	11 Deg.		11¼ Deg.		11½ Deg.		11¾ Deg.		Distance.
	Lat.	Dep.	Lat.	Dep.	Lat.	Dep.	Lat.	Dep.	
51	50·06	9·73	50·02	9·95	49·98	10·17	49·93	10·39	51
52	51·04	9·92	51·00	10·14	50·96	10·37	50·91	10·59	52
53	52·03	10·11	51·98	10·34	51·94	10·57	51·89	10·79	53
54	53·01	10·30	52·96	10·53	52·92	10·77	52·87	11·00	54
55	53·99	10·49	53·94	10·73	53·90	10·97	53·85	11·20	55
56	54·97	10·69	54·92	10·93	54·88	11·16	54·83	11·40	56
57	55·95	10·88	55·90	11·12	55·86	11·36	55·81	11·61	57
58	56·93	11·07	56·89	11·32	56·84	11·56	56·78	11·81	58
59	57·92	11·26	57·87	11·51	57·82	11·76	57·76	12·01	59
60	58·90	11·45	58·85	11·71	58·80	11·96	58·74	12·22	60
61	59·88	11·64	59·83	11·90	59·78	12·16	59·72	12·42	61
62	60·86	11·83	60·81	12·10	60·76	12·36	60·70	12·63	62
63	61·84	12·02	61·79	12·29	61·74	12·56	61·68	12·83	63
64	62·82	12·21	62·77	12·49	62·72	12·76	62·66	13·03	64
65	63·81	12·40	63·75	12·68	63·70	12·96	63·64	13·24	65
66	64·79	12·59	64·73	12·88	64·68	13·16	64·62	13·44	66
67	65·77	12·78	65·71	13·07	65·66	13·36	65·60	13·64	67
68	66·75	12·98	66·69	13·27	66·63	13·56	66·58	13·85	68
69	67·73	13·17	67·67	13·46	67·61	13·76	67·55	14·05	69
70	68·71	13·36	68·66	13·66	68·59	13·96	68·53	14·25	70
71	69·70	13·55	69·64	13·85	69·57	14·16	69·51	14·46	71
72	70·68	13·74	70·62	14·05	70·55	14·35	70·49	14·66	72
73	71·66	13·93	71·60	14·24	71·53	14·55	71·47	14·87	73
74	72·64	14·12	72·58	14·44	72·51	14·75	72·45	15·07	74
75	73·62	14·31	73·56	14·63	73·49	14·95	73·43	15·27	75
76	74·60	14·50	74·54	14·83	74·47	15·15	74·41	15·48	76
77	75·59	14·69	75·52	15·02	75·45	15·35	75·39	15·68	77
78	76·57	14·88	76·50	15·22	76·43	15·55	76·37	15·88	78
79	77·55	15·07	77·48	15·41	77·41	15·75	77·34	16·09	79
80	78·53	15·26	78·46	15·61	78·39	15·95	78·32	16·29	80
81	79·51	15·46	79·44	15·80	79·37	16·15	79·30	16·49	81
82	80·49	15·65	80·42	16·00	80·35	16·35	80·28	16·70	82
83	81·48	15·84	81·41	16·19	81·33	16·55	81·26	16·90	83
84	82·46	16·03	82·39	16·39	82·31	16·75	82·24	17·11	84
85	83·44	16·22	83·37	16·58	83·29	16·95	83·22	17·31	85
86	84·42	16·41	84·35	16·78	84·27	17·15	84·20	17·51	86
87	85·40	16·60	85·33	16·97	85·25	17·35	85·18	17·72	87
88	86·38	16·79	86·31	17·17	86·23	17·54	86·16	17·92	88
89	87·36	16·98	87·29	17·36	87·21	17·74	87·14	18·12	89
90	88·35	17·17	88·27	17·56	88·19	17·94	88·11	18·33	90
91	89·33	17·36	89·25	17·75	89·17	18·14	89·09	18·53	91
92	90·31	17·55	90·23	17·95	90·15	18·34	90·07	18·74	92
93	91·29	17·75	91·21	18·14	91·13	18·54	91·05	18·94	93
94	92·27	17·94	92·19	18·34	92·11	18·74	92·03	19·14	94
95	93·25	18·13	93·17	18·53	93·09	18·94	93·01	19·35	95
96	94·24	18·32	94·16	18·73	94·07	19·14	93·99	19·55	96
97	95·22	18·51	95·14	18·92	95·05	19·34	94·97	19·75	97
98	96·20	18·70	96·12	19·12	96·03	19·54	95·95	19·96	98
99	97·18	18·89	97·10	19·31	97·01	19·74	96·93	20·16	99
100	98·16	19·08	98·08	19·51	97·99	19·94	97·90	20·36	100
Distance.	Dep.	Lat.	Dep.	Lat.	Dep.	Lat.	Dep.	Lat.	Distance.
	79 Deg.		78¾ Deg.		78½ Deg.		78¼ Deg.		

Distance.	12 Deg.		12¼ Deg.		12½ Deg.		12¾ Deg.		Distance.
	Lat.	Dep.	Lat.	Dep.	Lat.	Dep.	Lat.	Dep.	
1	0·98	0·21	0·98	0·21	0·98	0·22	0·98	0·22	1
2	1·96	0·42	1·95	0·42	1·95	0·43	1·95	0·44	2
3	2·93	0·62	2·93	0·64	2·93	0·65	2·93	0·66	3
4	3·91	0·83	3·91	0·85	3·91	0·87	3·90	0·88	4
5	4·89	1·04	4·89	1·06	4·88	1·08	4·88	1·10	5
6	5·87	1·25	5·86	1·27	5·86	1·30	5·85	1·32	6
7	6·85	1·46	6·84	1·49	6·83	1·52	6·83	1·54	7
8	7·83	1·66	7·82	1·70	7·81	1·73	7·80	1·77	8
9	8·80	1·87	8·80	1·91	8·79	1·95	8·78	1·99	9
10	9·78	2·08	9·77	2·12	9·76	2·16	9·75	2·21	10
11	10·76	2·29	10·75	2·33	10·74	2·38	10·73	2·43	11
12	11·74	2·49	11·73	2·55	11·72	2·60	11·70	2·65	12
13	12·72	2·70	12·70	2·76	12·69	2·81	12·68	2·87	13
14	13·69	2·91	13·68	2·97	13·67	3·03	13·65	3·09	14
15	14·67	3·12	14·66	3·18	14·64	3·25	14·63	3·31	15
16	15·65	3·33	15·64	3·39	15·62	3·46	15·61	3·53	16
17	16·63	3·53	16·61	3·61	16·60	3·68	16·58	3·75	17
18	17·61	3·74	17·59	3·82	17·57	3·90	17·56	3·97	18
19	18·58	3·95	18·57	4·03	18·55	4·11	18·53	4·19	19
20	19·56	4·16	19·54	4·24	19·53	4·33	19·51	4·41	20
21	20·54	4·37	20·52	4·46	20·50	4·55	20·48	4·63	21
22	21·52	4·57	21·50	4·67	21·48	4·76	21·46	4·86	22
23	22·50	4·78	22·48	4·88	22·45	4·98	22·43	5·08	23
24	23·48	4·99	23·45	5·09	23·43	5·19	23·41	5·30	24
25	24·45	5·20	24·43	5·30	24·41	5·41	24·38	5·52	25
26	25·43	5·41	25·41	5·52	25·38	5·63	25·36	5·74	26
27	26·41	5·61	26·39	5·73	26·36	5·84	26·33	5·96	27
28	27·39	5·82	27·36	5·94	27·34	6·06	27·31	6·18	28
29	28·37	6·03	28·34	6·15	28·31	6·28	28·28	6·40	29
30	29·34	6·24	29·32	6·37	29·29	6·49	29·26	6·62	30
31	30·32	6·45	30·29	6·58	30·27	6·71	30·24	6·84	31
32	31·30	6·65	31·27	6·79	31·24	6·93	31·21	7·06	32
33	32·28	6·86	32·25	7·00	32·22	7·14	32·19	7·28	33
34	33·26	7·07	33·23	7·21	33·19	7·36	33·16	7·50	34
35	34·24	7·28	34·20	7·43	34·17	7·58	34·14	7·72	35
36	35·21	7·48	35·18	7·64	35·15	7·79	35·11	7·95	36
37	36·19	7·69	36·16	7·85	36·12	8·01	36·09	8·17	37
38	37·17	7·90	37·13	8·06	37·10	8·22	37·06	8·39	38
39	38·15	8·11	38·11	8·27	38·08	8·44	38·04	8·61	39
40	39·13	8·32	39·09	8·49	39·05	8·66	39·01	8·83	40
41	40·10	8·52	40·07	8·70	40·03	8·87	39·99	9·05	41
42	41·08	8·73	41·04	8·91	41·00	9·09	40·96	9·27	42
43	42·06	8·94	42·02	9·12	41·98	9·31	41·94	9·49	43
44	43·04	9·15	43·00	9·34	42·96	9·52	42·92	9·71	44
45	44·02	9·36	43·98	9·55	43·93	9·74	43.89	9·93	45
46	44·99	9·56	44·95	9·76	44·91	9·96	44·87	10·15	46
47	45·97	9·77	45·93	9·97	45·89	10·17	45·84	10·37	47
48	46·95	9·98	46·91	10·18	46·86	10·39	46·82	10·59	48
49	47·93	10·19	47·88	10·40	47·84	10·61	47·79	10·81	49
50	48·91	10·40	48·86	10·61	48·81	10·82	48·77	11·03	50
Distance.	Dep.	Lat.	Dep.	Lat.	Dep.	Lat.	Dep.	Lat.	Distance.
	78 Deg.		77¾ Deg.		77½ Deg.		77¼ Deg.		

Distance.	12 Deg.		12¼ Deg.		12½ Deg.		12¾ Deg.		Distance.
	Lat.	Dep.	Lat.	Dep.	Lat.	Dep.	Lat.	Dep.	
51	49·89	10·60	49·84	10·82	49·79	11·04	49·74	11·26	51
52	50·86	10·81	50·82	11·03	50·77	11·25	50·72	11·48	52
53	51·84	11·02	51·79	11·25	51·74	11·47	51·69	11·70	53
54	52·82	11·23	52·77	11·46	52·72	11·69	52·67	11·92	54
55	53·80	11·44	53·75	11·67	53·70	11·90	53·64	12·14	55
56	54·78	11·64	54·72	11·88	54·67	12·12	54·62	12·36	56
57	55·75	11·85	55·70	12·09	55·65	12·34	55·59	12·58	57
58	56·73	12·06	56·68	12·31	56·63	12·55	56·57	12·80	58
59	57·71	12·27	57·66	12·52	57·60	12·77	57·55	13·02	59
60	58·69	12·47	58·63	12·73	58·58	12·99	58·52	13·24	60
61	59·67	12·68	59·61	12·94	59·55	13·20	59·50	13·46	61
62	60·65	12·89	60·59	13·16	60·53	13·42	60·47	13·68	62
63	61·62	13·10	61·57	13·37	61·51	13·64	61·45	13·90	63
64	62·60	13·31	62·54	13·58	62·48	13·85	62·42	14·12	64
65	63·58	13·51	63·52	13·79	63·46	14·07	63·40	14·35	65
66	64·56	13·72	64·50	14·00	64·44	14·29	64·37	14·57	66
67	65·54	13·93	65·47	14·22	65·41	14·50	65·35	14·79	67
68	66·51	14·14	66·45	14·43	66·39	14·72	66·32	15·01	68
69	67·49	14·35	67·43	14·64	67·36	14·93	67·30	15·23	69
70	68·47	14·55	68·41	14·85	68·34	15·15	68·27	15·45	70
71	69·45	14·76	69.38	15·06	69·32	15·37	69·25	15·67	71
72	70·43	14·97	70·36	15·28	70·29	15·58	70·22	15·89	72
73	71·40	15·18	71·34	15·49	71·27	15·80	71·20	16·11	73
74	72·38	15·39	72·32	15·70	72·25	16·02	72·18	16·33	74
75	73·36	15·59	73·29	15·91	73·22	16·23	73·15	16·55	75
76	74·34	15·80	74·27	16·13	74·20	16·45	74·13	16·77	76
77	75·32	16·01	75·25	16·34	75·17	16·67	75·10	16·99	77
78	76·30	16·22	76·22	16·55	76·15	16·88	76·08	17·21	78
79	77·27	16·43	77·20	16·76	77·13	17·10	77·05	17·44	79
80	78·25	16·63	78·18	16·97	78·10	17·32	78·03	17·66	80
81	79·23	16·84	79·16	17·19	79·08	17·53	79·00	17·88	81
82	80·21	17·05	80·13	17·40	80·06	17·75	79·98	18·10	82
83	81·19	17·26	81·11	17·61	81·03	17·96	80·95	18·32	83
84	82·16	17·46	82·09	17·82	82·01	18·18	81·93	18·54	84
85	83·14	17·67	83·06	18·04	82·99	18·40	82·90	18·76	85
86	84·12	17·88	84·04	18·25	83·96	18·61	83·88	18·98	86
87	85·10	18·09	85·02	18·46	84·94	18·83	84·85	19·20	87
88	86·08	18·30	86·00	18·67	85·91	19·05	85·83	19·42	88
89	87·06	18·50	86·97	18·88	86·89	19·26	86·81	19·64	89
90	88·03	18·71	87·95	19·10	87·87	19·48	87·78	19·86	90
91	89·01	18·92	88·93	19·31	88·84	19·70	88·76	20·08	91
92	89·99	19·13	89·91	19·52	89·82	19·91	89·73	20·30	92
93	90·97	19·34	90·88	19·73	90·80	20·13	90·71	20·52	93
94	91·95	19·54	91·86	19·94	91·77	20·35	91·68	20·75	94
95	92·92	19·75	92·84	20·16	92·75	20·56	92·66	20·97	95
96	93·90	19·96	93·81	20·37	93·72	20·78	93·63	21·19	96
97	94·88	20·17	94·79	20·58	94·70	20·99	94·61	21·41	97
98	95·86	20·38	95·77	20·79	95·68	21·21	95·58	21·63	98
99	96·84	20·58	96·75	21·01	96·65	21·43	96·56	21·85	99
100	97·81	20·79	97·72	21·22	97·63	21·64	97·53	22·07	100
	Dep.	Lat.	Dep.	Lat.	Dep.	Lat.	Dep.	Lat.	
Distance.	78 Deg.		77¾ Deg.		77½ Deg.		77¼ Deg.		Distance.

Distance.	13 Deg.		13¼ Deg.		13½ Deg.		13¾ Deg.		Distance.
	Lat.	Dep.	Lat.	Dep.	Lat.	Dep.	Lat.	Dep.	
1	0·97	0·23	0·97	0·23	0·97	0·23	0·97	0·24	1
2	1·95	0·45	1·95	0·46	1·95	0·47	1·94	0·48	2
3	2·92	0·67	2·92	0·69	2·92	0·70	2·91	0·71	3
4	3·90	0·90	3·89	0·92	3·89	0·93	3·89	0·95	4
5	4·87	1·12	4·87	1·15	4·86	1·17	4·86	1·19	5
6	5·85	1·35	5·84	1·38	5·83	1·40	5·83	1·43	6
7	6·82	1·57	6·81	1·60	6·81	1·63	6·80	1·66	7
8	7·80	1·80	7·79	1·83	7·78	1·87	7·77	1·90	8
9	8·77	2·02	8·76	2·06	8·75	2·10	8·74	2·14	9
10	9.74	2·25	9·73	2·29	9·72	2·33	9·71	2·38	10
11	10·72	2·47	10·71	2·52	10·70	2·57	10·68	2·61	11
12	11·69	2·70	11·68	2·75	11·67	2·80	11·66	2·85	12
13	12·67	2·92	12·65	2·98	12·64	3·03	12·63	3·09	13
14	13·64	3·15	13·63	3·21	13·61	3·27	13·60	3·33	14
15	14·62	3·37	14·60	3·44	14·59	3·50	14·57	3·57	15
16	15·59	3·60	15·57	3·67	15·56	3·74	15·54	3·80	16
17	16·57	3·82	16·55	3·90	16·53	3·97	16·51	4·04	17
18	17·54	4·05	17·52	4·13	17·50	4·20	17·48	4·28	18
19	18·51	4·27	18·49	4·35	18·48	4·44	18·46	4·52	19
20	19·49	4·50	19·47	4·58	19·45	4·67	19·43	4·75	20
21	20·46	4·72	20·44	4·81	20·42	4·90	20·40	4·99	21
22	21·44	4·95	21·41	5·04	21·39	5·14	21·37	5·23	22
23	22·41	5·17	22·39	5·27	22·36	5·37	22·34	5·47	23
24	23·38	5·40	23·36	5·50	23·34	5·60	23·31	5·70	24
25	24·36	5·62	24·33	5·73	24·31	5·84	24·28	5·94	25
26	25·33	5·85	25·31	5·96	25·28	6·07	25·25	6·18	26
27	26·31	6·07	26·28	6·19	26·25	6·30	26·23	6·42	27
28	27·28	6·30	27·25	6·42	27·23	6·54	27·20	6·66	28
29	28·26	6·52	28·23	6·65	28·20	6·77	28·17	6·89	29
30	29·23	6·75	29·20	6·88	29·17	7·00	29·14	7·13	30
31	30·21	6·97	30·17	7·11	30·14	7·24	30·11	7·37	31
32	31·18	7·20	31·15	7·33	31·12	7·47	31·08	7·61	32
33	32·15	7·42	32·12	7·56	32·09	7·70	32·05	7·84	33
34	33·13	7·65	33·09	7·79	33·06	7·94	33·03	8·08	34
35	34·10	7·87	34·07	8·02	34·03	8·17	34·00	8·32	35
36	35·08	8·10	35·04	8·25	35·01	8·40	34·97	8·56	36
37	36·05	8·32	36·02	8·48	35·98	8·64	35·94	8·79	37
38	37·03	8·55	36·99	8·71	36·95	8·87	36·91	9·03	38
39	38·00	8·77	37·96	8·94	37·92	9·10	37·88	9·27	39
40	38·97	9·00	38·94	9·17	38·89	9·34	38·85	9·51	40
41	39·95	9·22	39·91	9·40	39·87	9·57	39·83	9·75	41
42	40·92	9·45	40·88	9·63	40·84	9·80	40·80	9·98	42
43	41·90	9·67	41·86	9·86	41·81	10·04	41·77	10·22	43
44	42·87	9·90	42·83	10·08	42·78	10·27	42·74	10·46	44
45	43·85	10·12	43·80	10·31	43·76	10·51	43·71	10·70	45
46	44·82	10·35	44·78	10·54	44·73	10·74	44·68	10·93	46
47	45·80	10·57	45·75	10·77	45·70	10·97	45·65	11·17	47
48	46·77	10·80	46·72	11·00	46·67	11·21	46·62	11·41	48
49	47·74	11·02	47·70	11·23	47·65	11·44	47·60	11·65	49
50	48·72	11·25	48·67	11·46	48·62	11·67	48·57	11·88	50
Distance.	Dep.	Lat.	Dep.	Lat.	Dep.	Lat.	Dep.	Lat.	Distance.
	77 Deg.		76¾ Deg.		76½ Deg.		76¼ Deg.		

Distance.	13 Deg.		13¼ Deg.		13½ Deg.		13¾ Deg.		Distance.
	Lat.	Dep.	Lat.	Dep.	Lat.	Dep.	Lat.	Dep.	
51	49.69	11·47	49·64	11·69	49·59	11·91	49·54	12·12	51
52	50·67	11·70	50·62	11·92	50·56	12·14	50·51	12·36	52
53	51·64	11·92	51·59	12·15	51·54	12·37	51·48	12·60	53
54	52·62	12·15	52·56	12·38	52·51	12·61	52·45	12·84	54
55	53·59	12·37	53·54	12·61	53·48	12·84	53·42	13·07	55
56	54·56	12·60	54·51	12·84	54·45	13·07	54·40	13·31	56
57	55·54	12·82	55·48	13·06	55·43	13·31	55·37	13·55	57
58	56·51	13·05	56·46	13·29	56·40	13·54	56·34	13·79	58
59	57·49	13·27	57·43	13·52	57·37	13·77	57·31	14·02	59
60	58·46	13·50	58·40	13·75	58·34	14·01	58·28	14·26	60
61	59·44	13·72	59·38	13·98	59·31	14·24	59·25	14·50	61
62	60·41	13·95	60·35	14·21	60·29	14·47	60·22	14·74	62
63	61·39	14·17	61·32	14·44	61·26	14·71	61·19	14·97	63
64	62·36	14·40	62·30	14·67	62·23	14·94	62·17	15·21	64
65	63·33	14·62	63·27	14·90	63·20	15·17	63·14	15·45	65
66	64·31	14·85	64·24	15·13	64·18	15·41	64·11	15·69	66
67	65·28	15·07	65·22	15·36	65·15	15·64	65·08	15·93	67
68	66·26	15·30	66·19	15·59	66·12	15·87	66·05	16·16	68
69	67·23	15·52	67·16	15·81	67·09	16·11	67·02	16·40	69
70	68·21	15·75	68·14	16·04	68·07	16·34	67·99	16·64	70
71	69·18	15·97	69·11	16·27	69·04	16·57	68·97	16·88	71
72	70·15	16·20	70·08	16·50	70·01	16·81	69·94	17·11	72
73	71·13	16·42	71·06	16·73	70·98	17·04	70·91	17·35	73
74	72·10	16·65	72·03	16·96	71·96	17·28	71·88	17·59	74
75	73·08	16·87	73·00	17·19	72·93	17·50	72·85	17·83	75
76	74·05	17·10	73·98	17·42	73·90	17·74	73·82	18·06	76
77	75·03	17·32	74·95	17·65	74·87	17·98	74·79	18·30	77
78	76·00	17·55	75·92	17·88	75·84	18·21	75·76	18·54	78
79	76·98	17·77	76·90	18·11	76·82	18·44	76·74	18·78	79
80	77·95	18·00	77·87	18·34	77·79	18·68	77·71	19·01	80
81	78·92	18·22	78·84	18·57	78·76	18·91	78·68	19·25	81
82	79·90	18·45	79·82	18·79	79·73	19·14	79·65	19·49	82
83	80·87	18·67	80·79	19·02	80·71	19·38	80·62	19·73	83
84	81·85	18·90	81·76	19·25	81·68	19·61	81·59	19·97	84
85	82·82	19·12	82·74	19·48	82·65	19·84	82·56	20·20	85
86	83·80	19·35	83·71	19·71	83·62	20·08	83·54	20·44	86
87	84·77	19·57	84·68	19·94	84·60	20·31	84·51	20·68	87
88	85·74	19·80	85·66	20·17	85·57	20·54	85·48	20·92	88
89	86·72	20·02	86·63	20·40	86·54	20·78	86·45	21·15	89
90	87·69	20·25	87·60	20·63	87·51	21·01	87·42	21·39	90
91	88·67	20·47	88·58	20·86	88·49	21·24	88·39	21·63	91
92	89·64	20·70	89·55	21·09	89·46	21·48	89·36	21·87	92
93	90·62	20·92	90·52	21·32	90·43	21·71	90·33	22·10	93
94	91·59	21·15	91·50	21·54	91·40	21·94	91·31	22·34	94
95	92·57	21·37	92·47	21·77	92·38	22·18	92·28	22·58	95
96	93·54	21·60	93·44	22·00	93·35	22·41	93·25	22·82	96
97	94·51	21·82	94·42	22·23	94·32	22·64	94·22	23·06	97
98	95·49	22·05	95·39	22·46	95·29	22·88	95·19	23·29	98
99	96·46	22·27	96·36	22·69	96·26	23·11	96·16	23·53	99
100	97·44	22·50	97·34	22·92	97·24	23·34	97·13	23·77	100
Distance.	Dep.	Lat.	Dep.	Lat.	Dep.	Lat.	Dep.	Lat.	Distance.
	77 Deg.		76¾ Deg.		76½ Deg.		76¼ Deg.		

Distance.	14 Deg.		14¼ Deg.		14½ Deg.		14¾ Deg.		Distance.
	Lat.	Dep.	Lat.	Dep.	Lat.	Dep.	Lat.	Dep.	
1	0·97	0·24	0·97	0·25	0·97	0·25	0·97	0·25	1
2	1·94	0·48	1·94	0·49	1·94	0·50	1·93	0·51	2
3	2·91	0·73	2·91	0·74	2·90	0·75	2·90	0·76	3
4	3·88	0·97	3·88	0·98	3·87	1·00	3·87	1·02	4
5	4·85	1·21	4·85	1·23	4·84	1·25	4·84	1·27	5
6	5·82	1·45	5·82	1·48	5·81	1·50	5·80	1·53	6
7	6·79	1·69	6·78	1·72	6·78	1·75	6·77	1·78	7
8	7·76	1·94	7·75	1·97	7·75	2·00	7·74	2·04	8
9	8·73	2·18	8·72	2·22	8·71	2·25	8·70	2·29	9
10	9·70	2·42	9·69	2·46	9·68	2·50	9·67	2·55	10
11	10·67	2·66	10·66	2·71	10·65	2·75	10·64	2·80	11
12	11·64	2·90	11·63	2·95	11·62	3·00	11·60	3·06	12
13	12·61	3·15	12·60	3·20	12·59	3·25	12·57	3·31	13
14	13·58	3·39	13·57	3·45	13·55	3·51	13·54	3·56	14
15	14·55	3·63	14·54	3·69	14·52	3·76	14·51	3·82	15
16	15·52	3·87	15·51	3·94	15·49	4·01	15·47	4·07	16
17	16·50	4·11	16·48	4·18	16·46	4·26	16·44	4·33	17
18	17·47	4·35	17·45	4·43	17·43	4·51	17·41	4·58	18
19	18·44	4·60	18·42	4·68	18·39	4·76	18·37	4·84	19
20	19·41	4·84	19·38	4·92	19·36	5·01	19·34	5·09	20
21	20·38	5·08	20·35	5·17	20·33	5·26	20·31	5·35	21
22	21·35	5·32	21·32	5·42	21·30	5·51	21·28	5·60	22
23	22·32	5·56	22·29	5·66	22·27	5·76	22·24	5·86	23
24	23·29	5·81	23·26	5·91	23·24	6·01	23·21	6·11	24
25	24·26	6·05	24·23	6·15	24·20	6·26	24·18	6·37	25
26	25·23	6·29	25·20	6·40	25·17	6·51	25·14	6·62	26
27	26·20	6·53	26·17	6·65	26·14	6·76	26·11	6·87	27
28	27·17	6·77	27·14	6·89	27·11	7·01	27·08	7·13	28
29	28·14	7·02	28·11	7·14	28·08	7·26	28·04	7·38	29
30	29·11	7·26	29·08	7·38	29·04	7·51	29·01	7·64	30
31	30·08	7·50	30·05	7·63	30·01	7·76	29·98	7·89	31
32	31·05	7·74	31·02	7·88	30·98	8·01	30·95	8·15	32
33	32·02	7·98	31·98	8·12	31·95	8·26	31·91	8·40	33
34	32·99	8·23	32·95	8·37	32·92	8·51	32·88	8·66	34
35	33·96	8·47	33·92	8·62	33·89	8·76	33·85	8·91	35
36	34·93	8·71	34·89	8·86	34·85	9·01	34·81	9·17	36
37	35·90	8·95	35·86	9·11	35·82	9·26	35·78	9·42	37
38	36·87	9·19	36·83	9·35	36·79	9·51	36·75	9·67	38
39	37·84	9·44	37·80	9·60	37·76	9·76	37·71	9·93	39
40	38·81	9·68	38·77	9·85	38·73	10·02	38·68	10·18	40
41	39·78	9·92	39·74	10·09	39·69	10·27	39·65	10·44	41
42	40·75	10·16	40·71	10·34	40·66	10·52	40·62	10·69	42
43	41·72	10·40	41·68	10·58	41·63	10·77	41·58	10·95	43
44	42·69	10·64	42·65	10·83	42·60	11·02	42·55	11·20	44
45	43·66	10·89	43·62	11·08	43·57	11·27	43·52	11·46	45
46	44·63	11·13	44·58	11·32	44·53	11·52	44·48	11·71	46
47	45·60	11·37	45·55	11·57	45·50	11·77	45·45	11·97	47
48	46·57	11·61	46·52	11·82	46·47	12·02	46·42	12·22	48
49	47·54	11·85	47·49	12·06	47·44	12·27	47·39	12·48	49
50	48·51	12·10	48·46	12·31	48·41	12·52	48·35	12·73	50
Distance.	Dep.	Lat.	Dep.	Lat.	Dep.	Lat.	Dep.	Lat.	Distance.
	76 Deg		75¾ Deg.		75½ Deg.		75¼ Deg.		

Distance.	14 Deg.		14¼ Deg.		14½ Deg.		14¾ Deg.		Distance.
	Lat.	Dep.	Lat.	Dep.	Lat.	Dep.	Lat.	Dep.	
51	49·49	12·34	49·43	12·55	49·38	12·77	49·32	12·98	51
52	50·46	12·58	50·40	12·80	50·34	13·02	50·29	13·24	52
53	51·43	12·82	51·37	13·05	51·31	13·27	51·25	13·49	53
54	52·40	13·06	52·34	13·29	52·28	13·52	52·22	13·75	54
55	53·37	13·31	53·31	13·54	53·25	13·77	53·19	14·00	55
56	54·34	13·55	54·28	13·78	54·22	14·02	54·15	14·26	56
57	55·31	13·79	55·25	14·03	55·18	14·27	55·12	14·51	57
58	56·28	14·03	56·22	14·28	56·15	14·52	56·09	14·77	58
59	57·25	14·27	57·18	14·52	57·12	14·77	57·06	15·02	59
60	58·22	14·52	58·15	14·77	58·09	15·02	58·02	15·28	60
61	59·19	14·76	59·12	15·02	59·06	15·27	58·99	15·53	61
62	60·16	15·00	60·09	15·26	60·03	15·52	59·96	15·79	62
63	61·13	15·24	61·06	15·51	60·99	15·77	60·92	16·04	63
64	62·10	15·48	62·03	15·75	61·96	16·02	61·89	16·29	64
65	63·07	15·72	63·00	16·00	62·93	16·27	62·86	16·55	65
66	64·04	15·97	63·97	16·25	63·90	16·53	63·83	16·80	66
67	65·01	16·21	64·94	16·49	64·87	16·78	64·79	17·06	67
68	65·98	16·45	65·91	16·74	65·83	17·03	65·76	17·31	68
69	66·95	16·69	66·88	16·98	66·80	17·28	66·73	17·57	69
70	67·92	16·93	67·85	17·23	67·77	17·53	67·69	17·82	70
71	68·89	17·18	68·82	17·48	68·74	17·78	68·66	18·08	71
72	69·86	17·42	69·78	17·72	69·71	18·03	69·63	18·33	72
73	70·83	17·66	70·75	17·97	70·67	18·28	70·59	18·59	73
74	71·80	17·90	71·72	18·22	71·64	18·53	71·56	18·84	74
75	72·77	18·14	72·69	18·46	72·61	18·78	72·53	19·10	75
76	73·74	18·39	73·66	18·71	73·58	19·03	73·50	19·35	76
77	74·71	18·63	74·63	18·95	74·55	19·28	74·46	19·60	77
78	75·68	18·87	75·60	19·20	75·52	19·53	75·43	19·86	78
79	76·65	19·11	76·57	19·45	76·48	19·78	76·40	20·11	79
80	77·62	19·35	77·54	19·69	77·45	20·03	77·36	20·37	80
81	78·59	19·60	78·51	19·94	78·42	20·28	78·33	20·62	81
82	79·56	19·84	79·48	20·18	79·39	20·53	79·30	20·88	82
83	80·53	20·08	80·45	20·43	80·36	20·78	80·26	21·13	83
84	81·50	20·32	81·42	20·68	81·32	21·03	81·23	21·39	84
85	82·48	20·56	82·38	20·92	82·29	21·28	82·20	21·64	85
86	83·45	20·81	83·35	21·17	83·26	21·53	83·17	21·90	86
87	84·42	21·05	84·32	21·42	84·23	21·78	84·13	22·15	87
88	85·39	21·29	85·29	21·66	85·20	22·03	85·10	22·41	88
89	86·36	21·53	86·26	21·91	86·17	22·28	86·07	22·66	89
90	87·33	21·77	87·23	22·15	87·13	22·53	87·03	22·91	90
91	88·30	22·01	88·20	22·40	88·10	22·78	88·00	23·17	91
92	89·27	22·26	89·17	22·65	89·07	23·04	88·97	23·42	92
93	90·24	22·50	90·14	22·89	90·04	23·29	89·94	23·68	93
94	91·21	22·74	91·11	23·14	91·01	23·54	90·90	23·93	94
95	92·18	22·98	92·08	23·38	91·97	23·79	91·87	24·19	95
96	93·15	23·22	93·05	23·63	92·94	24·04	92·84	24·44	96
97	94·12	23·47	94·02	23·88	93·91	24·29	93·80	24·70	97
98	95·09	23·71	94·98	24·12	94·88	24·54	94·77	24·95	98
99	96·06	23·95	95·95	24·37	95·85	24·79	95·74	25·21	99
100	97·03	24·19	96·92	24·62	96·81	25·04	96·70	25·46	100
	Dep.	Lat.	Dep.	Lat.	Dep.	Lat.	Dep.	Lat.	
Distance.	76 Deg.		75¾ Deg.		75½ Deg.		75¼ Deg.		Distance.

Distance.	15 Deg.		15¼ Deg.		15½ Deg.		15¾ Deg.		Distance.
	Lat.	Dep.	Lat.	Dep.	Lat.	Dep.	Lat.	Dep.	
1	0·97	0·26	0·96	0·26	0·96	0·27	0·96	0·27	1
2	1·93	0·52	1·93	0·53	1·93	0·53	1·92	0·54	2
3	2·90	0·78	2·89	0·79	2·89	0·80	2·89	0·81	3
4	3·86	1·04	3·86	1·05	3·85	1·07	3·85	1·09	4
5	4·83	1·29	4·82	1·32	4·82	1·34	4·81	1·36	5
6	5·80	1·55	5·79	1·58	5·78	1·60	5·77	1·63	6
7	6·76	1·81	6·75	1·84	6·75	1·87	6·74	1·90	7
8	7·73	2·07	7·72	2·10	7·71	2·14	7·70	2·17	8
9	8·69	2·33	8·68	2·37	8·67	2·41	8·66	2·44	9
10	9·66	2·59	9·65	2·63	9·64	2·67	9·62	2·71	10
11	10·63	2·85	10·61	2·89	10·60	2·94	10·59	2·99	11
12	11·59	3·11	11·58	3·16	11·56	3·21	11·55	3·26	12
13	12·56	3·36	12·54	3·42	12·53	3·47	12·51	3·53	13
14	13·52	3·62	13·51	3·68	13·49	3·74	13·47	3·80	14
15	14·49	3·88	14·47	3·95	14·45	4·01	14·44	4·07	15
16	15·45	4·14	15·44	4·21	15·42	4·28	15·40	4·34	16
17	16·42	4·40	16·40	4·47	16·38	4·54	16·36	4·61	17
18	17·39	4·66	17·37	4·73	17·35	4·81	17·32	4·89	18
19	18·35	4·92	18·33	5·00	18·31	5·08	18·29	5·16	19
20	19·32	5·18	19·30	5·26	19·27	5·34	19·25	5·43	20
21	20·28	5·44	20·26	5·52	20·24	5·61	20·21	5·70	21
22	21·25	5·69	21·23	5·79	21·20	5·88	21·17	5·97	22
23	22·22	5·95	22·19	6·05	22·16	6·15	22·14	6·24	23
24	23·18	6·21	23·15	6·31	23·13	6·41	23·10	6·51	24
25	24·15	6·47	24·12	6·58	24·09	6·68	24·06	6·79	25
26	25·11	6·73	25·08	6·84	25·05	6·95	25·02	7·06	26
27	26·08	6·99	26·05	7·10	26·02	7·22	25·99	7·33	27
28	27·05	7·25	27·01	7·36	26·98	7·48	26·95	7·60	28
29	28·01	7·51	27·98	7·63	27·95	7·75	27·91	7·87	29
30	28·98	7·76	28·94	7·89	28·91	8·02	28·87	8·14	30
31	29·94	8·02	29·91	8·15	29·87	8·28	29·84	8·41	31
32	30·91	8·28	30·87	8·42	30·84	8·55	30·80	8·69	32
33	31·88	8·54	31·84	8·68	31·80	8·82	31·76	8·96	33
34	32·84	8·80	32·80	8·94	32·76	9·09	32·72	9·23	34
35	33·81	9·06	33·77	9·21	33·73	9·35	33·69	9·50	35
36	34·77	9·32	34·73	9·47	34·69	9·62	34·65	9·77	36
37	35·74	9·58	35·70	9·73	35·65	9·89	35·61	10·04	37
38	36·71	9·84	36·66	10·00	36·62	10·16	36·57	10·31	38
39	37·67	10·09	37·63	10·26	37·58	10·42	37·54	10·59	39
40	38·64	10·35	38·59	10·52	38·55	10·69	38·50	10·86	40
41	39·60	10·61	39·56	10·78	39·51	10·96	39·46	11·13	41
42	40·57	10·87	40·52	11·05	40·47	11·22	40·42	11·40	42
43	41·53	11·13	41·49	11·31	41·44	11·49	41·39	11·67	43
44	42·50	11·39	42·45	11·57	42·40	11·76	42·35	11·94	44
45	43·47	11·65	43·42	11·84	43·36	12·03	43·31	12·21	45
46	44·43	11·91	44·38	12·10	44·33	12·29	44·27	12·49	46
47	45·40	12·16	45·35	12·36	45·29	12·56	45·24	12·76	47
48	46·36	12·42	46·31	12·63	46·25	12·83	46·20	13·03	48
49	47·33	12·68	47·27	12·89	47·22	13·09	47·16	13·30	49
50	48·30	12·94	48·24	13·15	48·18	13·36	48·12	13·57	50
Distance.	Dep.	Lat.	Dep.	Lat.	Dep.	Lat.	Dep.	Lat.	Distance.
	75 Deg.		74¾ Deg.		74½ Deg.		74¼ Deg.		

Distance.	15 Deg.		15¼ Deg.		15½ Deg.		15¾ Deg.		Distance.
	Lat.	Dep.	Lat.	Dep.	Lat.	Dep.	Lat.	Dep.	
51	49·26	13·20	49·20	13·41	49·15	13·63	49·09	13·84	51
52	50·23	13·46	50·17	13·68	50·11	13·90	50·05	14·11	52
53	51·19	13·72	51·13	13·94	51·07	14·16	51·01	14·39	53
54	52·16	13·98	52·10	14·20	52·04	14·43	51·97	14·66	54
55	53·13	14·24	53·06	14·47	53·00	14·70	52·94	14·93	55
56	54·09	14·49	54·03	14·73	53·96	14·97	53·90	15·20	56
57	55·06	14·75	54·99	14·99	54·93	15·23	54·86	15·47	57
58	56·02	15·01	55·96	15·26	55·89	15·50	55·82	15·74	58
59	56·99	15·27	56·92	15·52	56·85	15·77	56·78	16·01	59
60	57·96	15·53	57·89	15·78	57·82	16·03	57·75	16·29	60
61	58·92	15·79	58·85	16·04	58·78	16·30	58·71	16·56	61
62	59·89	16·05	59·82	16·31	59·75	16·57	59·67	16·83	62
63	60·85	16·31	60·78	16·57	60·71	16·84	60·63	17·10	63
64	61·82	16·56	61·75	16·83	61·67	17·10	61·60	17·37	64
65	62·79	16·82	62·71	17·10	62·64	17·37	62·56	17·64	65
66	63·75	17·08	63·68	17·36	63·60	17·64	63·52	17·92	66
67	64·72	17·34	64·64	17·62	64·56	17·90	64·48	18·19	67
68	65·68	17·60	65·61	17·89	65·53	18·17	65·45	18·46	68
69	66·65	17·86	66·57	18·15	66·49	18·44	66·41	18·73	69
70	67·61	18·12	67·54	18·41	67·45	18·71	67·37	19·00	70
71	68·58	18·38	68·50	18·68	68·42	18·97	68·33	19·27	71
72	69·55	18·63	69·46	18·94	69·38	19·24	69·30	19·54	72
73	70·51	18·89	70·43	19·20	70·35	19·51	70·26	19·82	73
74	71·48	19·15	71·39	19·46	71·31	19·78	71·22	20·09	74
75	72·44	19·41	72·36	19·73	72·27	20·04	72·18	20·36	75
76	73·41	19·67	73·32	19·99	73·24	20·31	73·15	20·63	76
77	74·38	19·93	74·29	20·25	74·20	20·58	74·11	20·90	77
78	75·34	20·19	75·25	20·52	75·16	20·84	75·07	21·17	78
79	76·31	20·45	76·22	20·78	76·13	21·11	76·03	21·44	79
80	77·27	20·71	77·18	21·04	77·09	21·38	77·00	21·72	80
81	78·24	20·96	78·15	21·31	78·05	21·65	77·96	21·99	81
82	79·21	21·22	79·11	21·57	79·02	21·91	78·92	22·26	82
83	80·17	21·48	80·08	21·83	79·98	22·18	79·88	22·53	83
84	81·14	21·74	81·04	22·09	80·94	22·45	80·85	22·80	84
85	82·10	22·00	82·01	22·36	81·91	22·72	81·81	23·07	85
86	83·07	22·26	82·97	22·62	82·87	22·98	82·77	23·34	86
87	84·04	22·52	83·94	22·88	83·84	23·25	83·73	23·62	87
88	85·00	22·78	84·90	23·15	84·80	23·52	84·70	23·89	88
89	85·97	23·03	85·87	23·41	85·76	23·78	85·66	24·16	89
90	86·93	23·29	86·83	23·67	86·73	24·05	86·62	24·43	90
91	87·90	23·55	87·80	23·94	87·69	24·32	87·58	24·70	91
92	88·87	23·81	88·76	24·20	88·65	24·59	88·55	24·97	92
93	89·83	24·07	89·73	24·46	89·62	24·85	89·51	25·24	93
94	90·80	24·33	90·69	24·72	90·58	25·12	90·47	25·52	94
95	91·76	24·59	91·65	24·99	91·54	25·39	91·43	25·79	95
96	92·73	24·85	92·62	25·25	92·51	25·65	92·40	26·06	96
97	93·69	25·11	93·58	25·51	93·47	25·92	93·36	26·33	97
98	94·66	25·36	94·55	25·78	94·44	26·19	94·32	26·60	98
99	95·63	25·62	95·51	26·04	95·40	26·46	95·28	26·87	99
100	96·59	25·88	96·48	26·30	96·36	26·72	96·25	27·14	100
Distance.	Dep.	Lat.	Dep.	Lat.	Dep.	Lat.	Dep.	Lat.	Distance.
	75 Deg.		74¾ Deg.		74½ Deg.		74¼ Deg.		

Distance.	16 Deg.		16¼ Deg.		16½ Deg.		16¾ Deg.		Distance.
	Lat.	Dep.	Lat.	Dep.	Lat.	Dep.	Lat.	Dep.	
1	0·96	0·28	0·96	0·28	0·96	0·28	0·96	0·29	1
2	1·92	0·55	1·92	0·56	1·92	0·57	1·92	0·58	2
3	2·88	0·83	2·88	0·84	2·88	0·85	2·87	0·86	3
4	3·85	1·10	3·84	1·12	3·84	1·14	3·83	1·15	4
5	4·81	1·38	4·80	1·40	4·79	1·42	4·79	1·44	5
6	5·77	1·65	5·76	1·68	5·75	1·70	5·75	1·73	6
7	6·73	1·93	6·72	1·96	6·71	1·99	6·70	2·02	7
8	7·69	2·21	7·68	2·24	7·67	2·27	7·66	2·31	8
9	8·65	2·48	8·64	2·52	8·63	2·56	8·62	2·59	9
10	9·61	2·76	9·60	2·80	9·59	2·84	9·58	2·88	10
11	10·57	3·03	10·56	3·08	10·55	3·12	10·53	3·17	11
12	11·54	3·31	11·52	3·36	11·51	3·41	11·49	3·46	12
13	12·50	3·58	12·48	3·64	12·46	3·69	12·45	3·75	13
14	13·46	3·86	13·44	3·92	13·42	3·98	13·41	4·03	14
15	14·42	4·13	14·40	4·20	14·38	4·26	14·36	4·32	15
16	15·38	4·41	15·36	4·48	15·34	4·54	15·32	4·61	16
17	16·34	4·69	16·32	4·76	16·30	4·83	16·28	4·90	17
18	17·30	4·96	17·28	5·04	17·26	5·11	17·24	5·19	18
19	18·26	5·24	18·24	5·32	18·22	5·40	18·19	5·48	19
20	19·23	5·51	19·20	5·60	19·18	5·68	19·15	5·76	20
21	20·19	5·79	20·16	5·88	20·14	5·96	20·11	6·05	21
22	21·15	6·06	21·12	6·16	21·09	6·25	21·07	6·34	22
23	22·11	6·34	22·08	6·44	22·05	6·53	22·02	6·63	23
24	23·07	6·62	23·04	6·72	23·01	6·82	22·98	6·92	24
25	24·03	6·89	24·00	7·00	23·97	7·10	23·94	7·20	25
26	24·99	7·17	24·96	7·28	24·93	7·38	24·90	7·49	26
27	25·95	7·44	25·92	7·56	25·89	7·67	25·85	7·78	27
28	26·92	7·72	26·88	7·84	26·85	7·95	26·81	8·07	28
29	27·88	7·99	27·84	8·11	27·81	8·24	27·77	8·36	29
30	28·84	8·27	28·80	8·39	28·76	8·52	28·73	8·65	30
31	29·80	8·54	29·76	8·67	29·72	8·80	29·68	8·93	31
32	30·76	8·82	30·72	8·95	30·68	9·09	30·64	9·22	32
33	31·72	9·10	31·68	9·23	31·64	9·37	31·60	9·51	33
34	32·68	9·37	32·64	9·51	32·60	9·66	32·56	9·80	34
35	33·64	9·65	33·60	9·79	33·56	9·94	33·51	10·09	35
36	34·61	9·92	34·56	10·07	34·52	10·22	34·47	10·38	36
37	35·57	10·20	35·52	10·35	35·48	10·51	35·43	10·66	37
38	36·53	10·47	36·48	10·63	36·44	10·79	36·39	10·95	38
39	37·49	10·75	37·44	10·91	37·39	11·08	37·35	11·24	39
40	38·45	11·03	38·40	11·19	38·35	11·36	38·30	11·53	40
41	39·41	11·30	39·36	11·47	39·31	11·64	39·26	11·82	41
42	40·37	11·58	40·32	11·75	40·27	11·93	40·22	12·10	42
43	41·33	11·85	41·28	12·03	41·23	12·21	41·18	12·39	43
44	42·30	12·13	42·24	12·31	42·19	12·50	42·13	12·68	44
45	43·26	12·40	43·20	12·59	43·15	12·78	43·09	12·97	45
46	44·22	12·68	44·16	12·87	44·11	13·06	44·05	13·26	46
47	45·18	12·95	45·12	13·15	45·06	13·35	45·01	13·55	47
48	46·14	13·23	46·08	13·43	46·02	13·63	45·96	13·83	48
49	47·10	13·51	47·04	13·71	46·98	13·92	46·92	14·12	49
50	48·06	13·78	48·00	13·99	47·94	14·20	47·88	14·41	50
	Dep.	Lat.	Dep.	Lat.	Dep.	Lat.	Dep.	Lat.	
Distance.	74 Deg.		73¾ Deg.		73½ Deg.		73¼ Deg.		Distance.

Distance.	16 Deg.		16¼ Deg.		16½ Deg.		16¾ Deg.		Distance.
	Lat.	Dep.	Lat.	Dep.	Lat.	Dep.	Lat.	Dep.	
51	49·02	14·06	48·96	14·27	48·90	14·48	48·84	14·70	51
52	49·99	14·33	49·92	14·55	49·86	14·77	49·79	14·99	52
53	50·95	14·61	50·88	14·83	50·82	15·05	50·75	15·27	53
54	51·91	14·88	51·84	15·11	51·78	15·34	51·71	15·56	54
55	52·87	15·16	52·80	15·39	52·74	15·62	52·67	15·85	55
56	53·83	15·44	53·76	15·67	53·69	15·90	53·62	16·14	56
57	54·79	15·71	54·72	15·95	54·65	16·19	54·58	16·43	57
58	55·75	15·99	55·68	16·23	55·61	16·47	55·54	16·72	58
59	56·71	16·26	56·64	16·51	56·57	16·76	56·50	17·00	59
60	57·68	16·54	57·60	16·79	57·53	17·04	57·45	17·29	60
61	58·64	16·81	58·56	17·07	58·49	17·32	58·41	17·58	61
62	59·60	17·09	59·52	17·35	59·45	17·61	59·37	17·87	62
63	60·56	17·37	60·48	17·63	60·41	17·89	60·33	18·16	63
64	61·52	17·64	61·44	17·91	61·36	18·18	61·28	18·44	64
65	62·48	17·92	62·40	18·19	62·32	18·46	62·24	18·73	65
66	63·44	18·19	63·36	18·47	63·28	18·74	63·20	19·02	66
67	64·40	18·47	64·32	18·75	64·24	19·03	64·16	19·31	67
68	65·37	18·74	65·28	19·03	65·20	19·31	65·11	19·60	68
69	66·33	19·02	66·24	19·31	66·16	19·60	66·07	19·89	69
70	67·29	19·29	67·20	19·59	67·12	19·88	67·03	20·17	70
71	68·25	19·57	68·16	19·87	68·08	20·17	67·99	20·46	71
72	69·21	19·85	69·12	20·15	69·03	20·45	68·95	20·75	72
73	70·17	20·12	70·08	20·43	69·99	20·73	69·90	21·04	73
74	71·13	20·40	71·04	20·71	70·95	21·02	70·86	21·33	74
75	72·09	20·67	72·00	20·99	71·91	21·30	71·82	21·61	75
76	73·06	20·95	72·96	21·27	72·87	21·59	72·78	21·90	76
77	74·02	21·22	73·92	21·55	73·83	21·87	73·73	22·19	77
78	74·98	21·50	74·88	21·83	74·79	22·15	74·69	22·48	78
79	75·94	21·78	75·84	22·11	75·75	22·44	75·65	22·77	79
80	76·90	22·05	76·80	22·39	76·71	22·72	76·61	23·06	80
81	77·86	22·33	77·76	22·67	77·66	23·01	77·56	23·34	81
82	78·82	22·60	78·72	22·95	78·62	23·29	78·52	23·63	82
83	79·78	22·88	79·68	23·23	79·58	23·57	79·48	23·92	83
84	80·75	23·15	80·64	23·51	80·54	23·86	80·44	24·21	84
85	81·71	23·43	81·60	23·79	81·50	24·14	81·39	24·50	85
86	82·67	23·70	82·56	24·07	82·46	24·43	82·35	24·78	86
87	83·63	23·98	83·52	24·35	83·42	24·71	83·31	25·07	87
88	84·59	24·26	84·48	24·62	84·38	24·99	84·27	25·36	88
89	85·55	24·53	85·44	24·90	85·33	25·28	85·22	25·65	89
90	86·51	24·81	86·40	25·18	86·29	25·56	86·18	25·94	90
91	87·47	25·08	87·36	25·46	87·25	25·85	87·14	26·23	91
92	88·44	25·36	88·32	25·74	88·21	26·13	88·10	26·51	92
93	89·40	25·63	89·28	26·02	89·17	26·41	89·05	26·80	93
94	90·36	25·91	90·24	26·30	90·13	26·70	90·01	27·09	94
95	91·32	26·19	91·20	26·58	91·09	26·98	90·97	27·38	95
96	92·28	26·46	92·16	26·86	92·05	27·27	91·93	27·67	96
97	93·24	26·74	93·12	27·14	93·01	27·55	92·88	27·95	97
98	94·20	27·01	94·08	27·42	93·96	27·83	93·84	28·24	98
99	95·16	27·29	95·04	27·70	94·92	28·12	94·80	28·53	99
100	96·13	27·56	96·00	27·98	95·88	28·40	95·76	28·82	100
	Dep.	Lat.	Dep.	Lat.	Dep.	Lat.	Dep.	Lat.	
Distance.	74 Deg.		73¾ Deg.		73½ Deg.		73¼ Deg.		Distance.

Distance.	17 Deg.		17¼ Deg.		17½ Deg.		17¾ Deg.		Distance.
	Lat.	Dep.	Lat.	Dep.	Lat.	Dep.	Lat.	Dep.	
1	0·96	0·29	0·95	0·30	0·95	0·30	0·95	0·30	1
2	1·91	0·58	1·91	0·59	1·91	0·60	1·90	0·61	2
3	2·87	0·88	2·87	0·89	2·86	0·90	2·86	0·91	3
4	3·83	1·17	3·82	1·19	3·81	1·20	3·81	1·22	4
5	4·78	1·46	4·78	1·48	4·77	1·50	4·76	1·52	5
6	5·74	1·75	5·73	1·78	5·72	1·80	5·71	1·83	6
7	6·69	2·05	6·69	2·08	6·68	2·10	6·67	2·13	7
8	7·65	2·34	7·64	2·37	7·63	2·41	7·62	2·44	8
9	8·61	2·63	8·60	2·67	8·58	2·71	8·57	2·74	9
10	9·56	2·92	9·55	2·97	9·54	3·01	9·52	3·05	10
11	10·52	3·22	10·51	3·26	10·49	3·31	10·48	3·35	11
12	11·48	3·51	11·46	3·56	11·44	3·61	11·43	3·66	12
13	12·43	3·80	12·42	3·85	12·40	3·91	12·38	3·96	13
14	13·39	4·09	13·37	4·15	13·35	4·21	13·33	4·27	14
15	14·34	4·39	14·33	4·45	14·31	4·51	14·29	4·57	15
16	15·30	4·68	15·28	4·74	15·26	4·81	15·24	4·88	16
17	16·26	4·97	16·24	5·04	16·21	5·11	16·19	5·18	17
18	17·21	5·26	17·19	5·34	17·17	5·41	17·14	5·49	18
19	18·17	5·56	18·15	5·63	18·12	5·71	18·10	5·79	19
20	19·13	5·85	19·10	5·93	19·07	6·01	19·05	6·10	20
21	20·08	6·14	20·06	6·23	20·03	6·31	20·00	6·40	21
22	21·04	6·43	21·01	6·52	20·98	6·62	20·95	6·71	22
23	21·99	6·72	21·97	6·82	21·94	6·92	21·91	7·01	23
24	22·95	7·02	22·92	7·12	22·89	7·22	22·86	7·32	24
25	23·91	7·31	23·88	7·41	23·84	7·52	23·81	7·62	25
26	24·86	7·60	24·83	7·71	24·80	7·82	24·76	7·93	26
27	25·82	7·89	25·79	8·01	25·75	8·12	25·71	8·23	27
28	26·78	8·19	26·74	8·30	26·70	8·42	26·67	8·54	28
29	27·73	8·48	27·70	8·60	27·66	8·72	27·62	8·84	29
30	28·69	8·77	28·65	8·90	28·61	9·02	28·57	9·15	30
31	29·65	9·06	29·61	9·19	29·57	9·32	29·52	9·45	31
32	30·60	9·36	30·56	9·49	30·52	9·62	30·48	9·76	32
33	31·56	9·65	31·52	9·79	31·47	9·92	31·43	10·06	33
34	32·51	9·94	32·47	10·08	32·43	10·22	32·38	10·37	34
35	33·47	10·23	33·43	10·38	33·38	10·52	33·33	10·67	35
36	34·43	10·53	34·38	10·68	34·33	10·83	34·29	10·98	36
37	35·38	10·82	35·34	10·97	35·29	11·13	35·24	11·28	37
38	36·34	11·11	36·29	11·27	36·24	11·43	36·19	11·58	38
39	37·30	11·40	37·25	11·57	37·19	11·73	37·14	11·89	39
40	38·25	11·69	38·20	11·86	38·15	12·03	38·10	12·19	40
41	39·21	11·99	39·16	12·16	39·10	12·33	39·05	12·50	41
42	40·16	12·28	40·11	12·45	40·06	12·63	40·00	12·80	42
43	41·12	12·57	41·07	12·75	41·01	12·93	40·95	13·11	43
44	42·08	12·86	42·02	13·05	41·96	13·23	41·91	13·41	44
45	43·03	13·16	42·98	13·34	42·92	13·53	42.86	13·72	45
46	43·99	13·45	43·93	13·64	43·87	13·83	43·81	14·02	46
47	44·95	13·74	44·89	13·94	44·82	14·13	44·76	14·33	47
48	45·90	14·03	45·84	14·23	45·78	14·43	45·71	14·63	48
49	46·86	14·33	46·80	14·53	46·73	14·73	46·67	14·94	49
50	47·82	14·62	47·75	14·83	47·69	15·04	47·62	15·24	50
Distance.	Dep.	Lat.	Dep.	Lat.	Dep.	Lat.	Dep.	Lat.	Distance.
	73 Deg.		72¾ Deg.		72½ Deg.		72¼ Deg.		

Distance.	17 Deg.		17¼ Deg.		17½ Deg.		17¾ Deg.		Distance.
	Lat.	Dep.	Lat.	Dep.	Lat.	Dep.	Lat.	Dep.	
51	48·77	14·91	48·71	15·12	48·64	15·34	48·57	15·55	51
52	49·73	15·20	49·66	15·42	49·59	15·64	49·52	15·85	52
53	50·68	15·50	50·62	15·72	50·55	15·94	50·48	16·16	53
54	51·64	15·79	51·57	16·01	51·50	16·24	51·43	16·46	54
55	52·60	16·08	52·53	16·31	52·45	16·54	52·38	16·77	55
56	53·55	16·37	53·48	16·61	53·41	16·84	53·33	17·07	56
57	54·51	16·67	54·44	16·90	54·36	17·14	54·29	17·38	57
58	55·47	16·96	55·39	17·20	55·32	17·44	55·24	17·68	58
59	56·42	17·25	56·35	17·50	56·27	17·74	56·10	17·99	59
60	57·38	17·54	57·30	17·79	57·22	18·04	57·14	18·29	60
61	58·33	17·83	58·26	18·09	58·18	18·34	58·10	18·60	61
62	59·29	18·13	59·21	18·39	59·13	18·64	59·05	18·90	62
63	60·25	18·42	60·17	18·68	60·08	18·94	60·00	19·21	63
64	61·20	18·71	61·12	18·98	61·04	19·25	60·95	19·51	64
65	62·16	19·00	62·08	19·28	61·99	19·55	61·91	19·82	65
66	63·12	19·30	63·03	19·57	62·95	19·85	62·86	20·12	66
67	64·07	19·59	63·99	19·87	63·90	20·15	63·81	20·43	67
68	65·03	19·88	64·94	20·16	64·85	20·45	64·76	20·73	68
69	65·99	20·17	65·90	20·46	65·81	20·75	65·72	21·04	69
70	66·94	20·47	66·85	20·76	66·76	21·05	66·67	21·34	70
71	67·90	20·76	67.81	21·05	67·71	21·35	67·62	21·65	71
72	68·85	21·05	68·76	21·35	68·67	21·65	68·57	21·95	72
73	69·81	21·34	69·72	21·65	69·62	21·95	69·52	22·26	73
74	70·77	21·64	70·67	21·94	70·58	22·25	70·48	22·56	74
75	71·72	21·93	71·63	22·24	71·53	22·55	71·43	22·86	75
76	72·68	22·22	72·58	22·54	72·48	22·85	72·38	23·17	76
77	73·64	22·51	73·54	22·83	73·44	23·15	73·33	23·47	77
78	74·59	22·80	74·49	23·13	74·39	23·46	74·29	23·78	78
79	75·55	23·10	75·45	23·43	75·34	23·76	75·24	24·08	79
80	76·50	23·39	76·40	23·72	76·30	24·06	76·19	24·39	80
81	77·46	23·68	77·36	24·02	77·25	24·36	77·14	24·69	81
82	78·42	23·97	78·31	24·32	78·20	24·66	78·10	25·00	82
83	79·37	24·27	79·27	24·61	79·16	25·96	79·05	25·30	83
84	80·33	24·56	80·22	24·91	80·11	25·26	80·00	25·61	84
85	81·29	24·85	81·18	25·21	81·07	25·56	80·95	25·91	85
86	82·24	25·14	82·13	25·50	82·02	25·86	81·91	26·22	86
87	83·20	25·44	83·09	25·80	82·97	26·16	82·86	26·52	87
88	84·15	25·73	84·04	26·10	83·93	26·46	83·81	26·83	88
89	85·11	26·02	85·00	26·39	84·88	26·76	84·76	27·13	89
90	86·07	26·31	85·95	26·69	85·83	27·06	85·72	27·44	90
91	87 02	26·61	86·91	26·99	86·79	27·36	86·67	27·74	91
92	87·98	26·90	87·86	27·28	87·74	27·66	87·62	28·05	92
93	88·94	27·19	88·82	27·58	88·70	27·97	88·57	28·35	93
94	89·89	27·48	89·77	27·87	89·65	28·27	89·53	28·66	94
95	90·85	27·78	90·73	28·17	90·60	28·57	90·48	28·96	95
96	91·81	28·07	91·68	28·47	91·56	28·87	91·43	29·27	96
97	92·76	28·36	92·64	28·76	92·51	29·17	92·38	29·57	97
98	93·72	28·65	93·59	29·06	93·46	29·47	93·33	29·88	98
99	94·67	28·94	94·55	29·36	94·42	29·77	94·20	30·18	99
100	95·63	29·24	95·50	29·65	95·37	30·07	95·24	30·49	100
	Dep.	Lat.	Dep.	Lat.	Dep.	Lat.	Dep.	Lat.	
Distance.	73 Deg.		72¾ Deg.		72½ Deg.		72¼ Deg.		Distance.

Distance.	18 Deg.		18¼ Deg.		18½ Deg.		18¾ Deg.		Distance.
	Lat.	Dep.	Lat.	Dep.	Lat.	Dep.	Lat.	Dep.	
1	0·95	0·31	0·95	0·31	0·95	0·32	0·95	0·32	1
2	1·90	0·62	1·90	0·63	1·90	0·63	1·89	0·64	2
3	2·85	0·93	2·85	0·94	2·84	0·95	2·84	0·96	3
4	3·80	1·24	3·80	1·25	3·79	1·27	3·79	1·29	4
5	4·76	1·55	4·75	1·57	4·74	1·59	4·73	1·61	5
6	5·71	1·85	5·70	1·88	5·69	1·90	5·68	1·93	6
7	6·66	2·16	6·65	2·19	6·64	2·22	6·63	2·25	7
8	7·61	2·47	7·60	2·51	7·59	2·54	7·58	2·57	8
9	8·56	2·78	8·55	2·82	8·53	2·86	8·52	2·89	9
10	9.51	3·09	9·50	3·13	9·48	3·17	9·47	3·21	10
11	10·46	3·40	10·45	3·44	10·43	3·49	10·42	3·54	11
12	11·41	3·71	11·40	3·76	11·38	3·81	11·36	3·86	12
13	12·36	4·02	12·35	4·07	12·33	4·12	12·31	4·18	13
14	13·31	4·33	13·30	4·38	13·28	4·44	13·26	4·50	14
15	14·27	4·64	14·25	4·70	14·22	4·76	14·20	4·82	15
16	15·22	4·94	15·20	5·01	15·17	5·08	15·15	5·14	16
17	16·17	5·25	16·14	5·32	16·12	5·39	16·10	5·46	17
18	17·12	5·56	17·09	5·64	17·07	5·71	17·04	5·79	18
19	18·07	5·87	18·04	5·95	18·02	6·03	17·99	6·11	19
20	19·02	6·18	18·99	6·26	18·97	6·35	18·94	6·43	20
21	19·97	6·49	19·94	6·58	19·91	6·66	19·89	6·75	21
22	20·92	6·80	20·89	6·89	20·86	6·98	20·83	7·07	22
23	21·87	7·11	21·84	7·20	21·81	7·30	21·78	7·39	23
24	22·83	7·42	22·79	7·52	22·76	7·62	22·73	7·71	24
25	23·78	7·73	23·74	7·83	23·71	7·93	23·67	8·04	25
26	24·73	8·03	24·69	8·14	24·66	8·25	24·62	8·36	26
27	25·68	8·34	25·64	8·46	25·60	8·57	25·57	8·68	27
28	26·63	8·65	26·59	8·77	26·55	8·88	26·51	9·00	28
29	27·58	8·96	27·54	9·08	27·50	9·20	27·46	9·32	29
30	28·53	9·27	28·49	9·39	28·45	9·52	28·41	9·64	30
31	29·48	9·58	29·44	9·71	29·40	9·84	29·35	9·96	31
32	30·43	9·89	30·39	10·02	30·35	10·15	30·30	10·29	32
33	31·38	10·20	31·34	10·33	31·29	10·47	31·25	10·61	33
34	32·34	10·51	32·29	10·65	32·24	10·79	32·20	10·93	34
35	33·29	10·82	33·24	10·96	33·19	11·11	33·14	11·25	35
36	34·24	11·12	34·19	11·27	34·14	11·42	34·09	11·57	36
37	35·19	11·43	35·14	11·59	35·09	11·74	35·04	11·89	37
38	36·14	11·74	36·09	11·90	36·04	12·06	35·98	12·21	38
39	37·09	12·05	37·04	12·21	36·98	12·37	36·93	12·54	39
40	38·04	12·36	37·99	12·53	37·93	12·69	37·88	12·86	40
41	38·99	12·67	38·94	12·84	38·88	13·01	38·82	13·18	41
42	39·94	12·98	39·89	13·15	39·83	13·33	39·77	13·50	42
43	40·90	13·29	40·84	13·47	40·78	13·64	40·72	13·82	43
44	41·85	13·60	41·79	13·78	41·73	13·96	41·66	14·14	44
45	42·80	13·91	42·74	14·09	42·67	14·28	42·61	14·46	45
46	43·75	14·21	43·69	14·41	43·62	14·60	43·56	14·79	46
47	44·70	14·52	44·64	14·72	44·57	14·91	44·51	15·11	47
48	45·65	14·83	45·59	15·03	45·52	15·23	45·45	15·43	48
49	46·60	15·14	46·54	15·35	46·47	15·55	46·40	15·75	49
50	47·55	15·45	47·48	15·66	47·42	15·87	47·35	16·07	50
	Dep.	Lat.	Dep.	Lat.	Dep.	Lat.	Dep.	Lat.	
Distance.	72 Deg.		71¾ Deg.		71½ Deg.		71¼ Deg.		Distance.

Distance.	18 Deg.		18¼ Deg.		18½ Deg.		18¾ Deg.		Distance.
	Lat.	Dep.	Lat.	Dep.	Lat.	Dep.	Lat.	Dep.	
51	48·50	15·76	48·43	15·97	48·36	16·18	48·29	16·39	51
52	49.45	16·07	49·38	16·28	49·31	16·50	49·24	16·71	52
53	50·41	16·38	50·33	16·60	50·26	16·82	50·19	17·04	53
54	51·36	16·69	51·28	16·91	51·21	17·13	51·13	17·36	54
55	52·31	17·00	52·23	17·22	52·16	17·45	52·08	17·68	55
56	53·26	17·30	53·18	17·54	53·11	17·77	53·03	18·00	56
57	54·21	17·61	54·13	17·85	54·05	18·09	53·98	18·32	57
58	55·16	17·92	55·08	18·16	55·00	18·40	54·92	18·64	58
59	56·11	18·23	56·03	18·48	55·95	18·72	55·87	18·96	59
60	57·06	18·54	56·98	18·79	56·90	19·04	56·82	19·29	60
61	58·01	18·85	57·93	19·10	57·85	19·36	57·76	19·61	61
62	58·97	19·16	58·88	19·42	58·80	19·67	58·71	19·93	62
63	59·92	19·47	59·83	19·73	59·74	19·99	59·66	20·25	63
64	60·87	19·78	60·78	20·04	60·69	20·31	60·60	20·57	64
65	61·82	20·09	61·73	20·36	61·64	20·62	61·55	20·89	65
66	62·77	20·40	62·68	20·67	62·59	20·94	62·50	21·22	66
67	63·72	20·70	63·63	20·98	63·54	21·26	63·44	21·54	67
68	64·67	21·01	64·58	21·30	64·49	21·58	64·39	21·86	68
69	65·62	21·32	65·53	21·61	65·43	21·89	65·34	22·18	69
70	66·57	21·63	66·48	21·92	66·38	22·21	66·29	22·50	70
71	67·53	21·94	67·43	22·23	67·33	22·53	67·23	22·82	71
72	68·48	22·25	68·38	22·55	68·28	22·85	68·18	23·14	72
73	69·43	22·56	69·33	22·86	69·23	23·16	69·13	23·47	73
74	70·38	22·87	70·28	23·17	70·18	23·48	70·07	23·79	74
75	71·33	23·18	71·23	23·49	71·12	23·80	71·02	24·11	75
76	72·28	23·49	72·18	23·80	72·07	24·12	71·97	24·43	76
77	73·23	23·79	73·13	24·11	73·02	24·43	72·91	24·75	77
78	74·18	24·10	74·08	24·43	73·97	24·75	73·86	25·07	78
79	75·13	24·41	75·03	24·74	74·92	25·07	74·81	25·39	79
80	76·08	24·72	75·98	25·05	75·87	25·38	75·75	25·72	80
81	77·04	25·03	76·93	25·37	76·81	25·70	76·70	26·04	81
82	77·99	25·34	77·88	25·68	77·76	26·02	77·65	26·36	82
83	78·94	25·65	78·83	25·99	78·71	26·34	78·60	26·68	83
84	79·89	25·96	79·77	26·31	79·66	26·65	79·54	27·00	84
85	80·84	26·27	80·72	26·62	80·61	26·97	80·49	27·32	85
86	81·79	26·58	81·67	26·93	81·56	27·29	81·44	27·64	86
87	82·74	26·88	82·62	27·25	82·50	27·61	82·38	27·97	87
88	83·69	27·19	83·57	27·56	83·45	27·92	83·33	28·29	88
89	84·64	27·50	84·52	27·87	84·40	28·24	84·28	28·61	89
90	85·60	27·81	85·47	28·18	85·35	28·56	85·22	28·93	90
91	86·55	28·12	86·42	28·50	86·30	28·87	86·17	29·25	91
92	87·50	28·43	87·37	28·81	87·25	29·19	87·12	29·57	92
93	88·45	28·74	88·32	29·12	88·19	29·51	88·06	29·89	93
94	89·40	29·05	89·27	29·44	89·14	29·83	89·01	30·22	94
95	90·35	29·36	90·22	29·75	90·09	30·14	89·96	30·54	95
96	91·30	29·67	91·17	30·06	91·04	30·46	90·91	30·86	96
97	92·25	29·97	92·12	30·38	91·99	30·78	91·85	31·18	97
98	93·20	30·28	93·07	30·69	92·94	31·10	92·80	31·50	98
99	94·15	30·59	94·02	31·00	93·88	31·41	93·75	31·82	99
100	95·11	30·90	94·97	31·32	94·83	31·73	94·69	32·14	100
Distance.	Dep.	Lat.	Dep.	Lat.	Dep.	Lat.	Dep.	Lat.	Distance.
	72 Deg.		71¾ Deg.		71½ Deg.		71¼ Deg.		

Distance.	19 Deg.		19¼ Deg.		19½ Deg.		19¾ Deg.		Distance.
	Lat.	Dep.	Lat.	Dep.	Lat.	Dep.	Lat.	Dep.	
1	0·95	0·33	0·94	0·33	0·94	0·33	0·94	0·34	1
2	1·89	0·65	1·89	0·66	1·89	0·67	1·88	0·68	2
3	2·84	0·98	2·83	0·99	2·83	1·00	2·82	1·01	3
4	3·78	1·30	3·78	1·32	3·77	1·34	3·76	1·35	4
5	4·73	1·63	4·72	1·65	4·71	1·67	4·71	1·69	5
6	5·67	1·95	5·66	1·98	5·66	2·00	5·65	2·03	6
7	6·62	2·28	6·61	2·31	6·60	2·34	6·59	2·37	7
8	7·56	2·60	7·55	2·64	7·54	2·67	7·53	2·70	8
9	8·51	2·93	8·50	2·97	8·48	3·00	8·47	3·04	9
10	9·46	3·26	9·44	3·30	9·43	3·34	9·41	3·38	10
11	10·40	3·58	10·38	3·63	10·37	3·67	10·35	3·72	11
12	11·35	3·91	11·33	3·96	11·31	4·01	11·29	4·06	12
13	12·29	4·23	12·27	4·29	12·25	4·34	12·24	4·39	13
14	13·24	4·56	13·22	4·62	13·20	4·67	13·18	4·73	14
15	14·18	4·88	14·16	4·95	14·14	5·01	14·12	5·07	15
16	15·13	5·21	15·11	5·28	15·08	5·34	15·06	5·41	16
17	16·07	5·53	16·05	5·60	16·02	5·67	16·00	5·74	17
18	17·02	5·86	16·99	5·93	16·97	6·01	16·94	6·08	18
19	17·96	6·19	17·94	6·26	17·91	6·34	17·88	6·42	19
20	18·91	6·51	18·88	6·59	18·85	6·68	18·82	6·76	20
21	19·86	6·84	19·83	6·92	19·80	7·01	19·76	7·10	21
22	20·80	7·16	20·77	7·25	20·74	7·34	20·71	7·43	22
23	21·75	7·49	21·71	7·58	21·68	7·68	21·65	7·77	23
24	22·69	7·81	22·66	7·91	22·62	8·01	22·59	8·11	24
25	23·64	8·14	23·60	8·24	23·57	8·35	23·53	8·45	25
26	24·58	8·46	24·55	8·57	24·51	8·68	24·47	8·79	26
27	25·53	8·79	25·49	8·90	25·45	9·01	25·41	9·12	27
28	26·47	9·12	26·43	9·23	26·39	9·35	26·35	9·46	28
29	27·42	9·44	27·38	9·56	27·34	9·68	27·29	9·80	29
30	28·37	9·77	28·32	9·89	28·28	10·01	28·24	10·14	30
31	29·31	10·09	29·27	10·22	29·22	10·35	29·18	10·48	31
32	30·26	10·42	30·21	10·55	30·16	10·68	30·12	10·81	32
33	31·20	10·74	31·15	10·88	31·11	11·02	31·06	11·15	33
34	32·15	11·07	32·10	11·21	32·05	11·35	32·00	11·49	34
35	33·09	11·39	33·04	11·54	32·99	11·68	32·94	11·83	35
36	34·04	11·72	33·99	11·87	33·94	12·02	33·88	12·17	36
37	34·98	12·05	34·93	12·20	34·88	12·35	34·82	12·50	37
38	35·93	12·37	35·88	12·53	35·82	12·68	35·76	12·84	38
39	36·88	12·70	36·82	12·86	36·76	13·02	36·71	13·18	39
40	37·82	13·02	37·76	13·19	37·71	13·35	37·65	13·52	40
41	38·77	13·35	38·71	13·52	38·65	13·69	38·59	13·85	41
42	39·71	13·67	39·65	13·85	39·59	14·02	39·53	14·19	42
43	40·66	14·00	40·60	14·18	40·53	14·35	40·47	14·53	43
44	41·60	14·32	41·54	14·51	41·48	14·69	41·41	14·87	44
45	42·55	14·65	42·48	14·84	42·42	15·02	42·35	15·21	45
46	43·49	14·98	43·43	15·17	43·36	15·36	43·29	15·54	46
47	44·44	15·30	44·37	15·50	44·30	15·69	44·24	15·88	47
48	45·38	15·63	45·32	15·83	45·25	16·02	45·18	16·22	48
49	46·33	15·95	46·26	16·15	46·19	16·36	46·12	16·56	49
50	47·28	16·28	47·20	16·48	47·13	16·69	47·06	16·90	50
	Dep.	Lat.	Dep.	Lat.	Dep.	Lat.	Dep.	Lat.	
Distance.	71 Deg.		70¾ Deg.		70½ Deg.		70¼ Deg.		Distance.

Distance.	19 Deg.		19¼ Deg.		19½ Deg.		19¾ Deg.		Distance.
	Lat.	Dep.	Lat.	Dep.	Lat.	Dep.	Lat.	Dep.	
51	48·22	16·60	48·15	16·81	48·07	17·02	48·00	17·23	51
52	49·17	16·93	49·09	17·14	49·02	17·36	48·94	17·57	52
53	50·11	17·26	50·04	17·47	49·96	17·69	49·88	17·91	53
54	51·06	17·58	50·98	17·80	50·90	18·03	50·82	18·25	54
55	52·00	17·91	51·92	18·13	51·85	18·36	51·76	18·59	55
56	52·95	18·23	52·87	18·46	52·79	18·69	52·71	18·92	56
57	53·89	18·56	53·81	18·79	53·73	19·03	53·65	19·26	57
58	54·84	18·88	54·76	19·12	54·67	19·36	54·59	19·60	58
59	55·79	19·21	55·70	19·45	55·62	19·69	55·53	19·94	59
60	56·73	19·53	56·65	19·78	56·56	20·03	56·47	20·27	60
61	57·68	19·86	57·59	20·11	57·50	20·36	57·41	20·61	61
62	58·62	20·19	58·53	20·44	58·44	20·70	58·35	20·95	62
63	59·57	20·51	59·48	20·77	59·39	21·03	59·29	21·29	63
64	60·51	20·84	60·42	21·10	60·33	21·36	60·24	21·63	64
65	61·46	21·16	61·37	21·43	61·27	21·70	61·18	21·96	65
66	62·40	21·49	62·31	21·76	62·21	22·03	62·12	22·30	66
67	63·35	21·81	63·25	22·09	63·16	22·37	63·06	22·64	67
68	64·30	22·14	64·20	22·42	64·10	22·70	64·00	22·98	68
69	65·24	22·46	65·14	22·75	65·04	23·03	64·94	23·32	69
70	66·19	22·79	66·09	23·08	65·98	23·37	65·88	23·65	70
71	67·13	23·12	67·03	23·41	66·93	23·70	66·82	23·99	71
72	68·08	23·44	67·97	23·74	67·87	24·03	67·76	24·33	72
73	69·02	23·77	68·92	24·07	68·81	24·37	68·71	24·67	73
74	69·97	24·09	69·86	24·40	69·76	24·70	69·65	25·01	74
75	70·91	24·42	70·81	24·73	70·70	25·04	70·59	25·34	75
76	71·86	24·74	71·75	25·06	71·64	25·37	71·53	25·68	76
77	72·80	25·07	72·69	25·39	72·58	25·70	72·47	26·02	77
78	73·75	25·39	73·64	25·72	73·53	26·04	73·41	26·36	78
79	74·70	25·72	74·58	26·05	74·47	26·37	74·35	26·70	79
80	75·64	26·05	75·53	26·38	75·41	26·70	75·29	27·03	80
81	76·59	26·37	76·47	26·70	76·35	27·04	76·24	27·37	81
82	77·53	26·70	77·42	27·03	77·30	27·37	77·18	27·71	82
83	78·48	27·02	78·36	27·36	78·24	27·71	78·12	28·05	83
84	79·42	27·35	79·30	27·69	79·18	28·04	79·06	28·39	84
85	80·37	27·67	80·25	28·02	80·12	28·37	80·00	28·72	85
86	81·31	28·00	81·19	28·35	81·07	28·71	80·94	29·06	86
87	82·26	28·32	82·14	28·68	82·01	29·04	81·88	29·40	87
88	83·21	28·65	83·08	29·01	82·95	29·37	82·82	29·74	88
89	84·15	28·98	84·02	29·34	83·90	29·71	83·76	30·07	89
90	85·10	29·30	84·97	29·67	84·84	30·04	84·71	30·41	90
91	86·04	29·63	85·91	30·00	85·78	30·38	85·65	30·75	91
92	86·99	29·95	86·86	30·33	86·72	30·71	86·59	31·09	92
93	87·93	30·28	87·80	30·66	87·67	31·04	87·53	31·43	93
94	88·88	30·60	88·74	30·99	88·61	31·38	88·47	31·76	94
95	89·82	30·93	89·69	31·32	89·55	31·71	89·41	32·10	95
96	90·77	31·25	90·63	31·65	90·49	32·05	90·35	32·44	96
97	91·72	31·58	91·58	31·98	91·44	32·38	91·29	32·78	97
98	92·66	31·91	92·52	32·31	92·38	32·71	92·24	33·12	98
99	93·61	32·23	93·46	32·64	93·32	33·05	93·18	33·45	99
100	94·55	32·56	94·41	32·97	94·26	33·38	94·12	33·79	100
Distance.	Dep.	Lat.	Dep.	Lat.	Dep.	Lat.	Dep.	Lat.	Distance.
	71 Deg.		70¾ Deg.		70½ Deg.		70¼ Deg.		

Distance.	20 Deg.		20¼ Deg.		20½ Deg.		20¾ Deg.		Distance.
	Lat.	Dep.	Lat.	Dep.	Lat.	Dep.	Lat.	Dep.	
1	0·94	0·34	0·94	0·35	0·94	0·35	0·94	0·35	1
2	1·88	0·68	1·88	0·69	1·87	0·70	1·87	0·71	2
3	2·82	1·03	2·81	1·04	2·81	1·05	2·81	1·06	3
4	3·76	1·37	3·75	1·38	3·75	1·40	3·74	1·42	4
5	4·70	1·71	4·69	1·73	4·68	1·75	4·68	1·77	5
6	5·64	2·05	5·63	2·08	5·62	2·10	5·61	2·13	6
7	6·58	2·39	6·57	2·42	6·56	2·45	6·55	2·48	7
8	7·52	2·74	7·51	2·77	7·49	2·80	7·48	2·83	8
9	8·46	3·08	8·44	3·12	8·43	3·15	8·42	3·19	9
10	9·40	3·42	9·38	3·46	9·37	3·50	9·35	3·54	10
11	10·34	3·76	10·32	3·81	10·30	3·85	10·29	3·90	11
12	11·28	4·10	11·26	4·15	11·24	4·20	11·22	4·25	12
13	12·22	4·45	12·20	4·50	12·18	4·55	12·16	4·61	13
14	13·16	4·79	13·13	4·85	13·11	4·90	13·09	4·96	14
15	14·10	5·13	14·07	5·19	14·05	5·25	14·03	5·31	15
16	15·04	5·47	15·01	5·54	14·99	5·60	14·96	5·67	16
17	15·97	5·81	15·95	5·88	15·92	5·95	15·90	6·02	17
18	16·91	6·16	16·89	6·23	16·86	6·30	16·83	6·38	18
19	17·85	6·50	17·83	6·58	17·80	6·65	17·77	6·73	19
20	18·79	6·84	18·76	6·92	18·73	7·00	18·70	7·09	20
21	19·73	7·18	19·70	7·27	19·67	7·35	19·64	7·44	21
22	20·67	7·52	20·64	7·61	20·61	7·70	20·57	7·79	22
23	21·61	7·87	21·58	7·96	21·54	8·05	21·51	8·15	23
24	22·55	8·21	22·52	8·31	22·48	8·40	22·44	8·50	24
25	23·49	8·55	23·45	8·65	23·42	8·76	23·38	8·86	25
26	24·43	8·89	24·39	9·00	24·35	9·11	24·31	9·21	26
27	25·37	9·23	25·33	9·35	25·29	9·46	25·25	9·57	27
28	26·31	9·58	26·27	9·69	26·23	9·81	26·18	9·92	28
29	27·25	9·92	27·21	10·04	27·16	10·16	27·12	10·27	29
30	28·19	10·26	28·15	10·38	28·10	10·51	28·05	10·63	30
31	29·13	10·60	29·08	10·73	29·04	10·86	28·99	10·98	31
32	30·07	10·94	30·02	11·08	29·97	11·21	29·92	11·34	32
33	31·01	11·29	30·96	11·42	30·91	11·56	30·86	11·69	33
34	31·95	11·63	31·90	11·77	31·85	11·91	31·79	12·05	34
35	32·89	11·97	32·84	12·11	32·78	12·26	32·73	12·40	35
36	33·83	12·31	33·77	12·46	33·72	12·61	33·66	12·75	36
37	34·77	12·65	34·71	12·81	34·66	12·96	34·60	13·11	37
38	35·71	13·00	35·65	13·15	35·59	13·31	35·54	13·46	38
39	36·65	13·34	36·59	13·50	36·53	13·66	36·47	13·82	39
40	37·59	13·68	37·53	13·84	37·47	14·01	37·41	14·17	40
41	38·53	14·02	38·47	14·19	38·40	14·36	38·34	14·53	41
42	39·47	14·36	39·40	14·54	39·34	14·71	39·28	14·88	42
43	40·41	14·71	40·34	14·88	40·28	15·06	40·21	15·23	43
44	41·35	15·05	41·28	15·23	41·21	15·41	41·15	15·59	44
45	42·29	15·39	42·22	15·58	42·15	15·76	42·08	15·94	45
46	43·23	15·73	43·16	15·92	43·09	16·11	43.02	16·30	46
47	44·17	16·07	44·09	16·27	44·02	16·46	43·95	16·65	47
48	45·11	16·42	45·03	16·61	44·96	16·81	44·89	17·01	48
49	46·04	16·76	45·97	16·96	45·90	17·16	45·82	17·36	49
50	46·98	17·10	46·91	17·31	46·83	17·51	46·76	17·71	50
	Dep.	Lat.	Dep.	Lat.	Dep.	Lat.	Dep.	Lat.	
Distance.	70 Deg.		69¾ Deg.		69½ Deg.		69¼ Deg.		Distance.

Distance.	20 Deg.		20¼ Deg.		20½ Deg.		20¾ Deg.		Distance.
	Lat.	Dep.	Lat.	Dep.	Lat.	Dep.	Lat.	Dep.	
51	47·92	17·44	47·85	17·65	47·77	17·36	47·69	18·07	51
52	48·86	17·79	48·79	18·00	48·71	18·21	48·63	18·42	52
53	49·80	18·13	49·72	18·34	49·64	18·56	49·56	18·78	53
54	50·74	18·47	50·66	18·69	50·58	18·91	50·50	19·13	54
55	51·68	18·81	51·60	19·04	51·52	19·26	51·43	19·49	55
56	52·62	19·15	52·54	19·38	52·45	19·61	52·37	19·84	56
57	53·56	19·50	53·48	19·73	53·39	19·96	53·30	20·19	57
58	54·50	19·84	54·42	20·07	54·33	20·31	54·24	20·55	58
59	55·44	20·18	55·35	20·42	55·26	20·66	55·17	20·90	59
60	56·38	20·52	56·29	20·77	56·20	21·01	56·11	21·26	60
61	57·32	20·86	57·23	21·11	57·14	21·36	57·04	21·61	61
62	58·26	21·21	58·17	21·46	58·07	21·71	57·98	21·97	62
63	59·20	21·55	59·11	21·81	59·01	22·06	58·91	22·32	63
64	60·14	21·89	60·04	22·15	59·95	22·41	59·85	22·67	64
65	61·08	22·23	60·98	22·50	60·88	22·76	60·78	23·03	65
66	62·02	22·57	61·92	22·84	61·82	23·11	61·72	23·38	66
67	62·96	22·92	62·86	23·19	62·76	23·46	62·65	23·74	67
68	63·90	23·26	63·80	23·54	63·69	23·81	63·59	24·09	68
69	64·84	23·60	64·74	23·88	64·63	24·16	64·52	24·45	69
70	65·78	23·94	65·67	24·23	65·57	24·51	65·46	24·80	70
71	66·72	24·28	66·61	24·57	66·50	24·86	66·39	25·15	71
72	67·66	24·63	67·55	24·92	67·44	25·21	67·33	25·51	72
73	68·60	24·97	68·49	25·27	68·38	25·57	68·26	25·86	73
74	69·54	25·31	69·43	25·61	69·31	25·92	69·20	26·22	74
75	70·48	25·65	70·36	25·96	70·25	26·27	70·14	26·57	75
76	71·42	25·99	71·30	26·30	71·19	26·62	71·07	26·93	76
77	72·36	26·34	72·24	26·65	72·12	26·97	72·01	27·28	77
78	73·30	26·68	73·18	27·00	73·06	27·32	72·94	27·63	78
79	74·24	27·02	74·12	27·34	74·00	27·67	73·88	27·99	79
80	75·18	27·36	75·06	27·69	74·93	28·02	74·81	28·34	80
81	76·12	27·70	75·99	28·04	75·87	28·37	75·75	28·70	81
82	77·05	28·05	76·93	28·38	76·81	28·72	76·68	29·05	82
83	77·99	28·39	77·87	28·73	77·74	29·07	77·62	29·41	83
84	78·93	28·73	78·81	29·07	78·68	29·42	78·55	29·76	84
85	79·87	29·07	79·75	29·42	79·62	29·77	79·49	30·11	85
86	80·81	29·41	80·68	29·77	80·55	30·12	80·42	30·47	86
87	81·75	29·76	81·62	30·11	81·49	30·47	81·36	30·82	87
88	82·69	30·10	82·56	30·46	82·43	30·82	82·29	31·18	88
89	83·63	30·44	83·50	30·80	83·36	31·17	83·23	31·53	89
90	84·57	30·78	84·44	31·15	84·30	31·52	84·16	31·89	90
91	85·51	31·12	85·38	31·50	85·24	31·87	85·10	32·24	91
92	86·45	31·47	86·31	31·84	86·17	32·22	86·03	32·59	92
93	87·39	31·81	87·25	32·19	87·11	32·57	86·97	32·95	93
94	88·33	32·15	88·19	32·54	88·05	32·92	87·90	33·30	94
95	89·27	32·49	89·13	32·88	88·98	33·27	88·84	33·66	95
96	90·21	32·83	90·07	33·23	89·92	33·62	89·77	34·01	96
97	91·15	33·18	91·00	33·57	90·86	33·97	90·71	34·37	97
98	92·09	33·52	91·94	33·92	91·79	34·32	91·64	34·72	98
99	93·03	33·86	92·88	34·27	92·73	34·67	92·58	35·07	99
100	93·97	34·20	93·82	34·61	93·67	35·02	93·51	35·43	100
Distance.	Dep.	Lat.	Dep.	Lat.	Dep.	Lat.	Dep.	Lat.	Distance.
	70 Deg.		69¾ Deg.		69½ Deg.		69¼ Deg.		

Distance.	21 Deg.		21¼ Deg.		21½ Deg.		21¾ Deg.		Distance.
	Lat.	Dep.	Lat.	Dep.	Lat.	Dep.	Lat.	Dep.	
1	0·93	0·36	0·93	0·36	0·93	0·37	0·93	0·37	1
2	1·87	0·72	1·86	0·72	1·86	0·73	1·86	0·74	2
3	2·80	1·08	2·80	1·09	2·79	1·10	2·79	1·11	3
4	3·73	1·43	3·73	1·45	3·72	1·47	3·72	1·48	4
5	4·67	1·79	4·66	1·81	4·65	1·83	4·64	1·85	5
6	5·60	2·15	5·59	2·17	5·58	2·20	5·57	2·22	6
7	6·54	2·51	6·52	2·54	6·51	2·57	6·50	2·59	7
8	7·47	2·87	7·46	2·90	7·44	2·93	7·43	2·96	8
9	8·40	3·23	8·39	3·26	8·37	3·30	8·36	3·34	9
10	9·34	3·58	9·32	3·62	9·30	3·67	9·29	3·71	10
11	10·27	3·94	10·25	3·99	10·23	4·03	10·22	4·08	11
12	11·20	4·30	11·18	4·35	11·17	4·40	11·15	4·45	12
13	12·14	4·66	12·12	4·71	12·10	4·76	12·07	4·82	13
14	13·07	5·02	13·05	5·07	13·03	5·13	13·00	5·19	14
15	14·00	5·38	13·98	5·44	13·96	5·50	13·93	5·56	15
16	14·94	5·73	14·91	5·80	14·89	5·86	14·86	5·93	16
17	15·87	6·09	15·84	6·16	15·82	6·23	15·79	6·30	17
18	16·80	6·45	16·78	6·52	16·75	6·60	16·72	6·67	18
19	17·74	6·81	17·71	6·89	17·68	6·96	17·65	7·04	19
20	18·67	7·17	18·64	7·25	18·61	7·33	18·58	7·41	20
21	19·61	7·53	19·57	7·61	19·54	7·70	19·50	7·78	21
22	20·54	7·88	20·50	7·97	20·47	8·06	20·43	8·15	22
23	21·47	8·24	21·44	8·34	21·40	8·43	21·36	8·52	23
24	22·41	8·60	22·37	8·70	22·33	8·80	22·29	8·89	24
25	23·34	8·96	23·30	9·06	23·26	9·16	23·22	9·26	25
26	24·27	9·32	24·23	9·42	24·19	9·53	24·15	9·63	26
27	25·21	9·68	25·16	9·79	25·12	9·90	25·08	10·01	27
28	26·14	10·03	26·10	10·15	26·05	10·26	26·01	10·38	28
29	27·07	10·39	27·03	10·51	26·98	10·63	26·94	10·75	29
30	28·01	10·75	27·96	10·87	27·91	11·00	27·86	11·12	30
31	28·94	11·11	28·89	11·24	28·84	11·36	28·79	11·49	31
32	29·87	11·47	29·82	11·60	29·77	11·73	29·72	11·86	32
33	30·81	11·83	30·76	11·96	30·70	12·09	30·65	12·23	33
34	31·74	12·18	31·69	12·32	31·63	12·46	31·58	12·60	34
35	32·68	12·54	32·62	12·69	32·56	12·83	32·51	12·97	35
36	33·61	12·90	33·55	13·05	33·50	13·19	33·44	13·34	36
37	34·54	13·26	34·48	13·41	34·43	13·56	34·37	13·71	37
38	35·48	13·62	35·42	13·77	35·36	13·93	35·29	14·08	38
39	36·41	13·98	36·35	14·14	36·29	14·29	36·22	14·45	39
40	37·34	14·33	37·28	14·50	37·22	14·66	37·15	14·82	40
41	38·28	14·69	38·21	14·86	38·15	15·03	38·08	15·19	41
42	39·21	15·05	39·14	15·22	39·08	15·39	39·01	15·56	42
43	40·14	15·41	40·08	15·58	40·01	15·76	39·94	15·93	43
44	41·08	15·77	41·01	15·95	40·94	16·13	40·87	16·30	44
45	42·01	16·13	41·94	16·31	41·87	16·49	41·80	16·68	45
46	42·94	16·48	42·87	16·67	42·80	16·86	42·73	17·05	46
47	43·88	16·84	43·80	17·03	43·73	17·23	43·65	17·42	47
48	44·81	17·20	44·74	17·40	44·66	17·59	44·58	17·79	48
49	45·75	17·56	45·67	17·76	45·59	17·96	45·51	18·16	49
50	46·68	17·92	46·60	18·12	46·52	18·33	46·44	18·53	50
Distance.	Dep.	Lat.	Dep.	Lat.	Dep.	Lat.	Dep.	Lat.	Distance.
	69 Deg.		68¾ Deg.		68½ Deg.		68¼ Deg.		

Distance.	21 Deg.		21¼ Deg.		21½ Deg.		21¾ Deg.		Distance.
	Lat.	Dep.	Lat.	Dep.	Lat.	Dep.	Lat.	Dep.	
51	47·61	18·28	47·53	18·48	47·45	18·69	47·37	18·90	51
52	48·55	18·64	48·46	18·85	48·38	19·06	48·30	19·27	52
53	49·48	18·99	49·40	19·21	49·31	19·42	49·23	19·64	53
54	50·41	19·35	50·33	19·57	50·24	19·79	50·16	20·01	54
55	51·35	19·71	51·26	19·93	51·17	20·16	51·08	20·38	55
56	52·28	20·07	52·19	20·30	52·10	20·52	52·01	20·75	56
57	53·21	20·43	53·12	20·66	53·03	20·89	52·94	21·12	57
58	54·15	20·79	54·06	21·02	53·96	21·26	53·87	21·49	58
59	55·08	21·14	54·99	21·38	54·89	21·62	54·80	21·86	59
60	56·01	21·50	55·92	21·75	55·83	21·99	55·73	22·23	60
61	56·95	21·86	56·85	22·11	56·76	22·36	56·66	22·60	61
62	57·88	22·22	57·78	22·47	57·69	22·72	57·59	22·97	62
63	58·82	22·58	58·72	22·83	58·62	23·09	58·52	23·35	63
64	59·75	22·94	59·65	23·20	59·55	23·46	59·44	23·72	64
65	60·68	23·29	60·58	23·56	60·48	23·82	60·37	24·09	65
66	61·62	23·65	61·51	23·92	61·41	24·19	61·30	24·46	66
67	62·55	24·01	62·44	24·28	62·34	24·56	62·23	24·83	67
68	63·48	24·37	63·38	24·65	63·27	24·92	63·16	25·20	68
69	64·42	24·73	64·31	25·01	64·20	25·29	64·09	25·57	69
70	65·35	25·09	65·24	25·37	65·13	25·66	65·02	25·94	70
71	66·28	25·44	66·17	25·73	66·06	26·02	65·95	26·31	71
72	67·22	25·80	67·10	26·10	66·99	26·39	66·87	26·68	72
73	68·15	26·16	68·04	26·46	67·92	26·75	67·80	27·05	73
74	69·08	26·52	68·97	26·82	68·85	27·12	68·73	27·42	74
75	70·02	26·88	69·90	27·18	69·78	27·49	69·66	27·79	75
76	70·95	27·24	70·83	27·55	70·71	27·85	70·59	28·16	76
77	71·89	27·59	71·76	27·91	71·64	28·22	71·52	28·53	77
78	72·82	27·95	72·70	28·27	72·57	28·59	72·45	28·90	78
79	73·75	28·31	73·63	28·63	73·50	28·95	73·38	29·27	79
80	74·69	28·67	74·56	29·00	74·43	29·32	74·30	29·64	80
81	75·62	29·03	75·49	29·36	75·36	29·69	75·23	30·02	81
82	76·55	29·39	76·42	29·72	76·29	30·05	76·16	30·39	82
83	77·49	29·74	77·36	30·08	77·22	30·42	77·09	30·76	83
84	78·42	30·10	78·29	30·44	78·16	30·79	78·02	31·13	84
85	79·35	30·46	79·22	30·81	79·09	31·15	78·95	31·50	85
86	80·29	30·82	80·15	31·17	80·02	31·52	79·88	31·87	86
87	81·22	31·18	81·08	31·53	80·95	31·89	80·81	32·24	87
88	82·16	31·54	82·02	31·89	81·88	32·25	81·74	32·61	88
89	83·09	31·89	82·95	32·26	82·81	32·62	82·66	32·98	89
90	84·02	32·25	83·88	32·62	83·74	32·99	83·59	33·35	90
91	84·96	32·61	84·81	32·98	84·67	33·35	84·52	33·72	91
92	85·89	32·97	85·74	33·34	85·60	33·72	85·45	34·09	92
93	86·82	33·33	86·68	33·71	86·53	34·08	86·38	34·46	93
94	87·76	33·69	87·61	34·07	87·46	34·45	87·31	34·83	94
95	88·69	34·04	88·54	34·43	88·39	34·82	88·24	35·20	95
96	89·62	34·40	89·47	34·79	89·32	35·18	89·17	35·57	96
97	90·56	34·76	90·40	35·16	90·25	35·55	90·09	35·94	97
98	91·49	35·12	91·34	35·52	91·18	35·92	91·02	36·31	98
99	92·42	35·48	92·27	35·88	92·11	36·28	91·95	36·69	99
100	93·36	35·84	93·20	36·24	93·04	36·65	92·88	37·06	100
	Dep.	Lat.	Dep.	Lat.	Dep.	Lat.	Dep.	Lat.	
Distance.	69 Deg.		68¾ Deg.		68 ½ Deg.		68¼ Deg.		Distance.

Distance.	22 Deg.		22¼ Deg.		22½ Deg.		22¾ Deg.		Distance.
	Lat.	Dep.	Lat.	Dep.	Lat.	Dep.	Lat.	Dep.	
1	0·93	0·37	0·93	0·38	0·92	0·38	0·92	0·39	1
2	1·85	0·75	1·85	0·76	1·85	0·77	1·84	0·77	2
3	2·78	1·12	2·78	1·14	2·77	1·15	2·77	1·16	3
4	3·71	1·50	3·70	1·51	3·70	1·53	3·69	1·55	4
5	4·64	1·87	4·63	1·89	4·62	1·91	4·61	1·93	5
6	5·56	2·25	5·55	2·27	5·54	2·30	5·53	2·32	6
7	6·49	2·62	6·48	2·65	6·47	2·68	6·46	2·71	7
8	7·42	3·00	7·40	3·03	7·39	3·06	7·38	3·09	8
9	8·34	3·37	8·33	3·41	8·31	3·44	8·30	3·48	9
10	9.27	3·75	9·26	3·79	9·24	3·83	9·22	3·87	10
11	10·20	4·12	10·18	4·17	10·16	4·21	10·14	4·25	11
12	11·13	4·50	11·11	4·54	11·09	4·59	11·07	4·64	12
13	12·05	4·87	12·03	4·92	12·01	4·97	11·99	5·03	13
14	12·98	5·24	12·96	5·30	12·93	5·36	12·91	5·41	14
15	13·91	5·62	13·88	5·68	13·86	5·74	13·83	5·80	15
16	14·83	5·99	14·81	6·06	14·78	6·12	14·76	6·19	16
17	15·76	6·37	15·73	6·44	15·71	6·51	15·68	6·57	17
18	16·69	6·74	16·66	6·82	16·63	6·89	16·60	6·96	18
19	17·62	7·12	17·59	7·19	17·55	7·27	17·52	7·35	19
20	18·54	7·49	18·51	7·57	18·48	7·65	18·44	7·73	20
21	19·47	7·87	19·44	7·95	19·40	8·04	19·37	8·12	21
22	20·40	8·24	20·36	8·33	20·33	8·42	20·29	8·51	22
23	21·33	8·62	21·29	8·71	21·25	8·80	21·21	8·89	23
24	22·25	8·99	22·21	9·09	22·17	9·18	22·13	9·28	24
25	23·18	9·37	23·14	9·47	23·10	9·57	23·05	9·67	25
26	24·11	9·74	24·06	9·84	24·02	9·95	23·98	10·05	26
27	25·03	10·11	24·99	10·22	24·94	10·33	24·90	10·44	27
28	25·96	10·49	25·92	10·60	25·87	10·72	25·82	10·83	28
29	26·89	10·86	26·84	10·98	26·79	11·10	26·74	11·21	29
30	27·82	11·24	27·77	11·36	27·72	11·48	27·67	11·60	30
31	28·74	11·61	28·69	11·74	28·64	11·86	28·59	11·99	31
32	29·67	11·99	29·62	12·12	29·56	12·25	29·51	12·37	32
33	30·60	12·36	30·54	12.50	30·49	12·63	30·43	12·76	33
34	31·52	12·74	31·47	12·87	31·41	13·01	31·35	13·15	34
35	32·45	13·11	32·39	13·25	32·34	13·39	32·28	13·53	35
36	33·38	13·49	33·32	13·63	33·26	13·78	33·20	13·92	36
37	34·31	13·86	34·24	14·01	34·18	14·16	34·12	14·31	37
38	35·23	14·24	35·17	14·39	35·11	14·54	35·04	14·70	38
39	36·16	14·61	36·10	14·77	36·03	14·92	35·97	15·08	39
40	37·09	14·98	37·02	15·15	36·96	15·31	36·89	15·47	40
41	38·01	15·36	37·95	15·52	37·88	15·69	37·81	15·86	41
42	38·94	15·73	38·87	15·90	38·80	16·07	38·73	16·24	42
43	39·87	16·11	39·80	16·28	39·73	16·46	39·65	16·63	43
44	40·80	16·48	40·72	16·66	40·65	16·84	40·58	17·02	44
45	41·72	16·86	41·65	17·04	41·57	17·22	41·50	17·40	45
46	42·65	17·23	42·57	17·42	42·50	17·60	42·42	17·79	46
47	43·58	17·61	43·50	17·80	43·42	17·99	43·34	18·18	47
48	44·50	17·98	44·43	18·18	44·35	18·37	44·27	18·56	48
49	45·43	18·36	45·35	18·55	45·27	18·75	45·19	18·95	49
50	46·36	18·73	46·28	18·93	46·19	19·13	46·11	19·34	50
Distance.	Dep.	Lat.	Dep.	Lat.	Dep.	Lat.	Dep.	Lat.	Distance.
	68 Deg.		67¾ Deg.		67½ Deg.		67¼ Deg.		

Distance.	22 Deg.		22¼ Deg.		22½ Deg.		22¾ Deg.		Distance.
	Lat.	Dep.	Lat.	Dep.	Lat.	Dep.	Lat.	Dep.	
51	47·29	19·10	47·20	19·31	47·12	19·52	47·03	19·72	51
52	48·21	19·48	48·13	19·69	48·04	19·90	47·95	20·11	52
53	49·14	19·85	49·05	20·07	48·97	20·28	48·88	20·50	53
54	50·07	20·23	49·98	20·45	49·89	20·66	49·80	20·88	54
55	51·00	20·60	50·90	20·83	50·81	21·05	50·72	21·27	55
56	51·92	20·98	51·83	21·20	51·74	21·43	51·64	21·66	56
57	52·85	21·35	52·76	21·58	52·66	21·81	52·57	22·04	57
58	53·78	21·73	53·68	21·96	53·59	22·20	53·49	22·43	58
59	54·70	22·10	54·61	22·34	54·51	22·58	54·41	22·82	59
60	55·63	22·48	55·53	22·72	55·43	22·96	55·33	23·20	60
61	56·56	22·85	56·47	23·10	56·36	23·34	56·25	23·59	61
62	57·49	23·23	57·38	23·48	57·28	23·73	57·18	23·98	62
63	58·41	23·60	58·31	23·85	58·20	24·11	58·10	24·36	63
64	59·34	23·97	59·23	24·23	59·13	24·49	59·02	24·75	64
65	60·27	24·35	60·16	24·61	60·05	24·87	59·94	25·14	65
66	61·19	24·72	61·09	24·99	60·98	25·26	60·87	25·52	66
67	62·12	25·10	62·01	25·37	61·90	25·64	61·79	25·91	67
68	63·05	25·47	62·94	25·75	62·82	26·02	62·71	26·30	68
69	63·98	25·85	63·86	26·13	63·75	26·41	63·63	26·68	69
70	64·90	26·22	64·79	26·51	64·67	26·79	64·55	27·07	70
71	65·83	26·60	65·71	26·88	65·60	27·17	65·48	27·46	71
72	66·76	26·97	66·64	27·26	66·52	27·55	66·40	27·84	72
73	67·68	27·35	67.56	27·64	67·44	27·94	67·32	28·23	73
74	68·61	27·72	68·49	28·02	68·37	28·32	68·24	28·62	74
75	69·54	28·10	69·42	28·40	69·29	28·70	69·17	29·00	75
76	70·47	28·47	70·34	28·78	70·21	29·08	70·09	29·39	76
77	71·39	28·84	71·27	29·16	71·14	29·47	71·01	29·78	77
78	72·32	29·22	72·19	29·53	72·06	29·85	71·93	30·16	78
79	73·25	29·59	73·12	29·91	72·99	30·23	72·85	30·55	79
80	74·17	29·97	74·04	30·29	73·91	30·61	73·78	30·94	80
81	75·10	30·34	74·97	30·67	74·83	31·00	74·70	31·32	81
82	76·03	30·72	75·89	31·05	75·76	31·38	75·62	31·71	82
83	76·96	31·09	76·82	31·43	76·68	31·76	76·54	32·10	83
84	77·88	31·47	77·75	31·81	77·61	32·15	77·46	32·48	84
85	78·81	31·84	78·67	32·19	78·53	32·53	78·39	32·87	85
86	79·74	32·22	79·60	32·56	79·45	32·91	79·31	33·26	86
87	80·66	32·59	80·52	32·94	80·38	33·29	80·23	33·64	87
88	81·59	32·97	81·45	33·32	81·30	33·68	81·15	34·03	88
89	82·52	33·34	82·37	33·70	82·23	34·06	82·08	34·42	89
90	83·45	33·71	83·30	34·08	83·15	34·44	83·00	34·80	90
91	84·37	34·09	84·22	34·46	84·07	34·82	83·92	35·19	91
92	85·30	34·46	85·15	34·84	85·00	35·21	84·84	35·58	92
93	86·23	34·84	86·08	35·21	85·92	35·59	85·76	35·96	93
94	87 16	35·21	87·00	35·59	86·84	35·97	86·69	36·35	94
95	88·08	35·59	87·93	35·97	87·77	36·35	87·61	36·74	95
96	89·01	35·96	88·85	36·35	88·69	36·74	88·53	37·12	96
97	89·94	36·34	89·78	36·73	89·62	37·12	89·45	37·51	97
98	90·86	36·71	90·70	37·11	90·54	37·50	90·38	37·90	98
99	91·79	37·09	91·63	37·49	91·46	37·89	91·30	38·28	99
100	92·72	37·46	92·55	37·86	92·39	38·27	92·22	38·67	100
Distance.	Dep.	Lat.	Dep.	Lat.	Dep.	Lat.	Dep.	Lat.	Distance.
	68 Deg.		67¾ Deg.		67½ Deg.		67¼ Deg.		

Distance.	23 Deg. Lat.	23 Deg. Dep.	23¼ Deg. Lat.	23¼ Deg. Dep.	23½ Deg. Lat.	23½ Deg. Dep.	23¾ Deg. Lat.	23¾ Deg. Dep.	Distance.
1	0·92	0·39	0·92	0·39	0·92	0·40	0·92	0·40	1
2	1·84	0·78	1·84	0·79	1·83	0·80	1·83	0·81	2
3	2·76	1·17	2·76	1·18	2·75	1·20	2·75	1·21	3
4	3·68	1·56	3·68	1·58	3·67	1·59	3·66	1·61	4
5	4·60	1·95	4·59	1·97	4·59	1·99	4·58	2·01	5
6	5·52	2·34	5·51	2·37	5·50	2·39	5·49	2·42	6
7	6·44	2·74	6·43	2·76	6·42	2·79	6·41	2·82	7
8	7·36	3·13	7·35	3·16	7·34	3·19	7·32	3·22	8
9	8·28	3·52	8·27	3·55	8·25	3·59	8·24	3·62	9
10	9·20	3·91	9·19	3·95	9·17	3·99	9·15	4·03	10
11	10·13	4·30	10·11	4·34	10·09	4·39	10·07	4·43	11
12	11·05	4·69	11·03	4·74	11·00	4·78	10·98	4·83	12
13	11·97	5·08	11·94	5·13	11·92	5·18	11·90	5·24	13
14	12·89	5·47	12·86	5·53	12·84	5·58	12·81	5·64	14
15	13·81	5·86	13·78	5·92	13·76	5·98	13·73	6·04	15
16	14·73	6·25	14·70	6·32	14·67	6·38	14·64	6·44	16
17	15·65	6·64	15·62	6·71	15·59	6·78	15·56	6·85	17
18	16·57	7·03	16·54	7·11	16·51	7·18	16·48	7·25	18
19	17·49	7·42	17·46	7·50	17·42	7·58	17·39	7·65	19
20	18·41	7·81	18·38	7·89	18·34	7·97	18·31	8·05	20
21	19·33	8·21	19·29	8·29	19·26	8·37	19·22	8·46	21
22	20·25	8·60	20·21	8·68	20·18	8·77	20·14	8·86	22
23	21·17	8·99	21·13	9·08	21·09	9·17	21·05	9·26	23
24	22·09	9·38	22·05	9·47	22·01	9·57	21·97	9·67	24
25	23·01	9·77	22·97	9·87	22·93	9·97	22·88	10·07	25
26	23·93	10·16	23·89	10·26	23·84	10·37	23·80	10·47	26
27	24·85	10·55	24·81	10·66	24·76	10·77	24·71	10·87	27
28	25·77	10·94	25·73	11·05	25·68	11·16	25·63	11·28	28
29	26·69	11·33	26·64	11·45	26·59	11·56	26·54	11·68	29
30	27·62	11·72	27·56	11·84	27·51	11·96	27·46	12·08	30
31	28·54	12·11	28·48	12·24	28·43	12·36	28·37	12·49	31
32	29·46	12·50	29·40	12·63	29·35	12·76	29·29	12·89	32
33	30·38	12·89	30·32	13·03	30·26	13·16	30·21	13·29	33
34	31·30	13·28	31·24	13·42	31·18	13·56	31·12	13·69	34
35	32·22	13·68	32·16	13·82	32·10	13·96	32·04	14·10	35
36	33·14	14·07	33·08	14·21	33·01	14·35	32·95	14·50	36
37	34·06	14·46	34·00	14·61	33·93	14·75	33·87	14·90	37
38	34·98	14·85	34·91	15·00	34·85	15·15	34·78	15·30	38
39	35·90	15·24	35·83	15·39	35·77	15·55	35·70	15·71	39
40	36·82	15·63	36·75	15·79	36·68	15·95	36·61	16.11	40
41	37·74	16·02	37·67	16·18	37·60	16·35	37·53	16·51	41
42	38·66	16·41	38·59	16·58	38·52	16·75	38·44	16·92	42
43	39·58	16·80	39·51	16·97	39·43	17·15	39·36	17·32	43
44	40·50	17·19	40·43	17·37	40·35	17·54	40·27	17·72	44
45	41·42	17·58	41·35	17·76	41·27	17·94	41·19	18·12	45
46	42·34	17·97	42·26	18·16	42·18	18·34	42·10	18·53	46
47	43·26	18·36	43·18	18·55	43·10	18·74	43.02	18·93	47
48	44·18	18·76	44·10	18·95	44·02	19·14	43·93	19·33	48
49	45·10	19·15	45·02	19·34	44·94	19·54	44·85	19·73	49
50	46·03	19·54	45·94	19·74	45·85	19·94	45·77	20·14	50
Distance.	Dep.	Lat.	Dep.	Lat.	Dep.	Lat.	Dep.	Lat.	Distance.
	67 Deg.		66¾ Deg.		66½ Deg.		66¼ Deg.		

Distance.	23 Deg.		23¼ Deg.		23½ Deg.		23¾ Deg.		Distance.
	Lat.	Dep.	Lat.	Dep.	Lat.	Dep.	Lat.	Dep.	
51	46·95	19·93	46·86	20·13	46·77	20·34	46·68	20·54	51
52	47·87	20·32	47·78	20·53	47·69	20·73	47·60	20·94	52
53	48·79	20·71	48·70	20·92	48·60	21·13	48·51	21·35	53
54	49·71	21·10	49·61	21·32	49·52	21·53	49·43	21·75	54
55	50·63	21·49	50·53	21·71	50·44	21·93	50·34	22·15	55
56	51·55	21·88	51·45	22·11	51·36	22·33	51·26	22·55	56
57	52·47	22·27	52·37	22·50	52·27	22·73	52·17	22·96	57
58	53·39	22·66	53·29	22·90	53·19	23·13	53·09	23·36	58
59	54·31	23·05	54·21	23·29	54·11	23·53	54·00	23·76	59
60	55·23	23·44	55·13	23·68	55·02	23·92	54·92	24·16	60
61	56·15	23·83	56·05	24·08	55·94	24·32	55·83	24·57	61
62	57·07	24·23	56·97	24·47	56·86	24·72	56·75	24·97	62
63	57·99	24·62	57·88	24·87	57·77	25·12	57·66	25·37	63
64	58·91	25·01	58·80	25·26	58·69	25·52	58·58	25·78	64
65	59·83	25·40	59·72	25·66	59·61	25·92	59·50	26·18	65
66	60·75	25·79	60·64	26·05	60·53	26·32	60·41	26·58	66
67	61·67	26·18	61·56	26·45	61·44	26·72	61·33	26·98	67
68	62·59	26·57	62·48	26·84	62·36	27·11	62·24	27·39	68
69	63·51	26·96	63·40	27·24	63·28	27·51	63·16	27·79	69
70	64·44	27·35	64·32	27·63	64·19	27·91	64·07	28·19	70
71	65·36	27·74	65·23	28·03	65·11	28·31	64·99	28·59	71
72	66·28	28·13	66·15	28·42	66·03	28·71	65·90	29·00	72
73	67·20	28·52	67·07	28·82	66·95	29·11	66·82	29·40	73
74	68·12	28·91	67·99	29·21	67·86	29·51	67·73	29·80	74
75	69·04	29·30	68·91	29·61	68·78	29·91	68·65	30·21	75
76	69·96	29·70	69·83	30·00	69·70	30·30	69·56	30·61	76
77	70·88	30·09	70·75	30·40	70·61	30·70	70·48	31·01	77
78	71·80	30·48	71·67	30·79	71·53	31·10	71·39	31·41	78
79	72·72	30·87	72·58	31·18	72·45	31·50	72·31	31·82	79
80	73·64	31·26	73·50	31·58	73·36	31·90	73·22	32·22	80
81	74·56	31·65	74·42	31·97	74·28	32·30	74·14	32·62	81
82	75·48	32·04	75·34	32·37	75·20	32·70	75·06	33·03	82
83	76·40	32·43	76·26	32·76	76·12	33·10	75·97	33·43	83
84	77·32	32·82	77·18	33·16	77·03	33·49	76·89	33·83	84
85	78·24	33·21	78·10	33·55	77·95	33·89	77·80	34·23	85
86	79·16	33·60	79·02	33·95	78·87	34·29	78·72	34·64	86
87	80·08	23·99	79·93	34·34	79·78	34·69	79·63	35·04	87
88	81·00	34·38	80·85	34·74	80·70	35·09	80·55	35·44	88
89	81·92	34·78	81·77	35·13	81·62	35·49	81·46	35·84	89
90	82·85	35·17	82·69	35·53	82·54	35·89	82·38	36·25	90
91	83·77	35·56	83·61	35·92	83·45	36·29	83·29	36·65	91
92	84·69	35·95	84·53	36·32	84·37	36·68	84·21	37·05	92
93	85·61	36·34	85·45	36·71	85·29	37·08	85·12	37·46	93
94	86·53	36·73	86·37	37·11	86·20	37·48	86·04	37·86	94
95	87·45	37·12	87·29	37·50	87·12	37·88	86·95	38·26	95
96	88·37	37·51	88·20	37·90	88·04	38·28	87·87	38·66	96
97	89·29	37·90	89·12	38·29	88·95	38·68	88·79	39·07	97
98	90·21	38·29	90·04	38·68	89·87	39·08	89·70	39·47	98
99	91·13	38·68	90·96	39·08	90·79	39·48	90·62	39·87	99
100	92·05	39·07	91·88	39·47	91·71	39·87	91·53	40·27	100
Distance.	Dep.	Lat.	Dep.	Lat.	Dep.	Lat.	Dep.	Lat.	Distance.
	67 Deg.		66¾ Deg.		66½ Deg.		66¼ Deg.		

Distance.	24 Deg.		24¼ Deg.		24½ Deg.		24¾ Deg.		Distance.
	Lat.	Dep.	Lat.	Dep.	Lat.	Dep.	Lat.	Dep.	
1	0·91	0·41	0·91	0·41	0·91	0·41	0·91	0·42	1
2	1·83	0·81	1·82	0·82	1·82	0·83	1·82	0·84	2
3	2·74	1·22	2·74	1·23	2·73	1·24	2·72	1·26	3
4	3·65	1·63	3·65	1·64	3·64	1·66	3·63	1·67	4
5	4·57	2·03	4·56	2·05	4·55	2·07	4·54	2·09	5
6	5·48	2·44	5·47	2·46	5·46	2·49	5·45	2·51	6
7	6·39	2·85	6·38	2·87	6·37	2·90	6·36	2·93	7
8	7·31	3·25	7·29	3·29	7·28	3·32	7·27	3·35	8
9	8·22	3·66	8·21	3·70	8·19	3·73	8·17	3·77	9
10	9·14	4·07	9·12	4·11	9·10	4·15	9·08	4·19	10
11	10·05	4·47	10·03	4·52	10·01	4·56	9·99	4·61	11
12	10·96	4·88	10·94	4·93	10·92	4·98	10·90	5·02	12
13	11·88	5·29	11·85	5·34	11·83	5·39	11·81	5·44	13
14	12·79	5·69	12·76	5·75	12·74	5·81	12·71	5·86	14
15	13·70	6·10	13·68	6·16	13·65	6·22	13·62	6·28	15
16	14·62	6·51	14·59	6·57	14·56	6·64	14·53	6·70	16
17	15·53	6·92	15·50	6·98	15·47	7·05	15·44	7·12	17
18	16·44	7·32	16·41	7·39	16·38	7·46	16·35	7·54	18
19	17·36	7·73	17·32	7·80	17·29	7·88	17·25	7·95	19
20	18·27	8·13	18·24	8·21	18·20	8·29	18·16	8·37	20
21	19·18	8·54	19·15	8·63	19·11	8·71	19·07	8·79	21
22	20·10	8·95	20·06	9·04	20·02	9·12	19·98	9·21	22
23	21·01	9·35	20·97	9·45	20·93	9·54	20·89	9·63	23
24	21·93	9·76	21·88	9·86	21·84	9·95	21·80	10·05	24
25	22·84	10·17	22·79	10·27	22·75	10·37	22·70	10·47	25
26	23·75	10·58	23·71	10·68	23·66	10·78	23·61	10·89	26
27	24·67	10·98	24·62	11·09	24·57	11·20	24·52	11·30	27
28	25·58	11·39	25·53	11·50	25·48	11·61	25·43	11·72	28
29	26·49	11·80	26·44	11·91	26·39	12·03	26·34	12·14	29
30	27·41	12·20	27·35	12·32	27·30	12·44	27·24	12·56	30
31	28·32	12·61	28·26	12·73	28·21	12·86	28·15	12·98	31
32	29·23	13·02	29·18	13·14	29·12	13·27	29·06	13·40	32
33	30·15	13·42	30·09	13·55	30·03	13·68	29·97	13·82	33
34	31·06	13·83	31·00	13·96	30·94	14·10	30·88	14·23	34
35	31·97	14·24	31·91	14·38	31·85	14·51	31·78	14·65	35
36	32·89	14·64	32·82	14·79	32·76	14·93	32·69	15·07	36
37	33·80	15·05	33·74	15·20	33·67	15·34	33·60	15·49	37
38	34·71	15·46	34·65	15·61	34·58	15·76	34·51	15·91	38
39	35·63	15·86	35·56	16·02	35·49	16·17	35·42	16·33	39
40	36·54	16·27	36·47	16·43	36·40	16·59	36·33	16·75	40
41	37·46	16·68	37·38	16·84	37·31	17·00	37·23	17·16	41
42	38·37	17·08	38·29	17·25	38·22	17·42	38·14	17·58	42
43	39·28	17·49	39·21	17·66	39·13	17·83	39·05	18·00	43
44	40·20	17·90	40·12	18·07	40·04	18·25	39·96	18·42	44
45	41·11	18·30	41·03	18·48	40·95	18·66	40·87	18·84	45
46	42·02	18·71	41·94	18·89	41·86	19·08	41·77	19·26	46
47	42·94	19·12	42·85	19·30	42·77	19·49	42·68	19·68	47
48	43·85	19·52	43·76	19·71	43·68	19·91	43·59	20·10	48
49	44·76	19·93	44·68	20·13	44·59	20·32	44·50	20·51	49
50	45·68	20·34	45·59	20·54	45·50	20·73	45·41	20·93	50
	Dep.	Lat.	Dep.	Lat.	Dep.	Lat.	Dep.	Lat.	
Distance.	66 Deg.		65¾ Deg.		65½ Deg.		65¼ Deg.		Distance.

Distance.	24 Deg.		24¼ Deg.		24½ Deg.		24¾ Deg.		Distance.
	Lat.	Dep.	Lat.	Dep.	Lat.	Dep.	Lat.	Dep.	
51	46·59	20·74	46·50	20·95	46·41	21·15	46·32	21·35	51
52	47·50	21·15	47·41	21·36	47·32	21·56	47·22	21·77	52
53	48·42	21·56	48·32	21·77	48·23	21·98	48·13	22·19	53
54	49·33	21·96	49·24	22·18	49·14	22·39	49·04	22·61	54
55	50·24	22·37	50·15	22·59	50·05	22·81	49·95	23·03	55
56	51·16	22·78	51·06	23·00	50·96	23·22	50·86	23·44	56
57	52·07	23·18	51·97	23·41	51·87	23·64	51·76	23·86	57
58	52·99	23·59	52·88	23·82	52·78	24·05	52·67	24·28	58
59	53·90	24·00	53·79	24·23	53·69	24·47	53·58	24·70	59
60	54·81	24·40	54·71	24·64	54·60	24·88	54·49	25·12	60
61	55·73	24·81	55·62	25·05	55·51	25·30	55·40	25·54	61
62	56·64	25·22	56·53	25·46	56·42	25·71	56·30	25·96	62
63	57·55	25·62	57·44	25·88	57·33	26·13	57·21	26·38	63
64	58·47	26·03	58·35	26·29	58·24	26·54	58·12	26·79	64
65	59·38	26·44	59·26	26·70	59·15	26·96	59·03	27·21	65
66	60·29	26·84	60·18	27·11	60·06	27·37	59·94	27·63	66
67	61·21	27·25	61·09	27·52	60·97	27·78	60·85	28·05	67
68	62·12	27·66	62·00	27·93	61·88	28·20	61·75	28·47	68
69	63·03	28·06	62·91	28·34	62·79	28·61	62·66	28·89	69
70	63·95	28·47	63·82	28·75	63·70	29·03	63·57	29·31	70
71	64·86	28·88	64·74	29·16	64·61	29·44	64·48	29·72	71
72	65·78	29·28	65·65	29·57	65·52	29·86	65·39	30·14	72
73	66·69	29·69	66·56	29·98	66·43	30·27	66·29	30·56	73
74	67·60	30·10	67·47	30·39	67·34	30·69	67·20	30·98	74
75	68·52	30·51	68·38	30·80	68·25	31·10	68·11	31·40	75
76	69·43	30·91	69·29	31·21	69·16	31·52	69·02	31·82	76
77	70·34	31·32	70·21	31·63	70·07	31·93	69·93	32·24	77
78	71·26	31·73	71·12	32·04	70·98	32·35	70·84	32·66	78
79	72·17	32·13	72·03	32·45	71·89	32·76	71·74	33·07	79
80	73·08	32·54	72·94	32·86	72·80	33·18	72·65	33·49	80
81	74·00	32·95	73·85	33·27	73·71	33·59	73·56	33·91	81
82	74·91	33·35	74·76	33·68	74·62	34·00	74·47	34·33	82
83	75·82	33·76	75·68	34·09	75·53	34·42	75·38	34·75	83
84	76·74	34·17	76·59	34·50	76·44	34·83	76·28	35·17	84
85	77·65	34·57	77·50	34·91	77·35	35·25	77·19	35·59	85
86	78·56	34·98	78·41	35·32	78·26	35·66	78·10	36·00	86
87	79·48	35·39	79·32	35·73	79·17	36·08	79·01	36·42	87
88	80·39	35·79	80·24	36·14	80·08	36·49	79·92	36·84	88
89	81·31	36·20	81·15	36·55	80·99	36·91	80·82	37·26	89
90	82·22	36·61	82·06	36·96	81·90	37·32	81·73	37·68	90
91	83·13	37·01	82·97	37·38	82·81	37·74	82·64	38·10	91
92	84·05	37·42	83·88	37·79	83·72	38·15	83·55	38·52	92
93	84·96	37·83	84·79	38·20	84·63	38·57	84·46	38·94	93
94	85·87	38·23	85·71	38·61	85·54	38·98	85·37	39·35	94
95	86·79	38·64	86·62	39·02	86·45	39·40	86·27	39·77	95
96	87·70	39·05	87·53	39·43	87·36	39·81	87·18	40·19	96
97	88·61	39·45	88·44	39·84	88·27	40·23	88·09	40·61	97
98	89·53	39·86	89·35	40·25	89·18	40·64	89·00	41·03	98
99	90·44	40·27	90·26	40·66	90·09	41·05	89·91	41·45	99
100	91·35	40·67	91·18	41·07	91·00	41·47	90·81	41·87	100
Distance.	Dep.	Lat.	Dep.	Lat.	Dep.	Lat.	Dep.	Lat.	Distance.
	66 Deg.		65¾ Deg.		65½ Deg.		65¼ Deg.		

Distance.	25 Deg.		25¼ Deg.		25½ Deg.		25¾ Deg.		Distance.
	Lat.	Dep.	Lat.	Dep.	Lat.	Dep.	Lat.	Dep.	
1	0·91	0·42	0·90	0·43	0·90	0·43	0·90	0·43	1
2	1·81	0·85	1·81	0·85	1·81	0·86	1·80	0·87	2
3	2·72	1·27	2·71	1·28	2·71	1·29	2·70	1·30	3
4	3·63	1·69	3·62	1·71	3·61	1·72	3·60	1·74	4
5	4·53	2·11	4·52	2·13	4·51	2·15	4·50	2·17	5
6	5·44	2·54	5·43	2·56	5·42	2·58	5·40	2·61	6
7	6·34	2·96	6·33	2·99	6·32	3·01	6·30	3·04	7
8	7·25	3·38	7·24	3·41	7·22	3·44	7·21	3·48	8
9	8·16	3·80	8·14	3·84	8·12	3·87	8·11	3·91	9
10	9·06	4·23	9·04	4·27	9·03	4·31	9·01	4·34	10
11	9·97	4·65	9·95	4·69	9·93	4·74	9·91	4·78	11
12	10·88	5·07	10·85	5·12	10·83	5·17	10·81	5·21	12
13	11·78	5·49	11·76	5·55	11·73	5·60	11·71	5·65	13
14	12·69	5·92	12·66	5·97	12·64	6·03	12·61	6·08	14
15	13·59	6·34	13·57	6·40	13·54	6·46	13·51	6·52	15
16	14·50	6·76	14·47	6·83	14·44	6·89	14·41	6·95	16
17	15·41	7·18	15·38	7·25	15·34	7·32	15·31	7·39	17
18	16·31	7·61	16·28	7·68	16·25	7·75	16·21	7·82	18
19	17·22	8·03	17·18	8·10	17·15	8·18	17·11	8·25	19
20	18·13	8·45	18·09	8·53	18·05	8·61	18·01	8·69	20
21	19·03	8·87	18·99	8·96	18·95	9·04	18·91	9·12	21
22	19·94	9·30	19·90	9·38	19·86	9·47	19·82	9·56	22
23	20·85	9·72	20·80	9·81	20·76	9·90	20·72	9·99	23
24	21·75	10·14	21·71	10·24	21·66	10·33	21·62	10·43	24
25	22·66	10·57	22·61	10·66	22·56	10·76	22·52	10·86	25
26	23·56	10·99	23·52	11·09	23·47	11·19	23·42	11·30	26
27	24·47	11·41	24·42	11·52	24·37	11·62	24·32	11·73	27
28	25·38	11·83	25·32	11·94	25·27	12·05	25·22	12·16	28
29	26·28	12·26	26·23	12·37	26·17	12·48	26·12	12·60	29
30	27·19	12·68	27·13	12·80	27·08	12·92	27·02	13·03	30
31	28·10	13·10	28·04	13·22	27·98	13·35	27·92	13·47	31
32	29·00	13·52	28·94	13·65	28·88	13·78	28·82	13·90	32
33	29·91	13·95	29·85	14·08	29·79	14·21	29·72	14·34	33
34	30·81	14·37	30·75	14·50	30·69	14·64	30·62	14·77	34
35	31·72	14·79	31·66	14·93	31·59	15·07	31·52	15·21	35
36	32·63	15·21	32·56	15·36	32·49	15·50	32·43	15·64	36
37	33·53	15·64	33·46	15·78	33·40	15·93	33·33	16·07	37
38	34·44	16·06	34·37	16·21	34·30	16·36	34·23	16·51	38
39	35·35	16·48	35·27	16·64	35·20	16·79	35·13	16·94	39
40	36·25	16·90	36·18	17·06	36·10	17·22	36·03	17·38	40
41	37·16	17·33	37·08	17·49	37·01	17·65	36·93	17·81	41
42	38·06	17·75	37·99	17·92	37·91	18·08	37·83	18·25	42
43	38·97	18·17	38·89	18·34	38·81	18·51	38·73	18·68	43
44	39·88	18·60	39·80	18·77	39·71	18·94	39·63	19·12	44
45	40·78	19·02	40·70	19·20	40·62	19·37	40·53	19·55	45
46	41·69	19·44	41·60	19·62	41·52	19·80	41·43	19·98	46
47	42·60	19·86	42·51	20·05	42·42	20·23	42·33	20·42	47
48	43·50	20·29	43·41	20·48	43·32	20·66	43·23	20·85	48
49	44·41	20·71	44·32	20·90	44·23	21·10	44·13	21·29	49
50	45·32	21·13	45·22	21·33	45·13	21·53	45·03	21·72	50
Distance.	Dep.	Lat.	Dep.	Lat.	Dep.	Lat.	Dep.	Lat.	Distance.
	65 Deg.		64¾ Deg.		64½ Deg.		64¼ Deg.		

Distance.	25 Deg. Lat.	25 Deg. Dep.	25¼ Deg. Lat.	25¼ Deg. Dep.	25½ Deg. Lat.	25½ Deg. Dep.	25¾ Deg. Lat.	25¾ Deg. Dep.	Distance.
51	46·22	21·55	46·13	21·75	46·03	21·96	45·94	22·16	51
52	47·13	21·98	47·03	22·18	46·93	22·39	46·84	22·59	52
53	48·03	22·40	47·94	22·61	47·84	22·82	47·74	23·03	53
54	48·94	22·82	48·84	23·03	48·74	23·25	48·64	23·46	54
55	49·85	23·24	49·74	23·46	49·64	23·68	49·54	23·89	55
56	50·75	23·67	50·65	23·89	50·54	24·11	50·44	24·33	56
57	51·66	24·09	51·55	24·31	51·45	24·54	51·34	24·76	57
58	52·57	24·51	52·46	24·74	52·35	24·97	52·24	25·20	58
59	53·47	24·93	53·36	25·17	53·25	25·40	53·14	25·63	59
60	54·38	25·36	54·27	25·59	54·16	25·83	54·04	26·07	60
61	55·28	25·78	55·17	26·02	55·06	26·26	54·94	26·50	61
62	56·19	26·20	56·08	26·45	55·96	26·69	55·84	26·94	62
63	57·10	26·62	56·98	26·87	56·86	27·12	56·74	27·37	63
64	58·00	27·05	57·89	27·30	57·77	27·55	57·64	27·80	64
65	58·91	27·47	58·79	27·73	58·67	27·98	58.55	28·24	65
66	59·82	27·89	59·69	28·15	59·57	28·41	59·45	28·67	66
67	60·72	28·32	60·60	28·58	60·47	28·84	60·35	29·11	67
68	61.63	28·74	61·50	29·01	61·38	29·27	61·25	29·54	68
69	62.54	29·16	62·41	29·43	62·28	29·71	62·15	29·98	69
70	63.44	29·58	63·31	29·86	63·18	30·14	63·05	30·41	70
71	64·35	30·01	64·22	30·29	64·08	30·57	63·95	30·85	71
72	65·25	30·43	65·12	30·71	64·99	31·00	64·85	31·28	72
73	66·16	30·85	66·03	31·14	65·89	31·43	65·75	31·71	73
74	67·07	31·27	66·93	31·57	66·79	31·86	66·65	32·15	74
75	67·97	31·70	67·83	31·99	67·69	32·29	67·55	32·58	75
76	68·88	32·12	68·74	32·42	68·60	32·72	68·45	33·02	76
77	69·79	32·54	69·64	32·85	69·50	33·15	69·35	33·45	77
78	70·69	32·96	70·55	33·27	70·40	33·58	70·25	33·89	78
79	71·60	33·39	71·45	33·70	71·30	34·01	71·16	34·32	79
80	72·50	33·81	72·36	34·13	72·21	34·44	72·06	34·76	80
81	73·41	34·23	73·26	34·55	73·11	34·87	72·96	35·19	81
82	74·32	34·65	74·17	34·98	74·01	35·30	73·86	35·62	82
83	75·22	35·08	75·07	35·41	74·91	35·73	74·76	36·06	83
84	76·13	35·50	75·97	35·83	75·82	36·16	75·66	36·49	84
85	77·04	35·92	76·88	36·26	76·72	36·59	76·56	36·93	85
86	77·94	36·35	77·78	36·68	77·62	37·02	77·46	37·36	86
87	78·85	36·77	78·69	37·11	78·52	37·45	78·36	37·80	87
88	79·76	37·19	79·59	37·54	79·43	37·88	79·26	38·23	88
89	80·66	37·61	80·50	37·96	80·33	38·32	80·16	38·67	89
90	81·57	38·04	81·40	38·39	81·23	38·75	81·06	39·10	90
91	82·47	38·46	82·31	38·82	82·14	39·18	81·96	39·53	91
92	83·38	38·88	83·21	39·24	83·04	39·61	82·86	39·97	92
93	84·29	39·30	84·11	39·67	83·94	40·04	83·76	40·40	93
94	85·19	39·73	85·02	40·10	84·84	40·47	84·67	40·84	94
95	86·10	40·15	85·92	40·52	85·75	40·90	85·57	41·27	95
96	87·01	40·57	86·83	40·95	86·65	41·33	86·47	41·71	96
97	87·91	40·99	87·73	41·38	87·55	41·76	87·37	42·14	97
98	88·82	41·42	88·64	41·80	88·45	42·19	88·27	42·58	98
99	89·72	41·84	89·54	42·23	89·36	42·62	89·17	43·01	99
100	90·63	42·26	90·45	42·66	90·26	43·05	90·07	43·44	100
Distance.	Dep.	Lat.	Dep.	Lat.	Dep.	Lat.	Dep.	Lat.	Distance.
	65 Deg.		64¾ Deg.		64½ Deg.		64¼ Deg.		

Distance.	26 Deg.		26¼ Deg.		26½ Deg.		26¾ Deg.		Distance.
	Lat.	Dep.	Lat.	Dep.	Lat.	Dep.	Lat.	Dep.	
1	0·90	0·44	0·90	0·44	0·89	0·45	0·89	0·45	1
2	1·80	0·88	1·79	0·88	1·79	0·89	1·79	0·90	2
3	2·70	1·32	2·69	1·33	2·68	1·34	2·68	1·35	3
4	3·60	1·75	3·59	1·77	3·58	1·78	3·57	1·80	4
5	4·49	2·19	4·48	2·21	4·47	2·23	4·46	2·25	5
6	5·39	2·63	5·38	2·65	5·37	2·68	5·36	2·70	6
7	6·29	3·07	6·28	3·10	6·26	3·12	6·25	3·15	7
8	7·19	3·51	7·17	3·54	7·16	3·57	7·14	3·60	8
9	8·09	3·95	8·07	3·98	8·05	4·02	8·04	4·05	9
10	8·99	4·38	8·97	4·42	8·95	4·46	8·93	4·50	10
11	9·89	4·82	9·87	4·87	9·84	4·91	9·82	4·95	11
12	10·79	5·26	10·76	5·31	10·74	5·35	10·72	5·40	12
13	11·68	5·70	11·66	5·75	11·63	5·80	11·61	5·85	13
14	12·58	6·14	12·56	6·19	12·53	6·25	12·50	6·30	14
15	13·48	6·58	13·45	6·63	13·42	6·69	13·39	6·75	15
16	14·38	7·01	14·35	7·08	14·32	7·14	14·29	7·20	16
17	15·28	7·45	15·25	7·52	15·21	7·59	15·18	7·65	17
18	16·18	7·89	16·14	7·96	16·11	8·03	16·07	8·10	18
19	17·08	8·33	17·04	8·40	17·00	8·48	16·97	8·55	19
20	17·98	8·77	17·94	8·85	17·90	8·92	17·86	9·00	20
21	18·87	9·21	18·83	9·29	18·79	9·37	18·75	9·45	21
22	19·77	9·64	19·73	9·73	19·69	9·82	19·65	9·90	22
23	20·67	10·08	20·63	10·17	20·58	10·26	20·54	10·35	23
24	21·57	10·52	21·52	10·61	21·48	10·71	21·43	10·80	24
25	22·47	10·96	22·42	11·06	22·37	11·15	22·32	11·25	25
26	23·37	11·40	23·32	11·50	23·27	11·60	23·22	11·70	26
27	24·27	11·84	24·22	11·94	24·16	12·05	24·11	12·15	27
28	25·17	12·27	25·11	12·38	25·06	12·49	25·00	12·60	28
29	26·06	12·71	26·01	12·83	25·95	12·94	25·90	13·05	29
30	26·96	13·15	26·91	13·27	26·85	13·39	26·79	13·50	30
31	27·86	13·59	27·80	13·71	27·74	13·83	27·68	13·95	31
32	28·76	14·03	28·70	14·15	28·64	14·28	28·58	14·40	32
33	29·66	14·47	29·60	14·60	29·53	14·72	29·47	14·85	33
34	30·56	14·90	30·49	15·04	30·43	15·17	30·36	15·30	34
35	31·46	15·34	31·39	15·48	31·32	15·62	31·25	15·75	35
36	32·36	15·78	32·29	15·92	32·22	16·06	32·15	16·20	36
37	33·26	16·22	33·18	16·36	33·11	16·51	33·04	16·65	37
38	34·15	16·66	34·08	16·81	34·01	16·96	33·93	17·10	38
39	35·05	17·10	34·98	17·25	34·90	17·40	34·83	17·55	39
40	35·95	17·53	35·87	17·69	35·80	17·85	35·72	18·00	40
41	36·85	17·97	36·77	18·13	36·69	18·29	36·61	18·45	41
42	37·75	18·41	37·67	18·58	37·59	18·74	37·51	18·90	42
43	38·65	18·85	38·57	19·02	38·48	19·19	38·40	19·35	43
44	39·55	19·29	39·46	19·46	39·38	19·63	39·29	19·80	44
45	40·45	19·73	40·36	19·90	40·27	20·08	40·18	20·25	45
46	41·34	20·17	41·26	20·35	41·17	20·53	41·08	20·70	46
47	42·24	20·60	42·15	20·79	42·06	20·97	41·97	21·15	47
48	43·14	21·04	43·05	21·23	42·96	21·42	42·86	21·60	48
49	44·04	21·48	43·95	21·67	43·85	21·86	43·76	22·05	49
50	44·94	21·92	44·84	22·11	44·75	22·31	44·65	22·50	50
Distance.	Dep.	Lat.	Dep.	Lat.	Dep.	Lat.	Dep.	Lat.	Distance.
	64 Deg.		63¾ Deg.		63½ Deg.		63¼ Deg.		

Distance.	26 Deg.		26¼ Deg.		26½ Deg.		26¾ Deg.		Distance.
	Lat.	Dep.	Lat.	Dep.	Lat.	Dep.	Lat.	Dep.	
51	45·84	22·36	45·74	22·56	45·64	22·76	45·54	22·96	51
52	46·74	22·80	46·64	23·00	46·54	23·20	46·43	23·41	52
53	47·64	23·23	47·53	23·44	47·43	23·65	47·33	23·86	53
54	48·53	23·67	48·43	23·88	48·33	24·09	48·22	24·31	54
55	49·43	24·11	49·33	24·33	49·22	24·54	49·11	24·76	55
56	50·33	24·55	50·22	24·77	50·12	24·99	50·01	25·21	56
57	51·23	24·99	51·12	25·21	51·01	25·43	50·00	25·66	57
58	52·13	25·43	52·02	25·65	51·91	25·88	51·79	26·11	58
59	53·03	25·86	52·92	26·09	52·80	26·33	52·69	26·56	59
60	53·93	26·30	53·81	26·54	53·70	26·77	53·58	27·01	60
61	54·83	26·74	54·71	26·98	54·59	27·22	54·47	27·46	61
62	55·73	27·18	55·61	27·42	55·49	27·66	55·36	27·91	62
63	56·62	27·62	56·50	27·86	56·38	28·11	56·26	28·36	63
64	57·52	28·06	57·40	28·31	57·28	28·56	57·15	28·81	64
65	58·42	28·49	58·30	28·75	58·17	29·00	58·04	29·26	65
66	59·32	28·93	59·19	29·19	59·07	29·45	58·94	29·71	66
67	60·22	29·37	60·09	29·63	59·96	29·90	59·83	30·16	67
68	61·12	29·81	60·99	30·08	60·86	30·34	60·72	30·61	68
69	62·02	30·25	61·88	30·52	61·75	30·79	61·62	31·06	69
70	62·92	30·69	62·78	30·96	62·65	31·23	62·51	31·51	70
71	63·81	31·12	63·68	31·40	63·54	31·68	63·40	31·96	71
72	64·71	31·56	64·57	31·84	64·44	32·13	64·29	32·41	72
73	65·61	32·00	65·47	32·29	65·33	32·57	65·19	32·86	73
74	66·51	32·44	66·37	32·73	66·23	33·02	66·08	33·31	74
75	67·41	32·88	67·27	33·17	67·12	33·46	66·97	33·76	75
76	68·31	33·32	68·16	33·61	68·01	33·91	67·87	34·21	76
77	69·21	33·75	69·06	34·06	68·91	34·36	68·76	34·66	77
78	70·11	34·19	69·96	34·50	69·80	34·80	69·65	35·11	78
79	71·00	34·63	70·85	34·94	70·70	35·25	70·55	35·56	79
80	71·90	35·07	71·75	35·38	71·59	35·70	71·44	36·01	80
81	72·80	35·51	72·65	35·83	72·49	36·14	72·33	36·46	81
82	73·70	35·95	73·54	36·27	73·38	36·59	73·22	36·91	82
83	74·60	36·38	74·44	36·71	74·28	37·03	74·12	37·36	83
84	75·50	36·82	75·34	37·15	75·17	37·48	75·01	37·81	84
85	76·40	37·26	76·23	37·59	76·07	37·93	75·90	38·26	85
86	77·30	37·70	77·13	38·04	76·96	38·37	76·80	38·71	86
87	78·20	38·14	78·03	38·48	77·86	38·82	77·69	39·16	87
88	79·09	38·58	78·92	38·92	78·75	39·27	78·58	39·61	88
89	79·99	39·01	79·82	39·36	79·65	39·71	79·48	40·06	89
90	80·89	39·45	80·72	39·81	80·54	40·16	80·37	40·51	90
91	81·79	39·89	81·62	40·25	81·44	40·60	81·26	40·96	91
92	82·69	40·33	82·51	40·69	82·33	41·05	82·15	41·41	92
93	83·59	40·77	83·41	41·13	83·23	41·50	83·05	41·86	93
94	84·49	41·21	84·31	41·58	84·12	41·94	83·94	42·31	94
95	85·39	41·65	85·20	42·02	85·02	42·39	84·83	42·76	95
96	86·28	42·08	86·10	42·46	85·91	42·83	85·73	43·21	96
97	87·18	42·52	87·00	42·90	86·81	43·28	86·62	43·66	97
98	88·08	42·96	87·89	43·34	87·70	43·73	87·51	44.11	98
99	88·98	43·40	88·79	43·79	88·60	44·17	88·40	44·56	99
100	89·88	43·84	89·69	44·23	89·49	44·62	89·30	45·01	100
	Dep.	Lat.	Dep.	Lat.	Dep.	Lat.	Dep.	Lat.	
Distance.	64 Deg.		63¾ Deg.		63½ Deg.		63¼ Deg.		Distance.

Distance.	27 Deg. Lat.	27 Deg. Dep.	27¼ Deg. Lat.	27¼ Deg. Dep.	27½ Deg. Lat.	27½ Deg. Dep.	27¾ Deg. Lat.	27¾ Deg. Dep.	Distance.
1	0·89	0·45	0·89	0·46	0·89	0·46	0·88	0·47	1
2	1·78	0·91	1·78	0·92	1·77	0·92	1·77	0·93	2
3	2·67	1·36	2·67	1·37	2·66	1·39	2·65	1·40	3
4	3·56	1·82	3·56	1·83	3·55	1·85	3·54	1·86	4
5	4·45	2·27	4·45	2·29	4·44	2·31	4·42	2·33	5
6	5·35	2·72	5·33	2·75	5·32	2·77	5·31	2·79	6
7	6·24	3·18	6·22	3·21	6·21	3·23	6·19	3·26	7
8	7·13	3·63	7·11	3·66	7·10	3·69	7·08	3·72	8
9	8·02	4·09	8·00	4·12	7·98	4·16	7·96	4·19	9
10	8·91	4·54	8·89	4·58	8·87	4·62	8·85	4·66	10
11	9·80	4·99	9·78	5·04	9·76	5·08	9·73	5·12	11
12	10·69	5·45	10·67	5·49	10·64	5·54	10·62	5·59	12
13	11·58	5·90	11·56	5·95	11·53	6·00	11·50	6·05	13
14	12·47	6·36	12·45	6·41	12·42	6·46	12·39	6·52	14
15	13·37	6·81	13·34	6·87	13·31	6·93	13·27	6·98	15
16	14·26	7·26	14·22	7·33	14·19	7·39	14·16	7·45	16
17	15·15	7·72	15·11	7·78	15·08	7·85	15·04	7·92	17
18	16·04	8·17	16·00	8·24	15·97	8·31	15·93	8·38	18
19	16·93	8·63	16·89	8·70	16·85	8·77	16·81	8·85	19
20	17·82	9·08	17·78	9·16	17·74	9·23	17·70	9·31	20
21	18·71	9·53	18·67	9·62	18·63	9·70	18·58	9·78	21
22	19·60	9·99	19·56	10·07	19·51	10·16	19·47	10·24	22
23	20·49	10·44	20·45	10·53	20·40	10·62	20·35	10·71	23
24	21·38	10·90	21·34	10·99	21·29	11·08	21·24	11·17	24
25	22·28	11·35	22·23	11·45	22·18	11·54	22·12	11·64	25
26	23·17	11·80	23·11	11·90	23·06	12·01	23·01	12·11	26
27	24·06	12·26	24·00	12·36	23·95	12·47	23·89	12·57	27
28	24·95	12·71	24·89	12·82	24·84	12·93	24·78	13·04	28
29	25·84	13·17	25·78	13·28	25·72	13·39	25·66	13·50	29
30	26·73	13·62	26·67	13·74	26·61	13·85	26·55	13·97	30
31	27·62	14·07	27·56	14·19	27·50	14·31	27·43	14·43	31
32	28·51	14·53	28·45	14·65	28·38	14·78	28·32	14·90	32
33	29·40	14·98	29·34	15·11	29·27	15·24	29·20	15·37	33
34	30·29	15·44	30·23	15·57	30·16	15·70	30·09	15·83	34
35	31·19	15·89	31·12	16·03	31·05	16·16	30·97	16·30	35
36	32·08	16·34	32·20	16·48	31·93	16·62	31·86	16·76	36
37	32·97	16·80	32·89	16·94	32·82	17·08	32·74	17·23	37
38	33·86	17·25	33·78	17·40	33·71	17·55	33·63	17·69	38
39	34·75	17·71	34·67	17·86	34·59	18·01	34·51	18·16	39
40	35·64	18·16	35·56	18·31	35·48	18·47	35·40	18.62	40
41	36·53	18·61	36·45	18·77	36·37	18·93	36·28	19·09	41
42	37·42	19·07	37·34	19·23	37·25	19·39	37·17	19·56	42
43	38·31	19·52	38·23	19·69	38·14	19·86	38·05	20·02	43
44	39·20	19·98	39·12	20·15	39·03	20·32	38·94	20·49	44
45	40·10	20·43	40·01	20·60	39·92	20·78	39·82	20·95	45
46	40·99	20·88	40·89	21·06	40·80	21·24	40·71	21·42	46
47	41·88	21·34	41·78	21·52	41·69	21·70	41,59	21·88	47
48	42·77	21·79	42·67	21·98	42·58	22·16	42·48	22·35	48
49	43·66	22·25	43·56	22·44	43·46	22·63	43·36	22·82	49
50	44·55	22·70	44·45	22·89	44·35	23·09	44·25	23·28	50
Distance.	Dep.	Lat.	Dep.	Lat.	Dep.	Lat.	Dep.	Lat.	Distance.
	63 Deg.		62¾ Deg.		62½ Deg.		62¼ Deg.		

Distance.	27 Deg.		27¼ Deg.		27½ Deg.		27¾ Deg.		Distance.
	Lat.	Dep.	Lat.	Dep.	Lat.	Dep.	Lat.	Dep.	
51	45·44	23·15	45·34	23·35	45·24	23·55	45·13	23·75	51
52	46·33	23·61	46·23	23·81	46·12	24·01	46·02	24·21	52
53	47·22	24·06	47·12	24·27	47·01	24·47	46·90	24·68	53
54	48·11	24·52	48·01	24·73	47·90	24·93	47·79	25·14	54
55	49·01	24·97	48·90	25·18	48·79	25·40	48·67	25·61	55
56	49·90	25·42	49·78	25·64	49·67	25·86	49·56	26·07	56
57	50·79	25·88	50·67	26·10	50·56	26·32	50·44	26·54	57
58	51·68	26·33	51·56	26·56	51·45	26·78	51·33	27·01	58
59	52·57	26·79	52·45	27·01	52·33	27·24	52·21	27·47	59
60	53·46	27·24	53·34	27·47	53·22	27·70	53·10	27·94	60
61	54·35	27·69	54·23	27·93	54·11	28·17	53·98	28·40	61
62	55·24	28·15	55·12	28·39	54·99	28·63	54·87	28·87	62
63	56·13	28·60	56·01	28·85	55·88	29·09	55·75	29·33	63
64	57·02	29·06	56·90	29·30	56·77	29·55	56·64	29·80	64
65	57·92	29·51	57·79	29·76	57·66	30·01	57·52	30·26	65
66	58·81	29·96	58·68	30·22	58·54	30·48	58·41	30·73	66
67	59·70	30·42	59·56	30·68	59·43	30·94	59·29	31·20	67
68	60·59	30·87	60·45	31·14	60·32	31·40	60·18	31·66	68
69	61·48	31·33	61·34	31·59	61·20	31·86	61·06	32·13	69
70	62·37	31·78	62·23	32·05	62·09	32·32	61·95	32·59	70
71	63·26	32·23	63·12	32·51	62·98	32·78	62·83	33·06	71
72	64·15	32·69	64·01	32·97	63·86	33·25	63·72	33·52	72
73	65·04	33·14	64·90	33·42	64·75	33·71	64·60	33·99	73
74	65·93	33·60	65·79	33·88	65·64	34·17	65·49	34·46	74
75	66·83	34·05	66·68	34·34	66·53	34·63	66·37	34·92	75
76	67·72	34·50	67·57	34·80	67·41	35·09	67·26	35·39	76
77	68·61	34·96	68·45	35·26	68·30	35·55	68·14	35·85	77
78	69·50	35·41	69·34	35·71	69·19	36·02	69·03	36·32	78
79	70·39	35·87	70·23	36·17	70·07	36·48	69·91	36·78	79
80	71·28	36·32	71·12	36·63	70·96	36·94	70·80	37·25	80
81	72·17	36·77	72·01	37·09	71·85	37·40	71·68	37·71	81
82	73·06	37·23	72·90	37·55	72·73	37·86	72·57	38·18	82
83	73·95	37·68	73·79	38·00	73·62	38·33	73·45	38·65	83
84	74·84	38·14	74·68	38·46	74·51	38·79	74·34	39·11	84
85	75·74	38·59	75·57	38·92	75·40	39·25	75·22	39·58	85
86	76·63	39·04	76·46	39·38	76·28	39·71	76·11	40·04	86
87	77·52	39·50	77·34	39·83	77·17	40·17	76·99	40·51	87
88	78·41	39·95	78·23	40·29	78·06	40·63	77·88	40·97	88
89	79·30	40·41	79·12	40·75	78·94	41·10	78·76	41·44	89
90	80·19	40·86	80·01	41·21	79·83	41·56	79·65	41·91	90
91	81·08	41·31	80·90	41·67	80·72	42·02	80·53	42·37	91
92	81·97	41·77	81·79	42·12	81·60	42·48	81·42	42·84	92
93	82·86	42·22	82·68	42·58	82·49	42·94	82·30	43·30	93
94	83·75	42·68	83·57	43·04	83·38	43·40	83·19	43·77	94
95	84·65	43·13	84·46	43·50	84·27	43·87	84·07	44·23	95
96	85·54	43·58	85·35	43·96	85·15	44·33	84·96	44·70	96
97	86·43	44·04	86·23	44·41	86·04	44·79	85·84	45·16	97
98	87·32	44·49	87·12	44·87	86·93	45·25	86·73	45·63	98
99	88·21	44·95	88·01	45·33	87·81	45·71	87·61	46·10	99
100	89·10	45·40	88·90	45·79	88·70	46·17	88·50	46·56	100
Distance.	Dep.	Lat.	Dep.	Lat.	Dep.	Lat.	Dep.	Lat.	Distance.
	63 Deg.		62¾ Deg.		62½ Deg.		62¼ Deg.		

Distance.	28 Deg.		28¼ Deg.		28½ Deg.		28¾ Deg.		Distance.
	Lat.	Dep.	Lat.	Dep.	Lat.	Dep.	Lat.	Dep.	
1	0·88	0·47	0·88	0·47	0·88	0·48	0·88	0·48	1
2	1·77	0·94	1·76	0·95	1·76	0·95	1·75	0·96	2
3	2·65	1·41	2·64	1·42	2·64	1·43	2·63	1·44	3
4	3·53	1·88	3·52	1·89	3·52	1·91	3·51	1·92	4
5	4·41	2·35	4·40	2·37	4·39	2·39	4·38	2·40	5
6	5·30	2·82	5·29	2·84	5·27	2·86	5·26	2·89	6
7	6·18	3·29	6·17	3·31	6·15	3·34	6·14	3·37	7
8	7·06	3·76	7·05	3·79	7·03	3·82	7·01	3·85	8
9	7·95	4·23	7·93	4·26	7·91	4·29	7·89	4·33	9
10	8.83	4·69	8·81	4·73	8·79	4·77	8·77	4·81	10
11	9·71	5·16	9·69	5·21	9·67	5·25	9·64	5·29	11
12	10·60	5·63	10·57	5·68	10·55	5·73	10·52	5·77	12
13	11·48	6·10	11·45	6·15	11·42	6·20	11·40	6·25	13
14	12·36	6·57	12·33	6·63	12·30	6·68	12·27	6·73	14
15	13·24	7·04	13·21	7·10	13·18	7·16	13·15	7·21	15
16	14·13	7·51	14·09	7·57	14·06	7·63	14·03	7·70	16
17	15·01	7·98	14·98	8·05	14·94	8·11	14·90	8·18	17
18	15·89	8·45	15·86	8·52	15·82	8·59	15·78	8·66	18
19	16·78	8·92	16·74	8·99	16·70	9·07	16·66	9·14	19
20	17·66	9·39	17·62	9·47	17·58	9·54	17·53	9·62	20
21	18·54	9·86	18·50	9·94	18·46	10·02	18·41	10·10	21
22	19·42	10·33	19·38	10·41	19·33	10·50	19·29	10·58	22
23	20·31	10·80	20·26	10·89	20·21	10·97	20·16	11·06	23
24	21·19	11·27	21·14	11·36	21·09	11·45	21·04	11·54	24
25	22·07	11·74	22·02	11·83	21·97	11·93	21·92	12·02	25
26	22·96	12·21	22·90	12·31	22·85	12·41	22·79	12·51	26
27	23·84	12·68	23·78	12.78	23·73	12·88	23·67	12·99	27
28	24·72	13·15	24·66	13·25	24·61	13·36	24·55	13·47	28
29	25·61	13·61	25·55	13·73	25·49	13·84	25·43	13·95	29
30	26·49	14·08	26·43	14·20	26·36	14·31	26·30	14·43	30
31	27·37	14·55	27·31	14·67	27·24	14·79	27·18	14·91	31
32	28·25	15·02	28·19	15·15	28·12	15·27	28·06	15·39	32
33	29·14	15·49	29·07	15·62	29·00	15·75	28·93	15·87	33
34	30·02	15·96	29·95	16·09	29·88	16·22	29·81	16·35	34
35	30·90	16·43	30·83	16·57	30·76	16·70	30·69	16·83	35
36	31·79	16·90	31·71	17·04	31·64	17·18	31·56	17·32	36
37	32·67	17·37	32·59	17·51	32·52	17·65	32·44	17·80	37
38	33·55	17·84	33·47	17·99	33·39	18·13	33·32	18·28	38
39	34·43	18·31	34·35	18·46	34·27	18·61	34·19	18·76	39
40	35·32	18·78	35·24	18·93	35·15	19·09	35·07	19·24	40
41	36·20	19·25	36·12	19·41	36·03	19·56	35·95	19·72	41
42	37·08	19·72	37·00	19·88	36·91	20·04	36·82	20·20	42
43	37·97	20·19	37·88	20·35	37·79	20·52	37·70	20·68	43
44	38·85	20·66	38·76	20·83	38·67	20·99	38·58	21·16	44
45	39·73	21·13	39·64	21·30	39·55	21·47	39·45	21·64	45
46	40·62	21·60	40·52	21·77	40·43	21·95	40·33	22·13	46
47	41·50	22·07	41·40	22·25	41·30	22·43	41·21	22·61	47
48	42·38	22·53	42·28	22·72	42·18	22·90	42·08	23·09	48
49	43·26	23·00	43·16	23·19	43·06	23·38	42·96	23·57	49
50	44·15	23·47	44·04	23·67	43·94	23·86	43·84	24·05	50
Distance.	Dep.	Lat.	Dep.	Lat.	Dep.	Lat.	Dep.	Lat.	Distance.
	62 Deg.		61¾ Deg.		61½ Deg.		61¼ Deg.		

Distance.	28 Deg.		28¼ Deg.		28½ Deg.		28¾ Deg.		Distance.
	Lat.	Dep.	Lat.	Dep.	Lat.	Dep.	Lat.	Dep.	
51	45·03	23·94	44·93	24·14	44·82	24·34	44·71	24·53	51
52	45·91	24·41	45·81	24·61	45·70	24.81	45·59	25·01	52
53	46·80	24·88	46·69	25·09	46·58	25·29	46·47	25·49	53
54	47·68	25·35	47·57	25·56	47·46	25·77	47·34	25·97	54
55	48·56	25·82	48·45	26·03	48·33	26·24	48·22	26·45	55
56	49·45	26·29	49·33	26·51	49·21	26·72	49·10	26·94	56
57	50·33	26·76	50·21	26·98	50·09	27·20	49·97	27·42	57
58	51·21	27·23	51·09	27·45	50·97	27·68	50·85	27·90	58
59	52·09	27·70	51·97	27·93	51·85	28·15	51·73	28·38	59
60	52·98	28·17	52·85	28·40	52·73	28·63	52·60	28·86	60
61	53·86	28·64	53·73	28·87	53·61	29·11	53·48	29·34	61
62	54·74	29·11	54·62	29·35	54·49	29·58	54·36	29·82	62
63	55·63	29·58	55·50	29·82	55·37	30·06	55·23	30·30	63
64	56·51	30·05	56·38	30·29	56·24	30·54	56·11	30·78	64
65	57·39	30·52	57·26	30·77	57·12	31·02	56·99	31·26	65
66	58·27	30·99	58·14	31·24	58·00	31·49	57·86	31·75	66
67	59·16	31·45	59·02	31·71	58·88	31·97	58·74	32·23	67
68	60·04	31·92	59·90	32·19	59·76	32·45	59·62	32·71	68
69	60·92	32.39	60·78	32·66	60·64	32·92	60·49	33·19	69
70	61·81	32·86	61·66	33·13	61·52	33·40	61·37	33·67	70
71	62·69	33·33	62·54	33·61	62·40	33·88	62·25	34·15	71
72	63·57	33·80	63·42	34·08	63·27	34·36	63·12	34·63	72
73	64·46	34·27	64·30	34·55	64·15	34·83	64·00	35·11	73
74	65·34	34·74	65·19	35·03	65·03	35·31	64·88	35·59	74
75	66·22	35·21	66·07	35·50	65·91	35·79	65·75	36·07	75
76	67·10	35·68	66·95	35·97	66·79	36·26	66·63	36·56	76
77	67·99	36·15	67.83	36·45	67·67	36·74	67·51	37·04	77
78	68·87	36·62	68·71	36·92	68·55	37·22	68·38	37·52	78
79	69·75	37·09	69·59	37·39	69·43	37·70	69·26	38·00	79
80	70·64	37·56	70·47	37·87	70·31	38·17	70·14	38·48	80
81	71·52	38·03	71·35	38·34	71·18	38·65	71·01	38·96	81
82	72·40	38·50	72·23	38·81	72·06	39·13	71·89	39·44	82
83	73·28	38·97	73·11	39·29	72·94	39·60	72·77	39·92	83
84	74·17	39·44	73·99	39·76	73·82	40·08	73·64	40·40	84
85	75·05	39·91	74·88	40·23	74·70	40·56	74·52	40·88	85
86	75·93	40·37	75·76	40·71	75·58	41·04	75·40	41·36	86
87	76·82	40·84	76·64	41·18	76·46	41·51	76·28	41·85	87
88	77·70	41·31	77·52	41·65	77·34	41·99	77·15	42·33	88
89	78·58	41·78	78·40	42·13	78·21	42·47	78·03	42·81	89
90	79·47	42·25	79·28	42·60	79·09	42·94	78·91	43·29	90
91	80·35	42·72	80·16	43·07	79·97	43·42	79·78	43·77	91
92	81·23	43·19	81·04	43·55	80·85	43·90	80·66	44·25	92
93	82·11	43·66	81·92	44·02	81·73	44·38	81·54	44·73	93
94	83·00	44.13	82·80	44·49	82·61	44·85	82·41	45·21	94
95	83·88	44·60	83·68	44·97	83·49	45·33	83·29	45·69	95
96	84·76	45·07	84·57	45·44	84·37	45·81	84·17	46·17	96
97	85·65	45·54	85·45	45·91	85·25	46·28	85·04	46·66	97
98	86·53	46·01	86·33	46·39	86·12	46·76	85·92	47·14	98
99	87 41	46·48	87·21	46·86	87·00	47·24	86·80	47·62	99
100	88·29	46·95	88.09	47·33	87·88	47·72	87·67	48·10	100
Distance.	Dep.	Lat.	Dep.	Lat.	Dep.	Lat.	Dep.	Lat.	Distance.
	62 Deg.		61¾ Deg.		61½ Deg.		61¼ Deg.		

Distance.	29 Deg.		29¼ Deg.		29½ Deg.		29¾ Deg.		Distance.
	Lat.	Dep.	Lat.	Dep.	Lat.	Dep.	Lat.	Dep.	
1	0·87	0·48	0·87	0·49	0·87	0·49	0·87	0·50	1
2	1·75	0·97	1·74	0·98	1·74	0·98	1·74	0·99	2
3	2·62	1·45	2·62	1·47	2·61	1·48	2·60	1·49	3
4	3·50	1·94	3·49	1·95	3·48	1·97	3·47	1·98	4
5	4·37	2·42	4·36	2·44	4·35	2·46	4·34	2·48	5
6	5·25	2·91	5·23	2·93	5·22	2·95	5·21	2·98	6
7	6·12	3·39	6·11	3·42	6·09	3·45	6·08	3·47	7
8	7·00	3·88	6·98	3·91	6·96	3·94	6·95	3·97	8
9	7·87	4·36	7·85	4·40	7·83	4·43	7·81	4·47	9
10	8·75	4·85	8·72	4·89	8·70	4·92	8·68	4·96	10
11	9·62	5·33	9·60	5·37	9·57	5·42	9·55	5·46	11
12	10·50	5·82	10·47	5·86	10·44	5·91	10·42	5·95	12
13	11·37	6·30	11·34	6·35	11·31	6·40	11·29	6·45	13
14	12·24	6·79	12·21	6·84	12·18	6·89	12·15	6·95	14
15	13·12	7·27	13·09	7·33	13·06	7·39	13·02	7·44	15
16	13·99	7·76	13·96	7·82	13·93	7·88	13·89	7·94	16
17	14·87	8·24	14·83	8·31	14·80	8·37	14·76	8·44	17
18	15·74	8·73	15·70	8·80	15·67	8·86	15·63	8·93	18
19	16·62	9·21	16·58	9·28	16·54	9·36	16·50	9·43	19
20	17·49	9·70	17·45	9·77	17·41	9·85	17·36	9·92	20
21	18·37	10·18	18·32	10·26	18·28	10·34	18·23	10·42	21
22	19·24	10·67	19·19	10·75	19·15	10·83	19·10	10·92	22
23	20·12	11·15	20·07	11·24	20·02	11·33	19·97	11·41	23
24	20·99	11·64	20·94	11·73	20·89	11·82	20·84	11·91	24
25	21·87	12·12	21·81	12·22	21·76	12·31	21·70	12·41	25
26	22·74	12·60	22·68	12·70	22·63	12·80	22·57	12·90	26
27	23·61	13·09	23·56	13·19	23·50	13·30	23·44	13·40	27
28	24·49	13·57	24·43	13·68	24·37	13·79	24·31	13·89	28
29	25·36	14·06	25·30	14·17	25·24	14·28	25·18	14·39	29
30	26·24	14·54	26·17	14·66	26·11	14·77	26·05	14·89	30
31	27·11	15·03	27·05	15·15	26·98	15·27	26·91	15·38	31
32	27·99	15·51	27·92	15·64	27·85	15·76	27·78	15·88	32
33	28·86	16·00	28·79	16·12	28·72	16·25	28·65	16·38	33
34	29·74	16·48	29·66	16·61	29·59	16·74	29·52	16·87	34
35	30·61	16·97	30·54	17·10	30·46	17·23	30·39	17·37	35
36	31·49	17·45	31·41	17·59	31·33	17·73	31·26	17·86	36
37	32·36	17·94	32·28	18·08	32·20	18·22	32·12	18·36	37
38	33·24	18·42	33·15	18·57	33·07	18·71	32·99	18·86	38
39	34·11	18·91	34·03	19·06	33·94	19·20	33·86	19·35	39
40	34·98	19·39	34·90	19·54	34·81	19·70	34·73	19·85	40
41	35·86	19·88	35·77	20·03	35·68	20·19	35·60	20·34	41
42	36·73	20·36	36·64	20·52	36·55	20·68	36·46	20·84	42
43	37·61	20·85	37·52	21·01	37·43	21·17	37·33	21·34	43
44	38·48	21·33	38·39	21·50	38·30	21·67	38·20	21·83	44
45	39·36	21·82	39·26	21·99	39·17	22·16	39·07	22·33	45
46	40·23	22·30	40·13	22·48	40·04	22·65	39·94	22·83	46
47	41·11	22·79	41·01	22·97	40·91	23·14	40·81	23·32	47
48	41·98	23·27	41·88	23·45	41·78	23·63	41·67	23·82	48
49	42·86	23·76	42·75	23·94	42·65	24·13	42·54	24·31	49
50	43·73	24·24	43·62	24·43	43·52	24·62	43·41	24·81	50
Distance.	Dep.	Lat.	Dep.	Lat.	Dep.	Lat.	Dep.	Lat.	Distance.
	61 Deg.		60¾ Deg.		60½ Deg.		60¼ Deg.		

Distance.	29 Deg.		29¼ Deg.		29½ Deg.		29¾ Deg.		Distance.
	Lat.	Dep.	Lat.	Dep.	Lat.	Dep.	Lat.	Dep.	
51	44·61	24·73	44·50	24·92	44·39	25·11	44·28	25·31	51
52	45·48	25·21	45·37	25·41	45·26	25·61	45·15	25·80	52
53	46·35	25·69	46·24	25·90	46·13	26·10	46·01	26·30	53
54	47·23	26·18	47·11	26·39	47·00	26·59	46·88	26·80	54
55	48·10	26·66	47·99	26·87	47·87	27·08	47·75	27·29	55
56	48·98	27·15	48·86	27·36	48·74	27·58	48·62	27·79	56
57	49·85	27·63	49·73	27·85	49·61	28·07	49·49	28·28	57
58	50·73	28·12	50·60	28·34	50·48	28·56	50·36	28·78	58
59	51·60	28·60	51·48	28·83	51·35	29·05	51·22	29·28	59
60	52·48	29·09	52·35	29·32	52·22	29·55	52·09	29·77	60
61	53·35	29·57	53·22	29·81	53·09	30·04	52·96	30·27	61
62	54·23	30·06	54·09	30·29	53·96	30·53	53·83	30·77	62
63	55·10	30·54	54·97	30·78	54·83	31·02	54·70	31·26	63
64	55·98	31·03	55·84	31·27	55·70	31·52	55·56	31·76	64
65	56·85	31·51	56·71	31·76	56·57	32·01	56·43	32·25	65
66	57·72	32·00	57·58	32·25	57·44	32·50	57·30	32·75	66
67	58·60	32·48	58·46	32·74	58·31	32·99	58·17	33·25	67
68	59·47	32·97	59·33	33·23	59·18	33·48	59·04	33·74	68
69	60·35	33·45	60·20	33·71	60·05	33·98	59·91	34·24	69
70	61·22	33·94	61·07	34·20	60·92	34·47	60·77	34·74	70
71	62·10	34·42	61·95	34·69	61·80	34·96	61·64	35·23	71
72	62·97	34·91	62·82	35·18	62·67	35·45	62·51	35·73	72
73	63·85	35·39	63·69	35·67	63·54	35·95	63·38	36·22	73
74	64·72	35·88	64·56	36·16	64·41	36·44	64·25	36·72	74
75	65·60	36·36	65·44	36·65	65·28	36·93	65·11	37·22	75
76	66·47	36·85	66·31	37·14	66·15	37·42	65·98	37·71	76
77	67·35	37·33	67·18	37·62	67·02	37·92	66·85	38·21	77
78	68·22	37·82	68·05	38·11	67·89	38·41	67·72	38·70	78
79	69·09	38·30	68·93	38·60	68·76	38·90	68·59	39·20	79
80	69·97	38·78	69·80	39·09	69·63	39·39	69·46	39·70	80
81	70·84	39·27	79·67	39·58	70·50	39·89	70·32	40·19	81
82	71·72	39·75	71·54	40·07	71·37	40·38	71·19	40·69	82
83	72·59	40·24	72·42	40·56	72·24	40·87	72·06	41·19	83
84	73·47	40·72	73·29	41.04	73·11	41·36	72·93	41.68	84
85	74·34	41·21	74·16	41·53	73·98	41·86	73·80	42·18	85
86	75·22	41·69	75·03	42·02	74·85	42·35	74·67	42·67	86
87	76·09	42·18	75·91	42·51	75·72	42·84	75·53	43·17	87
88	76·97	42·66	76·78	43·00	76·59	43·33	76·40	43·67	88
89	77·84	43·15	77·65	43·49	77·46	43·83	77·27	44·16	89
90	78·72	43·63	78·52	43·98	78·33	44·32	78·14	44·66	90
91	79·59	44·12	79·40	44·46	79·20	44·81	79·01	45·16	91
92	80·46	44·60	80·27	44·95	80·07	45·30	79·87	45·65	92
93	81·34	45·09	81·14	45·44	80·94	45·80	80·74	46·15	93
94	82·21	45·57	82·01	45·93	81·81	46·29	81·61	46·64	94
95	83·09	46·06	82·89	46·42	82·68	46·78	82·48	47·14	95
96	83·96	46·54	83·76	46·91	83·55	47·27	83·35	47·64	96
97	84·84	47·03	84·63	47·40	84·42	47·77	84·22	48·13	97
98	85·71	47·51	85·50	47·88	85·20	48·26	85·08	48·63	98
99	86·59	48·00	86·38	48·37	86·17	48·75	85·95	49·13	99
100	87·46	48·48	87·25	48·86	87·04	49·24	86·82	49·62	100
	Dep.	Lat.	Dep.	Lat.	Dep.	Lat.	Dep.	Lat.	
Distance.	61 Deg.		60¾ Deg.		60½ Deg.		60¼ Deg.		Distance.

Distance.	30 Deg.		30¼ Deg.		30½ Deg.		30¾ Deg.		Distance.
	Lat.	Dep.	Lat.	Dep.	Lat.	Dep.	Lat.	Dep.	
1	0·87	0·50	0·86	0·50	0·86	0·51	0·86	0·51	1
2	1·73	1·00	1·73	1·01	1·72	1·02	1·72	1·02	2
3	2·60	1·50	2·59	1·51	2·58	1·52	2·58	1·53	3
4	3·46	2·00	3·46	2·02	3·45	2·03	3·44	2·05	4
5	4·33	2·50	4·32	2·52	4·31	2·54	4·30	2·56	5
6	5·20	3·00	5·18	3·02	5·17	3·05	5·16	3·07	6
7	6·06	3·50	6·05	3·53	6·03	3·55	6·02	3·58	7
8	6·93	4·00	6·91	4·03	6·89	4·06	6·88	4·09	8
9	7·79	4·50	7·77	4·53	7·75	4·57	7·73	4·60	9
10	8·66	5·00	8·64	5·04	8·62	5·08	8·59	5·11	10
11	9·53	5·50	9·50	5·54	9·48	5·58	9·45	5·62	11
12	10·39	6·00	10·37	6·05	10·34	6·09	10·31	6·14	12
13	11·26	6·50	11·23	6·55	11·20	6·60	11·17	6·65	13
14	12·12	7·00	12·09	7·05	12·06	7·11	12·03	7·16	14
15	12·99	7·50	12·96	7·56	12·92	7·61	12·89	7·67	15
16	13·86	8·00	13·82	8·06	13·79	8·12	13·75	8·18	16
17	14·72	8·50	14·69	8·56	14·65	8·63	14·61	8·69	17
18	15·59	9·00	15·55	9·07	15·51	9·14	15·47	9·20	18
19	16·45	9·50	16·41	9·57	16·37	9·64	16·33	9·71	19
20	17·32	10·00	17·28	10·08	17·23	10·15	17·19	10·23	20
21	18·19	10·50	18·14	10·58	18·09	10·66	18·05	10·74	21
22	19·05	11·00	19·00	11·08	18·96	11·17	18·91	11·25	22
23	19·92	11·50	19·87	11·59	19·82	11·67	19·77	11·76	23
24	20·78	12·00	20·73	12·09	20·68	12·18	20·63	12·27	24
25	21·65	12·50	21·60	12·59	21·54	12·69	21·49	12·78	25
26	22·52	13·00	22·46	13·10	22·40	13·20	22·34	13·29	26
27	23·38	13·50	23·32	13·60	23·26	13·70	23·20	13·80	27
28	24·25	14·00	24·19	14·11	24·13	14·21	24·06	14·32	28
29	25·11	14·50	25·05	14·61	24·99	14·72	24·92	14·83	29
30	25·98	15·00	25·92	15·11	25·85	15·23	25·78	15·34	30
31	26·85	15·50	26·78	15·62	26·71	15·73	26·64	15·85	31
32	27·71	16·00	27·64	16·12	27·57	16·24	27·50	16·36	32
33	28·58	16·50	28·51	16·62	28·43	16·75	28·36	16·87	33
34	29·44	17·00	29·37	17·13	29·30	17·26	29·22	17·38	34
35	30·31	17·50	30·23	17·63	30·16	17·76	30·08	17·90	35
36	31·18	18·00	31·10	18·14	31·02	18·27	30·94	18·41	36
37	32·04	18·50	31·96	18·64	31·88	18·78	31·80	18·92	37
38	32·91	19·00	32·83	19·14	32·74	19·29	32·66	19·43	38
39	33·77	19·50	33·69	19·65	33·60	19·79	33·52	19·94	39
40	34·64	20·00	34·55	20·15	34·47	20·30	34·38	20·45	40
41	35·51	20·50	35·42	20·65	35·33	20·81	35·24	20·96	41
42	36·37	21·00	36·28	21·16	36·19	21·32	36·10	21·47	42
43	37·24	21·50	37·14	21·66	37·05	21·82	36·95	21·99	43
44	38·11	22·00	38·01	22·17	37·91	22·33	37·81	22·50	44
45	38·97	22·50	38·87	22·67	38·77	22·84	38·67	23·01	45
46	39·84	23·00	39·74	23·17	39·63	23·35	39·53	23·52	46
47	40·70	23·50	40·60	23·68	40·50	23·85	40·39	24·03	47
48	41·57	24·00	41·46	24·18	41·36	24·36	41·25	24·54	48
49	42·44	24·50	42·33	24·68	42·22	24·87	42·11	25·05	49
50	43·30	25·00	43·19	25·19	43·08	25·38	42·97	25·56	50
Distance.	Dep.	Lat.	Dep.	Lat.	Dep.	Lat.	Dep.	Lat.	Distance.
	60 Deg.		59¾ Deg.		59½ Deg.		59¼ Deg.		

Distance.	30 Deg.		30¼ Deg.		30½ Deg.		30¾ Deg.		Distance.
	Lat.	Dep.	Lat.	Dep.	Lat.	Dep.	Lat.	Dep.	
51	44·17	25·50	44·06	25·69	43·94	25·88	43·83	26·08	51
52	45·03	26·00	44·92	26·20	44·80	26·39	44·69	26·59	52
53	45·90	26·50	45·78	26·70	45·67	26·90	45·55	27·10	53
54	46·77	27·00	46·65	27·20	46·53	27·41	46·41	27·61	54
55	47·63	27·50	47·51	27·71	47·39	27·91	47·27	28·12	55
56	48·50	28·00	48·37	28·21	48·25	28·42	48·13	28·63	56
57	49·36	28·50	49·24	28·72	49·11	28·93	48·99	29·14	57
58	50·23	29·00	50·10	29·22	49·97	29·44	49·85	29·65	58
59	51·10	29·50	50·97	29·72	50·84	29·94	50·70	30·17	59
60	51·96	30·00	51·83	30·23	51·70	30·45	51·56	30·68	60
61	52·83	30·50	52·69	30·73	52·56	30·96	52·42	31·19	61
62	53·69	31·00	53·56	31·23	53·42	31·47	53·28	31·70	62
63	54·56	31·50	54·42	31·74	54·28	31·97	54·14	32·21	63
64	55·43	32·00	55·29	32·24	55·14	32·48	55·00	32·72	64
65	56·29	32·50	56·15	32·75	56·01	32·99	55·86	33·23	65
66	57·16	33·00	57·01	33·25	56·87	33·50	56·72	33·75	66
67	58·02	33·50	57·88	33·75	57·73	34·01	57·58	34·26	67
68	58·89	34·00	58·74	34·26	58·59	34·51	58·44	34·77	68
69	59·76	34·50	59·60	34·76	59·45	35·02	59·30	35·28	69
70	60·62	35·00	60·47	35·26	60·31	35·53	60·16	35·79	70
71	61.49	35·50	61·33	35·77	61·18	36·04	61·02	36·30	71
72	62·35	36·00	62·20	36·27	62·04	36·54	61·88	36·81	72
73	63·22	36·50	63·06	36·78	62·90	37·05	62·74	37·32	73
74	64·09	37·00	63·92	37·28	63·76	37·56	63·60	37·84	74
75	64·95	37·50	64·79	37·78	64·62	38·07	64·46	38·35	75
76	65·82	38·00	65·65	38·29	65·48	38·57	65·31	38·86	76
77	66·68	38·50	66·52	38·79	66·35	39·08	66·17	39·37	77
78	67·55	39·00	67·38	39·29	67·21	39·59	67·03	39·88	78
79	68·42	39·50	68·24	39·80	68·07	40·10	67·89	40·39	79
80	69·28	40·00	69·11	40·30	68·93	40·60	68·75	40·90	80
81	70·15	40·50	69·97	40·81	69·79	41·11	69·61	41·41	81
82	71·01	41·00	70·83	41·31	70·65	41·62	70·47	41·93	82
83	71·88	41·50	71·70	41·81	71·52	42·13	71·33	42·44	83
84	72·75	42·00	72·56	42·32	72·38	42·63	72·19	42·95	84
85	73·61	42·50	73·43	42·82	73·24	43·14	73·05	43·46	85
86	74·48	43·00	74·29	43·32	74·10	43·65	73·91	43·97	86
87	75·34	43·50	75·15	43·83	74·96	44·16	74·77	44·48	87
88	76·21	44·00	76·02	44·33	75·82	44·66	75·63	44·99	88
89	77·08	44·50	76·88	44·84	76·68	45·17	76·49	45·51	89
90	77·94	45·00	77·75	45·34	77·55	45·68	77·35	46·02	90
91	78·81	45·50	78·61	45·84	78·41	46·19	78·21	46·53	91
92	79·67	46·00	79·47	46·35	79·27	46·69	79·07	47·04	92
93	80·54	46·50	80·34	46·85	80·13	47·20	79·92	47·55	93
94	81·41	47·00	81·20	47·35	80·99	47·71	80·78	48·06	94
95	82·27	47·50	82·06	47·86	81·85	48·22	81·64	48·57	95
96	83·14	48·00	82·93	48·36	82·72	48·72	82·50	49·08	96
97	84·00	48·50	83·79	48·87	83·58	49·23	83·36	49·60	97
98	84·87	49·00	84·66	49·37	84·44	49·74	84·22	50·11	98
99	85·74	49·50	85·52	49·87	85·30	50·25	85·08	50·62	99
100	86·60	50·00	86·38	50·38	86·16	50·75	85·94	51·13	100
	Dep.	Lat.	Dep.	Lat.	Dep.	Lat.	Dep.	Lat.	
Distance.	60 Deg.		59¾ Deg.		59½ Deg.		59¼ Deg.		Distance.

Distance.	31 Deg.		31¼ Deg.		31½ Deg.		31¾ Deg.		Distance.
	Lat.	Dep.	Lat.	Dep.	Lat.	Dep.	Lat.	Dep.	
1	0·86	0·51	0·85	0·52	0·85	0·52	0·85	0·53	1
2	1·71	1·03	1·71	1·04	1·71	1·04	1·70	1·05	2
3	2·57	1·55	2·56	1·56	2·56	1·57	2·55	1·58	3
4	3·43	2·06	3·42	2·08	3·41	2·09	3·40	2·10	4
5	4·29	2·58	4·27	2·59	4·26	2·61	4·25	2·63	5
6	5·14	3·09	5·13	3·11	5·12	3·13	5·10	3·16	6
7	6·00	3·61	5·98	3·63	5·97	3·66	5·95	3·68	7
8	6·86	4·12	6·84	4·15	6·82	4·18	6·80	4·21	8
9	7·71	4·64	7·69	4·67	7·67	4·70	7·65	4·74	9
10	8·57	5·15	8·55	5·19	8·53	5·22	8·50	5·26	10
11	9·43	5·67	9·40	5·71	9·38	5·75	9·35	5·79	11
12	10·29	6·18	10·26	6·23	10·23	6·27	10·20	6·31	12
13	11·14	6·70	11·11	6·74	11·08	6·79	11·05	6·84	13
14	12·00	7·21	11·97	7·26	11·94	7·31	11·90	7·37	14
15	12·86	7·73	12·82	7·78	12·79	7·84	12·76	7·89	15
16	13·71	8·24	13·68	8·30	13·64	8·36	13·61	8·42	16
17	14·57	8·76	14·53	8·82	14·49	8·88	14·46	8·95	17
18	15·43	9·27	15·39	9·34	15·35	9·40	15·31	9·47	18
19	16·29	9·79	16·24	9·86	16·20	9·93	16·16	10·00	19
20	17·14	10·30	17·10	10·38	17·05	10·45	17·01	10·52	20
21	18·00	10·82	17·95	10·89	17·91	10·97	17·86	11·05	21
22	18·86	11·33	18·81	11·41	18·76	11·49	18·71	11·58	22
23	19·71	11·85	19·66	11·93	19·61	12·02	19·56	12·10	23
24	20·57	12·36	20·52	12·45	20·46	12·54	20·41	12·63	24
25	21·43	12·88	21·37	12·97	21·32	13·06	21·26	13·16	25
26	22·29	13·39	22·23	13·49	22·17	13·58	22·11	13·68	26
27	23·14	13·91	23·08	14·01	23·02	14·11	22·96	14·21	27
28	24·00	14·42	23·94	14·53	23·87	14·63	23·81	14·73	28
29	24·86	14·94	24·79	15·04	24·73	15·15	24·66	15·26	29
30	25·71	15·45	25·65	15·56	25·58	15·67	25·51	15·79	30
31	26·57	15·97	26·50	16·08	26·43	16·20	26·36	16·31	31
32	27·43	16·48	27·36	16·60	27·28	16·72	27·21	16·84	32
33	28·29	17·00	28·21	17·12	28·14	17·24	28·06	17·37	33
34	29·14	17·51	29·07	17·64	28·99	17·76	28·91	17·89	34
35	30·00	18·03	29·92	18·16	29·84	18·29	29·76	18·42	35
36	30·86	18·54	30·78	18·68	30·70	18·81	30·61	18·94	36
37	31·72	19·06	31·63	19·19	31·55	19·33	31·46	19·47	37
38	32·57	19·57	32·49	19·71	32·40	19·85	32·31	20·00	38
39	33·43	20·09	33·34	20·23	33·25	20·38	33·16	20·52	39
40	34·29	20·60	34·20	20·75	34·11	20·90	34·01	21·05	40
41	35·14	21·12	35·05	21·27	34·96	21·42	34·86	21·57	41
42	36·00	21·63	35·91	21·79	35·81	21·94	35·71	22·10	42
43	36·86	22·15	36·76	22·31	36·66	22·47	36·57	22·63	43
44	37·72	22·66	37·62	22·83	37·52	22·99	37·42	23·15	44
45	38·57	23·18	38·47	23·34	38·37	23·51	38·27	23·68	45
46	39·43	23·69	39·33	23·86	39·22	24·03	39·12	24·21	46
47	40·29	24·21	40·18	24·38	40·07	24·56	39·97	24·73	47
48	41·14	24·72	41·04	24·90	40·93	25·08	40·82	25·26	48
49	42·00	25·24	41·89	25·42	41·78	25·60	41·67	25·78	49
50	42·86	25·75	42·75	25·94	42·63	26·12	42·52	26·31	50
	Dep.	Lat.	Dep.	Lat.	Dep.	Lat.	Dep.	Lat.	
Distance.	59 Deg.		58¾ Deg.		58½ Deg.		58¼ Deg.		Distance.

Distance.	31 Deg.		31¼ Deg.		31½ Deg.		31¾ Deg.		Distance.
	Lat.	Dep.	Lat.	Dep.	Lat.	Dep.	Lat.	Dep.	
51	43·72	26·27	43·60	26·46	43·48	26·65	43·37	26·84	51
52	44·57	26·78	44·46	26·98	44·34	27·17	44·22	27·36	52
53	45·43	27·30	45·31	27·49	45·19	27·69	45·07	27·89	53
54	46·29	27·81	46·17	28·01	46·04	28·21	45·92	28·42	54
55	47·14	28·33	47·02	28·53	46·90	28·74	46·77	28·94	55
56	48·00	28·84	47·88	29·05	47·75	29·26	47·62	29·47	56
57	48·86	29·36	48·73	29·57	48·60	29·78	48·47	29·99	57
58	49·72	29·87	49·58	30·09	49·45	30·30	49·32	30·52	58
59	50·57	30·39	50·44	30·61	50·31	30·83	50·17	31·05	59
60	51·43	30·90	51·29	31·13	51·16	31·35	51·02	31·57	60
61	52·29	31·42	52·15	31·65	52·01	31·87	51·87	32·10	61
62	53·14	31·93	53·00	32·16	52·86	32·39	52·72	32·63	62
63	54·00	32·45	53·86	32·68	53·72	32·92	53·57	33·15	63
64	54·86	32·96	54·71	33·20	54·57	33·44	54·42	33·68	64
65	55·72	33·48	55·57	33·72	55·42	33·96	55·27	34·20	65
66	56·57	33·99	56·42	34·24	56·27	34·48	56·12	34·73	66
67	57·43	34·51	57·28	34·76	57·13	35·01	56·98	35·26	67
68	58·29	35·02	58·13	35·28	57·98	35·53	57·82	35·78	68
69	59·14	35·54	58·99	35·80	58·83	36·05	58·67	36·31	69
70	60·00	36·05	59·84	36·31	59·68	36·57	59·52	36·83	70
71	60·86	36·57	60·70	36·83	60·54	37·10	60·37	37·36	71
72	61·72	37·08	61·55	37·35	61·39	37·62	61·23	37·89	72
73	62·57	37·60	62·41	37·87	62·24	38·14	62·08	38·41	73
74	63·43	38·11	63·26	38·39	63·10	38·66	62·93	38·94	74
75	64·29	38·63	64·12	38·91	63·95	39·19	63·78	39·47	75
76	65·14	39·14	64·97	39·43	64·80	39·71	64·63	39·99	76
77	66·00	39·66	65·83	39·95	65·65	40·23	65·48	40·52	77
78	66·86	40·17	66·68	40·46	66·51	40·75	66·33	41·04	78
79	67·72	40·69	67·54	40·98	67·36	41·28	67·18	41·57	79
80	68·57	41·20	68·39	41·50	68·21	41·80	68·03	42·10	80
81	69·43	41·72	69·25	42·02	69·06	42·32	68·88	42·62	81
82	70·29	42·23	70·10	42·54	69·92	42·84	69·73	43·15	82
83	71·14	42·75	70·96	43·06	70·77	43·37	70·58	43·68	83
84	72·00	43·26	71·81	43·58	71·62	43·89	71·43	44·20	84
85	72·86	43·78	72·67	44·10	72·47	44·41	72·28	44·73	85
86	73·72	44·29	73·52	44·61	73·33	44·93	73·13	45·25	86
87	74·57	44·81	74·38	45·13	74·18	45·46	73·98	45·78	87
88	75·43	45·32	75·23	45·65	75·03	45·98	74·83	46·31	88
89	76·29	45·84	76·09	46·17	75·88	46·50	75·68	46·83	89
90	77·15	46·35	76·94	46·69	76·74	47·02	76·53	47·36	90
91	78·00	46·87	77·80	47·21	77·59	47·55	77·38	47·89	91
92	78·86	47·38	78·65	47·73	78·44	48·07	78·23	48·41	92
93	79·72	47·90	79·51	48·25	79·30	48·59	79·08	48·94	93
94	80·57	48·41	80·36	48·76	80·15	49·11	79·93	49·47	94
95	81·43	48·93	81·22	49·28	81·00	49·64	80·78	49·99	95
96	82·29	49·44	82·07	49·80	81·85	50·16	81·63	50·52	96
97	83·15	49·96	82·93	50·32	82·71	50·68	82·48	51·04	97
98	84·00	50·47	83·78	50·84	83·56	51·20	83·33	51·57	98
99	84·86	50·99	84·64	51·36	84·41	51·73	84·18	52·10	99
100	85·72	51·50	85·49	51·88	85·26	52·25	85·04	52·62	100
Distance.	Dep.	Lat.	Dep.	Lat.	Dep.	Lat.	Dep.	Lat.	Distance.
	59 Deg.		58¾ Deg.		58½ Deg.		58¼ Deg.		

Distance.	32 Deg.		32¼ Deg.		32½ Deg.		32¾ Deg.		Distance.
	Lat.	Dep.	Lat.	Dep.	Lat.	Dep.	Lat.	Dep.	
1	0·85	0·53	0·85	0·53	0·84	0·54	0·84	0·54	1
2	1·70	1·06	1·69	1·07	1·69	1·07	1·68	1·08	2
3	2·54	1·59	2·54	1·60	2·53	1·61	2·52	1·62	3
4	3·39	2·12	3·38	2·13	3·37	2·15	3·36	2·16	4
5	4·24	2·65	4·23	2·67	4·22	2·69	4·21	2·70	5
6	5·09	3·18	5·07	3·20	5·06	3·22	5·05	3·25	6
7	5·94	3·71	5·92	3·74	5·90	3·76	5·89	3·79	7
8	6·78	4·24	6·77	4·27	6·75	4·30	6·73	4·33	8
9	7·63	4·77	7·61	4·80	7·59	4·84	7·57	4·87	9
10	8·48	5·30	8·46	5·34	8·43	5·37	8·41	5·41	10
11	9·33	5·83	9·30	5·87	9·28	5·91	9·25	5·95	11
12	10·18	6·36	10·15	6·40	10·12	6·45	10·09	6·49	12
13	11·02	6·89	10·99	6·94	10·96	6·98	10·93	7·03	13
14	11·87	7·42	11·84	7·47	11·81	7·52	11·77	7·57	14
15	12·72	7·95	12·69	8·00	12·65	8·06	12·62	8·11	15
16	13·57	8·48	13·53	8·54	13·49	8·60	13·46	8·66	16
17	14·42	9·01	14·38	9·07	14·34	9·13	14·30	9·20	17
18	15·26	9·54	15·22	9·61	15·18	9·67	15·14	9·74	18
19	16·11	10·07	16·07	10·14	16·02	10·21	15·98	10·28	19
20	16·96	10·60	16·91	10·67	16·87	10·75	16·82	10·82	20
21	17·81	11·13	17·76	11·21	17·71	11·28	17·66	11·36	21
22	18·66	11·66	18·61	11·74	18·55	11·82	18·50	11·90	22
23	19·51	12·19	19·45	12·27	19·40	12·36	19·34	12·44	23
24	20·35	12·72	20·30	12·81	20·24	12·90	20·18	12·98	24
25	21·20	13·25	21·14	13·34	21·08	13·43	21·03	13·52	25
26	22·05	13·78	21·99	13·87	21·93	13·97	21·87	14·07	26
27	22·90	14·31	22·83	14·41	22·77	14·51	22·71	14·61	27
28	23·75	14·84	23·68	14·94	23·61	15·04	23·55	15·15	28
29	24·59	15·37	24·53	15·47	24·46	15·58	24·39	15·69	29
30	25·44	15·90	25·37	16·01	25·30	16·12	25·23	16·23	30
31	26·29	16·43	26·22	16·54	26·15	16·66	26·07	16·77	31
32	27·14	16·96	27·06	17·08	26·99	17·19	26·91	17·31	32
33	27·99	17·49	27·91	17·61	27·83	17·73	27·75	17·85	33
34	28·83	18·02	28·75	18·14	28·68	18·27	28·60	18·39	34
35	29·68	18·55	29·60	18·68	29·52	18·81	29·44	18·93	35
36	30·53	19·08	30·45	19·21	30·36	19·34	30·28	19·48	36
37	31·38	19·61	31·29	19·74	31·21	19·88	31·12	20·02	37
38	32·23	20·14	32·14	20·28	32·05	20·42	31·96	20·56	38
39	33·07	20·67	32·98	20·81	32·89	20·95	32·80	21·10	39
40	33·92	21·20	33·83	21·34	33·74	21·49	33·64	21·64	40
41	34·77	21·73	34·67	21·88	34·58	22·03	34·48	22·18	41
42	35·62	22·26	35·52	22·41	35·42	22·57	35·32	22·72	42
43	36·47	22·79	36·37	22·95	36·27	23·10	36·16	23·26	43
44	37·31	23·32	37·21	23·48	37·11	23·64	37·01	23·80	44
45	38·16	23·85	38·06	24·01	37·95	24·18	37·85	24·34	45
46	39·01	24·38	38·90	24·55	38·80	24·72	38·69	24·88	46
47	39·86	24·91	39·75	25·08	39·64	25·25	39.53	25·43	47
48	40·71	25·44	40·59	25·61	40·48	25·79	40·37	25·97	48
49	41·55	25·97	41·44	26·15	41·33	26·33	41·21	26·51	49
50	42·40	26·50	42·29	26·68	42·17	26·86	42·05	27·05	50
	Dep.	Lat.	Dep.	Lat.	Dep.	Lat.	Dep.	Lat.	
Distance.	58 Deg.		57¾ Deg.		57½ Deg.		57¼ Deg.		Distance.

Distance.	32 Deg.		32¼ Deg.		32½ Deg.		32¾ Deg.		Distance.
	Lat.	Dep.	Lat.	Dep.	Lat.	Dep.	Lat.	Dep.	
51	43.25	27·03	43·13	27·21	43·01	27·40	42·89	27·59	51
52	44.10	27·56	43·98	27·75	43·86	27·94	43·73	28·13	52
53	44·95	28·09	44·82	28·28	44·70	28·48	44·58	28·67	53
54	45·79	28·62	45·67	28·82	45·54	29·01	45·42	29·21	54
55	46·64	29·15	46·51	29·35	46·39	29·55	46·26	29·75	55
56	47·49	29·68	47·36	29·88	47·23	30·09	47·10	30·29	56
57	48·34	30·21	48·21	30·42	48·07	30·63	47·94	30·84	57
58	49·19	30·74	49·05	30·95	48·92	31·16	48·78	31·38	58
59	50·03	31·27	49·90	31·48	49·76	31·70	49·62	31·92	59
60	50·88	31·80	50·74	32·02	50·60	32·24	50·46	32·46	60
61	51·73	32·33	51·59	32·55	51·45	32·78	51·30	33·00	61
62	52·58	32·85	52·44	33·08	52·29	33·31	52·14	33·54	62
63	53·43	33·38	53·28	33·62	53·13	33·85	52·99	34·08	63
64	54·28	33·91	54·13	34·15	53·98	34·39	53·83	34·62	64
65	55·12	34·44	54·97	34·68	54·82	34·92	54·67	35·16	65
66	55·97	34·97	55·82	35·22	55·66	35·46	55·51	35·70	66
67	56·82	35·50	56·66	35·75	56·51	36·00	56·35	36·25	67
68	57·67	36·03	57·51	36·29	57·35	36·54	57·19	36·79	68
69	58·52	36·56	58·36	36·82	58·19	37·07	58·03	37·33	69
70	59·36	37·09	59·20	37·35	59·04	37·61	58·87	37·87	70
71	60·21	37·62	60·05	37·89	59·88	38·15	59·71	38·41	71
72	61·06	38·15	60·89	38·42	60·72	38·69	60·55	38·95	72
73	61·91	38·68	61·74	38·95	61·57	39·22	61·40	39·49	73
74	62·76	39·21	62·58	39·49	62·41	39·76	62·24	40·03	74
75	63·60	39·74	63·43	40·02	63·25	40·30	63·08	40·57	75
76	64·45	40·27	64·28	40·55	64·10	40·83	63·92	41·11	76
77	65·30	40·80	65·12	41·09	64·94	41·37	64·76	41·65	77
78	66·15	41·33	65·97	41·62	65·78	41·91	65·60	42·20	78
79	67·00	41·86	66·81	42·16	66·63	42·45	66·44	42·74	79
80	67·84	42·39	67·66	42·69	67·47	42·98	67·28	43·28	80
81	68·69	42·92	68·50	43·22	68·31	43·52	68·12	43·82	81
82	69·54	43·45	69·35	43·76	69·16	44·06	68·97	44·36	82
83	70·39	43·98	70·20	44·29	70·00	44·60	69·81	44·90	83
84	71·24	44·51	71·04	44·82	70·84	45·13	70·65	45·44	84
85	72·08	45·04	71·89	45·36	71·69	45·67	71·49	45·98	85
86	72·93	45·57	72·73	45·89	72·53	46·21	72·33	46·52	86
87	73·78	46·10	73·58	46·42	73·38	46·75	73·17	47·06	87
88	74·63	46·63	74·42	46·96	74·22	47·28	74·01	47·61	88
89	75·48	47·16	75·27	47·49	75·06	47·82	74·85	48·15	89
90	76·32	47·69	76·12	48·03	75·91	48·36	75·69	48·69	90
91	77·17	48·22	76·96	48·56	76·75	48·89	76·53	49·23	91
92	78·02	48·75	77·81	49·09	77·59	49·43	77·38	49·77	92
93	78·87	49·28	78·65	49.63	78·44	49·97	78·22	50·31	93
94	79·72	49·81	79·50	50.16	79·28	50·51	79·06	50·85	94
95	80·56	50·34	80·34	50.69	80·12	51·04	79·90	51·39	95
96	81·41	50·87	81·19	51.23	80·97	51·58	80·74	51·93	96
97	82·26	51·40	82·04	51.76	81·81	52·12	81·58	52·47	97
98	83.11	51.93	82·88	52·29	82·65	52·66	82·42	53·02	98
99	83·96	52.46	83·73	52·83	83·50	53·19	83·26	53·56	99
100	84·80	52.99	84·57	53·36	84·34	53·73	84·10	54·10	100
Distance.	Dep.	Lat.	Dep.	Lat.	Dep.	Lat.	Dep.	Lat.	Distance.
	58 Deg.		57¾ Deg.		57½ Deg.		57¼ Deg.		

Distance.	33 Deg.		33¼ Deg.		33½ Deg.		33¾ Deg.		Distance.
	Lat.	Dep.	Lat.	Dep.	Lat.	Dep.	Lat.	Dep.	
1	0·84	0·54	0·84	0·55	0·83	0·55	0·83	0·56	1
2	1·68	1·09	1·67	1·10	1·67	1·10	1·66	1·11	2
3	2·52	1·63	2·51	1·64	2·50	1·66	2·49	1·67	3
4	3·35	2·18	3·35	2·19	3·34	2·21	3·33	2·22	4
5	4·19	2·72	4·18	2·74	4·17	2·76	4·16	2·78	5
6	5·03	3·27	5·02	3·29	5·00	3·31	4·99	3·33	6
7	5·87	3·81	5·85	3·84	5·84	3·86	5·82	3·89	7
8	6·71	4·36	6·69	4·39	6·67	4·42	6·65	4·44	8
9	7·55	4·90	7·53	4·93	7·50	4·97	7·48	5·00	9
10	8·39	5·45	8·36	5·48	8·34	5·52	8·31	5·56	10
11	9·23	5·99	9·20	6·03	9·17	6·07	9·15	6·11	11
12	10·06	6·54	10·04	6·58	10·01	6·62	9·98	6·67	12
13	10·90	7·08	10·87	7·13	10·84	7·18	10·81	7·22	13
14	11·74	7·62	11·71	7·68	11·67	7·73	11·64	7·78	14
15	12·58	8·17	12·54	8·22	12·51	8·28	12·47	8·33	15
16	13·42	8·71	13·38	8·77	13·34	8·83	13·30	8·89	16
17	14·26	9·26	14·22	9·32	14·18	9·38	14·13	9·44	17
18	15·10	9·80	15·05	9·87	15·01	9·93	14·97	10·00	18
19	15·93	10·35	15·89	10·42	15·84	10·49	15·80	10·56	19
20	16·77	10·89	16·73	10·97	16·68	11·04	16·63	11·11	20
21	17·61	11·44	17·56	11·51	17·51	11·59	17·46	11·67	21
22	18·45	11·98	18·40	12·06	18·35	12·14	18·29	12·22	22
23	19·29	12·53	19·23	12·61	19·18	12·69	19·12	12·78	23
24	20·13	13·07	20·07	13·16	20·01	13·25	19·96	13·33	24
25	20·97	13·62	20·91	13·71	20·85	13·80	20·79	13·89	25
26	21·81	14·16	21·74	14·26	21·68	14·35	21·62	14·44	26
27	22·64	14·71	22·58	14·80	22·51	14·90	22·45	15·00	27
28	23·48	15·25	23·42	15·35	23·35	15·45	23·28	15·56	28
29	24·32	15·79	24·25	15·90	24·18	16·01	24·11	16·11	29
30	25·16	16·34	25·09	16·45	25·02	16·56	24·94	16·67	30
31	26·00	16·88	25·92	17·00	25·85	17·11	25·78	17·22	31
32	26·84	17·43	26·76	17·55	26·68	17·66	26·61	17·78	32
33	27·68	17·97	27·60	18·09	27·52	18·21	27·44	18·33	33
34	28·51	18·52	28·43	18·64	28·35	18·77	28·27	18·89	34
35	29·35	19·06	29·27	19·19	29·19	19·32	29·10	19·44	35
36	30·19	19·61	30·11	19·74	30·02	19·87	29·93	20·00	36
37	31·03	20·15	30·94	20·29	30·85	20·42	30·76	20·56	37
38	31·87	20·70	31·78	20·84	31·69	20·97	31·60	21·11	38
39	32·71	21·24	32·62	21·38	32·52	21·53	32·43	21·67	39
40	33·55	21·79	33·45	21·93	33·36	22·08	33·26	22·22	40
41	34·39	22·33	34·29	22·48	34·19	22·63	34·09	22·78	41
42	35·22	22·87	35·12	23·03	35·02	23·18	34·92	23·33	42
43	36·06	23·42	35·96	23·58	35·86	23·73	35·75	23·89	43
44	36·90	23·96	36·80	24·12	36·69	24·29	36·58	24·45	44
45	37·74	24·51	37·63	24·67	37·52	24·84	37·42	25·00	45
46	38·58	25·05	38·47	25·22	38·36	25·39	38·25	25·56	46
47	39·42	25·60	39·31	25·77	39·19	25·94	39·08	26·11	47
48	40·26	26·14	40·14	26·32	40·03	26·49	39·91	26·67	48
49	41·09	26·69	40·98	26·87	40·86	27·04	40·74	27·22	49
50	41·93	27·23	41·81	27·41	41·69	27·60	41·57	27·78	50
	Dep.	Lat.	Dep.	Lat.	Dep.	Lat.	Dep.	Lat.	
Distance.	57 Deg.		56¾ Deg.		56½ Deg.		56¼ Deg.		Distance.

Distance.	33 Deg.		33¼ Deg.		33½ Deg.		33¾ Deg.		Distance.
	Lat.	Dep.	Lat.	Dep.	Lat.	Dep.	Lat.	Dep.	
51	42·77	27·78	42·65	27·96	42·53	28·15	42·40	28·33	51
52	43·61	28·32	43·40	28·51	43·36	28·70	43·24	28·89	52
53	44·45	28·87	44·32	29·06	44·20	29·25	44·07	29·45	53
54	45·29	29·41	45·16	29·61	45·03	29·80	44·90	30·00	54
55	46·13	29·96	46·00	30·16	45·86	30·36	45·73	30·56	55
56	46·97	30·50	46·83	30·70	46·70	30·91	46·56	31·11	56
57	47·80	31·04	47·67	31·25	47·53	31·46	47·39	31·67	57
58	48·64	31·59	48·50	31·80	48·37	32·01	48·23	32·22	58
59	49·48	32.13	49·34	32·35	49·20	32·56	49·06	32·78	59
60	50·32	32·68	50·18	32·90	50·03	33·12	49·89	33·33	60
61	51·16	33·22	51·01	33·45	50·87	33·67	50·72	33·89	61
62	52·00	33·77	51·85	33·99	51·70	34·22	51·55	34·45	62
63	52·84	34·31	52·69	34·54	52·53	34·77	52·38	35·00	63
64	53·67	34·86	53·52	35·09	53·37	35·32	53·21	35·56	64
65	54·51	35·40	54·36	35·64	54·20	35·88	54·05	36·11	65
66	55·35	35·95	55·19	36·19	55·04	36·43	54·88	36·67	66
67	56·19	36·49	56·03	36·74	55·87	36·98	55·71	37·22	67
68	57·03	37·04	56·87	37·28	56·70	37·53	56·54	37·78	68
69	57·87	37·58	57·70	37·83	57·54	38·08	57·37	38·33	69
70	58·71	38·12	58·54	38·38	58·37	38·64	58·20	38·89	70
71	59·55	38·67	59·38	38·93	59·21	39·19	59·03	39·45	71
72	60·38	39·21	60·21	39·48	60·04	39·74	59·87	40·00	72
73	61·22	39·76	61·05	40·03	60·87	40·29	60·70	40·56	73
74	62·06	40·30	61·89	40·57	61·71	40·84	61·53	41·11	74
75	62·90	40·85	62·72	41·12	62·54	41·40	62·36	41·67	75
76	63·74	41·39	63·56	41·67	63·38	41·95	63·19	42·22	76
77	64·58	41·94	64·39	42·22	64·21	42·50	64·02	42·78	77
78	65·42	42·48	65·23	42·77	65·04	43·05	64·85	43·33	78
79	66·25	43·03	66·07	43·32	65·88	43·60	65·69	43·89	79
80	67·09	43·57	66·90	43·86	66·71	44·15	66·52	44·45	80
81	67·93	44·12	67·74	44·41	67·54	44·71	67·35	45·00	81
82	68·77	44·66	68·58	44·96	68·38	45·26	68·18	45·56	82
83	69·61	45·20	69·41	45·51	69·21	45·81	69·01	46·11	83
84	70·45	45·75	70·25	46·06	70·05	46·36	69·84	46·67	84
85	71·29	46·29	71·08	46·60	70·88	46·91	70·67	47·22	85
86	72·13	46·84	71·92	47·15	71·71	47·47	71·51	47·78	86
87	72·96	47·38	72·76	47·70	72·55	48·02	72·34	48·33	87
88	73·80	47·93	73·59	48·25	73·38	48·57	73·17	48·89	88
89	74·64	48·47	74·43	48·80	74·22	49·12	74·00	49·45	89
90	75·48	49·02	75·27	49·35	75·05	49·67	74·83	50·00	90
91	76·32	49·56	76·10	49·89	75·88	50·23	75·66	50·56	91
92	77·16	50·11	76·94	50·44	76·72	50·78	76·50	51·11	92
93	78·00	50·65	77·77	50·99	77·55	51·33	77·33	51·67	93
94	78 83	51·20	78·61	51·54	78·39	51·88	78·16	52·22	94
95	79·67	51·74	79·45	52·09	79·22	52·43	78·99	52·78	95
96	80·51	52·29	80·28	52·64	80·05	52·99	79·82	53·33	96
97	81·35	52·83	81·12	53·18	80·89	53·54	80·65	53·89	97
98	82·19	53·37	81·96	53·73	81·72	54·09	81·48	54·45	98
99	83·03	53·92	82·79	54·28	82·55	54·64	82·32	55·00	99
100	83·87	54·46	83·63	54·83	83·39	55·19	83·15	55·56	100
Distance.	Dep.	Lat.	Dep.	Lat.	Dep.	Lat.	Dep.	Lat.	Distance.
	57 Deg.		56¾ Deg.		56½ Deg.		56¼ Deg.		

Distance.	34 Deg.		34¼ Deg.		34½ Deg.		34¾ Deg.		Distance.
	Lat.	Dep.	Lat.	Dep.	Lat.	Dep.	Lat.	Dep.	
1	0·83	0·56	0·83	0·56	0·82	0·57	0·82	0·57	1
2	1·66	1·12	1·65	1·13	1·65	1·13	1·64	1·14	2
3	2·49	1·68	2·48	1·69	2·47	1·70	2·46	1·71	3
4	3·32	2·24	3·31	2·25	3·30	2·27	3·29	2·28	4
5	4·15	2·80	4·13	2·81	4·12	2·83	4·11	2·85	5
6	4·97	3·36	4·96	3·38	4·94	3·40	4.93	3·42	6
7	5·80	3·91	5·79	3·94	5·77	3·96	5·75	3·99	7
8	6·63	4·47	6·61	4·50	6·59	4·53	6·57	4·56	8
9	7·46	5·03	7·44	5·07	7·42	5·10	7·39	5·13	9
10	8·29	5·59	8·27	5·63	8·24	5·66	8·22	5·70	10
11	9·12	6·15	9·09	6·19	9·07	6·23	9·04	6·27	11
12	9·95	6·71	9·92	6·75	9·89	6·80	9·86	6·84	12
13	10·78	7·27	10·75	7·32	10·71	7·36	10·68	7·41	13
14	11·61	7·83	11·57	7·88	11·54	7·93	11·50	7·98	14
15	12·44	8·39	12·40	8·44	12·36	8·50	12·32	8·55	15
16	13·26	8·95	13·23	9·00	13·19	9·06	13·15	9·12	16
17	14·09	9·51	14.05	9·57	14·01	9·63	13·97	9·69	17
18	14·92	10·07	14·88	10·13	14·83	10·20	14·79	10·26	18
19	15·75	10·62	15·71	10·69	15·66	10·76	15·61	10·83	19
20	16·58	11·18	16·53	11·26	16·48	11·33	16·43	11·40	20
21	17·41	11·74	17·36	11·82	17·31	11·89	17·25	11·97	21
22	18·24	12·30	18·18	12·38	18·13	12·46	18·08	12·54	22
23	19·07	12·86	19·01	12·94	18·95	13·03	18·90	13·11	23
24	19·90	13·42	19·84	13·51	19·78	13·59	19·72	13·68	24
25	20·73	13·98	20·66	14·07	20·60	14·16	20·54	14·25	25
26	21·55	14·54	21·49	14·63	21·43	14·73	21·36	14·82	26
27	22·38	15·10	22·32	15·20	22·25	15·29	22·18	15·39	27
28	23·21	15·66	23·14	15·76	23·08	15·86	23·01	15·96	28
29	24·04	16·22	23·97	16·32	23·90	16·43	23·83	16·53	29
30	24·87	16·78	24·80	16·88	24·72	16·99	24·65	17·10	30
31	25·70	17·33	25·62	17·45	25·55	17·56	25·47	17·67	31
32	26·53	17.89	26·45	18·01	26·37	18·12	26·29	18·24	32
33	27·36	18·45	27·28	18·57	27·20	18·69	27·11	18·81	33
34	28·19	19·01	28·10	19·14	28·02	19·26	27·94	19·38	34
35	29·02	19·57	28·93	19·70	28·84	19·82	28·76	19·95	35
36	29·85	20·13	29·76	20·26	29·67	20·39	29·58	20·52	36
37	30·67	20·69	30·58	20·82	30·49	20·96	30·40	21·09	37
38	31·50	21·25	31·41	21·39	31·32	21·52	31·22	21·66	38
39	32·33	21·81	32·24	21·95	32·14	22·09	32·04	22·23	39
40	33·16	22·37	33·06	22·51	32·97	22·66	32·87	22·80	40
41	33·99	22·93	33·89	23·07	33·79	23·22	33·69	23·37	41
42	34·82	23·49	34·72	23·64	34·61	23·79	34·51	23·94	42
43	35·65	24·05	35·54	24·20	35·44	24·36	35·33	24·51	43
44	36·48	24·60	36·37	24·76	36·26	24·92	36·15	25·08	44
45	37·31	25·16	37·20	25·33	37·09	25·49	36·97	25·65	45
46	38·14	25·72	38·02	25·89	37·91	26·05	37·80	26·22	46
47	38·96	26·28	38·85	26·45	38·73	26·62	38·62	26·79	47
48	39·79	26·84	39·68	27·01	39·56	27·19	39·44	27·36	48
49	40·62	27·40	40·50	27·58	40·38	27·75	40·26	27·93	49
50	41·45	27·96	41·33	28·14	41·21	28·32	41·08	28·50	50
Distance.	Dep.	Lat.	Dep.	Lat.	Dep.	Lat.	Dep.	Lat.	Distance.
	56 Deg.		55¾ Deg.		55½ Deg.		55¼ Deg.		

Distance.	34 Deg.		34¼ Deg.		34½ Deg.		34¾ Deg.		Distance.
	Lat.	Dep.	Lat.	Dep.	Lat.	Dep.	Lat.	Dep.	
51	42·28	28·52	42·16	28·70	42·03	28·89	41·90	29·07	51
52	43·11	29·08	42·98	29·27	42·85	29·45	42·73	29·64	52
53	43·94	29·64	43·81	29·83	43·68	30·02	43·55	30·21	53
54	44·77	30·20	44·64	30·39	44·50	30·59	44·37	30·78	54
55	45·60	30·76	45·46	30·95	45·33	31·15	45·19	31·35	55
56	46·43	31·31	46·29	31·52	46·15	31·72	46·01	31·92	56
57	47·26	31·87	47·12	32·08	46·98	32·29	46·83	32·49	57
58	48·08	32·43	47·94	32·64	47·80	32·85	47·66	33·06	58
59	48·91	32·99	48·77	33·21	48·62	33·42	48·48	33·63	59
60	49·74	33·55	49·60	33·77	49·45	33·98	49·30	34·20	60
61	50·57	34·11	50·42	34·33	50·27	34·55	50·12	34·77	61
62	51·40	34·67	51·25	34·89	51·10	35·12	50·94	35·34	62
63	52·23	35·23	52·08	35·46	51·92	35·68	51·76	35·91	63
64	53·06	35·79	52·90	36·02	52·74	36·25	52·59	36·48	64
65	53·89	36·35	53·73	36·58	53·57	36·82	53·41	37·05	65
66	54·72	36·91	54·55	37·15	54·39	37·38	54·23	37·62	66
67	55·55	37·46	55·38	37·71	55·22	37·95	55·05	38·19	67
68	56·37	38·03	56·21	38·27	56·04	38·52	55·87	38·76	68
69	57·20	38·58	57·03	38·83	56·86	39·08	56·69	39·33	69
70	58·03	39·14	57·86	39·40	57·69	39·65	57·52	39·90	70
71	58.86	39·70	58·69	39·96	58·51	40·21	58·34	40·47	71
72	59·69	40·26	59·51	40·52	59·34	40·78	59·16	41·04	72
73	60·52	40·82	60·34	41·08	60·16	41·35	59·98	41·61	73
74	61·35	41·38	61·17	41·65	60·99	41·91	60·80	42·18	74
75	62·18	41·94	61·99	42·21	61·81	42·48	61·62	42·75	75
76	63·01	42·50	62·82	42·77	62·63	43·05	62·45	43·32	76
77	63·84	43·06	63·65	43·34	63·46	43·61	63·27	43·89	77
78	64·66	43·62	64·47	43·90	64·28	44·18	64·09	44·46	78
79	65·49	44·18	65·30	44·46	65·11	44·75	64·91	45·03	79
80	66·32	44·74	66·13	45·02	65·93	45·31	65·73	45·60	80
81	67·15	45·29	66·95	45·59	66·75	45·88	66·55	46·17	81
82	67·98	45·85	67·78	46·15	67·58	46·45	67·37	46·74	82
83	68·81	46·41	68·61	46·71	68·40	47·01	68·20	47·31	83
84	69·64	46·97	69·43	47·28	69·23	47·58	69·02	47·88	84
85	70·47	47·53	70·26	47·84	70·05	48·14	69·84	48·45	85
86	71·30	48·09	71·09	48·40	70·87	48·71	70·66	49·02	86
87	72·13	48·65	71·91	48·96	71·70	49·28	71·48	49·59	87
88	72·96	49·21	72·74	49·53	72·52	49·84	72·30	50·16	88
89	73·78	49·77	73·57	50·09	73·35	50·41	73·13	50·73	89
90	74·61	50·33	74·39	50·65	74·17	50·98	73·95	51·30	90
91	75·44	50·89	75·22	51·22	75·00	51·54	74·77	51·87	91
92	76·27	51·45	76·05	51·78	75·82	52·11	75·59	52·44	92
93	77·10	52·00	76·87	52·34	76·64	52·68	76·41	53·01	93
94	77·93	52·56	77·70	52·90	77·47	53·24	77·23	53·58	94
95	78·76	53·12	78·53	53·47	78·29	53·81	78·06	54·15	95
96	79·59	53·68	79·35	54·03	79·12	54·37	78·88	54·72	96
97	80·42	54·24	80·18	54·59	79·94	54·94	79·70	55·29	97
98	81·25	54·80	81·01	55·15	80·76	55·51	80·52	55·86	98
99	82·07	55·36	81·83	55·72	81·59	56·07	81·34	56·43	99
100	82·90	55·92	82·66	56·28	82·41	56·64	82·16	57·00	100
Distance.	Dep.	Lat.	Dep.	Lat.	Dep.	Lat.	Dep.	Lat.	Distance.
	56 Deg.		55¾ Deg.		55½ Deg.		55¼ Deg.		

Distance.	35 Deg.		35¼ Deg.		35½ Deg.		35¾ Deg.		Distance.
	Lat.	Dep.	Lat.	Dep.	Lat.	Dep.	Lat.	Dep.	
1	0·82	0·57	0·82	0·58	0·81	0·58	0·81	0·58	1
2	1·64	1·15	1·63	1·15	1·63	1·16	1·62	1·17	2
3	2·46	1·72	2·45	1·73	2·44	1·74	2·43	1·75	3
4	3·28	2·29	3·27	2·31	3·26	2·32	3·25	2·34	4
5	4·10	2·87	4·08	2·89	4·07	2·90	4·06	2·92	5
6	4·91	3·44	4·90	3·46	4·88	3·48	4·87	3·51	6
7	5·73	4·01	5·72	4·04	5·70	4·06	5·68	4·09	7
8	6·55	4·59	6·53	4·62	6·51	4·65	6·49	4·67	8
9	7·37	5·16	7·35	5·19	7·33	5·23	7·30	5·26	9
10	8·19	5·74	8·17	5·77	8·14	5·81	8·12	5·84	10
11	9·01	6·31	8·98	6·35	8·96	6·39	8·93	6·43	11
12	9·83	6·88	9·80	6·93	9·77	6·97	9·74	7·01	12
13	10·65	7·46	10·62	7·50	10·58	7·55	10·55	7·60	13
14	11·47	8·03	11·43	8·08	11·40	8·13	11·36	8·18	14
15	12·29	8·60	12·25	8·66	12·21	8·71	12·17	8·76	15
16	13·11	9·18	13·07	9·23	13·03	9·29	12·99	9·35	16
17	13·93	9·75	13·88	9·81	13·84	9·87	13·80	9·93	17
18	14·74	10·32	14·70	10·39	14·65	10·45	14·61	10·52	18
19	15·56	10·90	15·52	10·97	15·47	11·03	15·42	11·10	19
20	16·38	11·47	16·33	11·54	16·28	11·61	16·23	11·68	20
21	17·20	12·05	17·15	12·12	17·10	12·19	17·04	12·27	21
22	18·02	12·62	17·97	12·70	17·01	12·78	17·85	12·85	22
23	18·84	13·19	18·78	13·27	18·72	13·36	18·67	13·44	23
24	19·66	13·77	19·60	13·85	19·54	13·94	19·48	14·02	24
25	20·48	14·34	20·42	14·43	20·35	14·52	20·29	14·61	25
26	21·30	14·91	21·23	15·01	21·17	15·10	21·10	15·19	26
27	22·12	15·49	22·05	15·58	21·98	15·68	21·91	15·77	27
28	22·94	16·06	22·87	16·16	22·80	16·26	22·72	16·36	28
29	23·76	16·63	23·68	16·74	23·61	16·84	23·54	16·94	29
30	24·57	17·21	24·50	17·31	24·42	17·42	24·35	17·53	30
31	25·39	17·78	25·32	17·89	25·24	18·00	25·16	18·11	31
32	26·21	18·35	26·13	18·47	26·05	18·58	25·97	18·70	32
33	27·03	18·93	26·95	19·05	26·87	19·16	26·78	19·28	33
34	27·85	19·50	27·77	19.62	27·68	19·74	27·59	19·86	34
35	28·67	20·08	28·58	20·20	28·49	20·32	28·41	20·45	35
36	29·49	20·65	29·40	20·78	29·31	20·91	29·22	21·03	36
37	30·31	21·22	30·22	21·35	30·12	21·49	30·03	21·62	37
38	31·13	21·80	31·03	21·93	30·94	22·07	30·84	22·20	38
39	31·95	22·37	31·85	22·51	31·75	22·65	31·65	22·79	39
40	32·77	22·94	32·67	23·09	32·56	23·23	32·46	23·37	40
41	33·59	23·52	33·48	23·66	33·38	23·81	33·27	23·95	41
42	34·40	24·09	34·30	24·24	34·19	24·39	34·09	24·54	42
43	35·22	24·66	35·12	24·82	35·01	24·97	34·90	25·12	43
44	36·04	25·24	35·93	25·39	35·82	25·55	35·71	25.71	44
45	36·86	25·81	36·75	25·97	36·64	26·13	36·52	26·29	45
46	37·68	26·38	37·57	26·55	37·45	26·71	37·33	26·88	46
47	38·50	26·96	38·38	27·13	38·26	27·29	38·14	27·46	47
48	39·32	27·53	39·20	27·70	39·08	27·87	38·96	28·04	48
49	40·14	28·11	40·02	28·28	39·89	28·45	39·77	28·63	49
50	40·96	28·68	40·83	28·86	40·71	29·04	40·58	29·21	50
	Dep.	Lat.	Dep.	Lat.	Dep.	Lat.	Dep.	Lat.	
Distance.	55 Deg.		54¾ Deg.		54½ Deg.		54¼ Deg.		Distance.

Distance.	35 Deg.		35¼ Deg.		35½ Deg.		35¾ Deg.		Distance.
	Lat.	Dep.	Lat.	Dep.	Lat.	Dep.	Lat.	Dep.	
51	41·78	29·25	41·65	29·43	41·52	29·62	41·39	29·80	51
52	42·60	29·83	42·47	30·01	42·33	30·20	42·20	30·38	52
53	43·42	30·40	43·28	30·59	43·15	30·78	43·01	30·97	53
54	44·23	30·97	44·10	31·17	43·96	31·36	43·82	31·55	54
55	45·05	31·55	44·92	31·74	44·78	31·94	44·64	32·13	55
56	45·87	32·12	45·73	32·32	45·59	32·52	45·45	32·72	56
57	46·69	32·69	46·55	32·90	46·40	33·10	46·26	33·30	57
58	47·51	33·27	47·37	33·47	47·22	33·68	47·07	33·89	58
59	48·33	33·84	48·18	34·05	48·03	34·26	47·88	34·47	59
60	49·15	34·41	49·00	34·63	48·85	34·84	48·69	35·05	60
61	49·97	34·99	49·82	35·21	49·66	35·42	49·51	35·64	61
62	50·79	35·56	50·63	35·78	50·48	36·00	50·32	36·22	62
63	51·61	36·14	51·45	36·36	51·29	36·58	51·13	36·81	63
64	52·43	36·71	52·27	36·94	52·10	37·16	51·94	37·39	64
65	53·24	37·28	53·08	37·51	52·92	37·75	52·75	37·98	65
66	54·06	37·86	53·90	38·09	53·73	38·33	53·56	38·56	66
67	54·88	38·43	54·71	38·67	54·55	38·91	54·38	39·14	67
68	55·70	39·00	55·53	39·25	55·36	39·49	55·19	39·73	68
69	56·52	39·58	56·35	39·82	56·17	40·07	56·00	40·31	69
70	57·34	40·15	57·16	40·40	56·99	40·65	56·81	40·90	70
71	58·16	40·72	57·98	40·98	57·80	41·23	57·62	41·48	71
72	58·98	41·30	58·80	41·55	58·62	41·81	58·43	42·07	72
73	59·80	41·87	59·61	42·13	59·43	42·39	59·24	42·65	73
74	60·62	42·44	60·43	42·71	60·24	42·97	60·06	43·23	74
75	61·44	43·02	61·25	43·29	61·06	43·55	60·87	43·82	75
76	62·26	43·59	62·06	43·86	61·87	44·13	61·68	44·40	76
77	63·07	44·17	62·88	44·44	62·69	44·71	62·49	44·99	77
78	63·89	44·74	63·70	45·02	63·50	45·29	63·30	45·57	78
79	64·71	45·31	64·51	45·59	64·32	45·88	64·11	46·16	79
80	65·53	45·89	65·33	46·17	65·13	46·46	64·93	46·74	80
81	66·35	46·46	66·15	46·75	65·94	47·04	65·74	47·32	81
82	67·17	47·03	66·96	47·33	66·76	47·62	66·55	47·91	82
83	67·99	47·61	67·78	47·90	67·57	48·20	67·36	48·49	83
84	68·81	48·18	68·60	48·48	68·39	48·78	68·17	49·08	84
85	69·63	48·75	69·41	49·06	69·20	49·36	68·98	49·66	85
86	70·45	49·33	70·23	49·63	70·01	49·94	69·80	50·25	86
87	71·27	49·90	71·05	50·21	70·83	50·52	70·61	50·83	87
88	72·09	50·47	71·86	50·79	71·64	51·10	71·42	51·41	88
89	72·90	51·05	72·68	51·37	72·46	51·68	72·23	52·00	89
90	73·72	51·62	73·50	51·94	73·27	52·26	73·04	52·58	90
91	74·54	52·20	74·31	52·52	74·08	52·84	73·85	53·17	91
92	75·36	52·77	75·13	53·10	74·90	53.42	74·66	53·75	92
93	76·18	53·34	75·95	53·67	75·71	54·01	75·48	54·34	93
94	77·00	53·92	76·76	54·25	76·53	54·59	76·29	54·92	94
95	77·82	54·49	77·58	54·83	77·34	55·17	77·10	55·50	95
96	78·64	55·06	78·40	55·41	78·16	55·75	77·91	56·09	96
97	79·46	55·64	79·21	55·98	78·97	56·33	78·72	56·67	97
98	80·28	56·21	80·03	56·56	79·78	56·91	79·53	57·26	98
99	81·10	56·78	80·85	57·14	80·60	57·49	80·35	57·84	99
100	81·92	57·36	81·66	57·71	81·41	58·07	81·16	58·42	100
Distance.	Dep.	Lat.	Dep.	Lat.	Dep.	Lat.	Dep.	Lat.	Distance.
	55 Deg.		54¾ Deg.		54½ Deg.		54¼ Deg.		

Distance.	36 Deg.		36¼ Deg.		36½ Deg.		36¾ Deg.		Distance.
	Lat.	Dep.	Lat.	Dep.	Lat.	Dep.	Lat.	Dep.	
1	0·81	0·59	0·81	0·59	0·80	0·59	0·80	0·60	1
2	1·62	1·18	1·61	1·18	1·61	1·19	1·60	1·20	2
3	2·43	1·76	2·42	1·77	2·41	1·78	2·40	1·79	3
4	3·24	2·35	3·23	2·37	3·22	2·38	3·20	2·39	4
5	4·05	2·94	4·03	2·96	4·02	2·97	4·01	2·99	5
6	4·85	3·53	4·84	3·55	4·82	3·57	4·81	3·59	6
7	5·66	4·11	5·65	4·14	5·63	4·16	5·61	4·19	7
8	6·47	4·70	6·45	4·73	6·43	4·76	6·41	4·79	8
9	7·28	5·29	7·26	5·32	7·23	5·35	7·21	5·38	9
10	8·09	5·88	8·06	5·91	8·04	5·95	8·01	5·98	10
11	8·90	6·47	8·87	6·50	8·84	6·54	8·81	6·58	11
12	9·71	7·05	9·68	7·10	9·65	7·14	9·61	7·18	12
13	10·52	7·64	10·48	7·69	10·45	7·73	10·42	7·78	13
14	11·33	8·23	11·29	8·28	11·25	8·33	11·22	8·38	14
15	12·14	8·82	12·10	8·87	12·06	8·92	12·02	8·97	15
16	12·94	9·40	12·90	9·46	12·86	9·52	12·82	9·57	16
17	13·75	9·99	13·71	10·05	13·67	10·11	13·62	10·17	17
18	14·56	10·58	14·52	10·64	14·47	10·71	14·42	10·77	18
19	15·37	11·17	15·32	11·23	15·27	11·30	15·22	11·37	19
20	16·18	11·76	16·13	11·83	16·08	11·90	16·03	11·97	20
21	16·99	12·34	16·94	12·42	16·88	12·49	16·83	12·56	21
22	17·80	12·93	17·74	13·01	17·68	13·09	17·63	13·16	22
23	18·61	13·52	18·55	13·60	18·49	13·68	18·43	13·76	23
24	19·42	14·11	19·35	14·19	19·29	14·28	19·23	14·36	24
25	20·23	14·69	20·16	14·78	20·10	14·87	20·03	14·96	25
26	21·03	15·28	20·97	15·37	20·90	15·47	20·83	15·56	26
27	21·84	15·87	21·77	15·97	21·70	16·06	21·63	16·15	27
28	22·65	16·46	22·58	16·56	22·51	16·65	22·44	16·75	28
29	23·46	17·05	23·39	17·15	23·31	17·25	23·24	17·35	29
30	24·27	17·63	24·19	17·74	24·12	17·84	24·04	17·95	30
31	25·08	18·22	25·00	18·33	24·92	18·44	24·84	18·55	31
32	25·89	18·81	25·81	18·92	25·72	19·03	25·64	19·15	32
33	26·70	19·40	26·61	19·51	26·53	19·63	26·44	19·74	33
34	27·51	19·98	27·42	20·10	27·33	20·22	27·24	20·34	34
35	28·32	20·57	28·23	20·70	28·13	20·82	28·04	20·94	35
36	29·12	21·16	29·03	21·29	28·94	21·41	28·85	21·54	36
37	29·93	21·75	29·84	21·88	29·74	22·01	29·65	22·14	37
38	30·74	22·34	30·64	22·47	30·55	22·60	30·45	22·74	38
39	31·55	22·92	31·45	23·06	31·35	23·20	31·25	23·33	39
40	32·36	23·51	32·26	23·65	32·15	23.79	32·05	23·93	40
41	33·17	24·10	33·06	24·24	32·96	24·39	32·85	24·53	41
42	33·98	24·69	33·87	24·83	33·76	24·98	33·65	25·13	42
43	34·79	25·27	34·68	25·43	34·57	25·58	34·45	25·73	43
44	35·60	25·86	35·48	26·02	35·37	26·17	35·26	26·33	44
45	36·41	26·45	36·29	26·61	36·17	26·77	36.06	26·92	45
46	37·21	27·04	37·10	27·20	36·98	27·36	36·86	27·52	46
47	38·02	27·63	37·90	27·79	37·78	27·96	37·66	28·12	47
48	38·83	28·21	38·71	28·38	38·59	28·55	38·46	28·72	48
49	39·64	28·80	39·52	28·97	39·39	29·15	39·26	29·32	49
50	40·45	29·39	40·32	29·57	40·19	29·74	40·06	29·92	50
Distance.	Dep.	Lat.	Dep.	Lat.	Dep.	Lat.	Dep.	Lat.	Distance.
	54 Deg.		53¾ Deg.		53½ Deg.		53¼ Deg.		

Distance.	36 Deg.		36¼ Deg.		36½ Deg.		36¾ Deg.		Distance.
	Lat.	Dep.	Lat.	Dep.	Lat.	Dep.	Lat.	Dep.	
51	41.26	29·98	41·13	30·16	41·00	30·34	40·86	30·51	51
52	42.07	30·56	41·94	30·75	41·80	30·93	41·67	31·11	52
53	42·88	31·15	42·74	31·34	42·60	31·53	42·47	31·71	53
54	43·69	31·74	43·55	31·93	43·41	32·12	43·27	32·31	54
55	44·50	32·33	44·35	32·52	44·21	32·72	44·07	32·91	55
56	45·30	32·92	45·16	33·11	45·02	33·31	44·87	33·51	56
57	46·11	33·50	45·97	33·70	45·82	33·90	45·67	34·10	57
58	46·92	34·09	46·77	34·30	46·62	34·50	46·47	34·70	58
59	47·73	34·68	47·58	34·89	47·43	35·09	47·27	35·30	59
60	48·54	35·27	48·39	35·48	48·23	35·69	48·08	35·90	60
61	49·35	35·85	49·19	36·07	49·04	36·28	48·88	36·50	61
62	50·16	36·44	50·00	36·66	49·84	36·88	49·68	37·10	62
63	50·97	37·03	50·81	37·25	50·64	37·47	50·48	37·69	63
64	51·78	37·62	51·61	37·84	51·45	38·07	51·28	38·29	64
65	52·59	38·21	52·42	38·44	52·25	38·66	52·08	38·89	65
66	53·40	38·79	53·23	39·03	53·05	39·26	52·88	39·49	66
67	54·20	39·38	54·03	39·62	53·86	39·85	53·68	40·09	67
68	55·01	39·97	54·84	40·21	54·66	40·45	54·49	40·69	68
69	55·82	40·56	55·64	40·80	55·47	41·04	55·29	41·28	69
70	56·63	41·14	56·45	41·39	56·27	41·64	56·09	41·88	70
71	57·44	41·73	57·26	41·98	57·07	42·23	56·89	42·48	71
72	58·25	42·32	58·06	42·57	57·88	42·83	57·69	43·08	72
73	59·06	42·91	58·87	43·17	58·68	43·42	58·49	43·68	73
74	59·87	43·50	59·68	43·76	59·49	44·02	59·29	44·28	74
75	60·68	44·08	60·48	44·35	60·29	44·61	60·09	44·87	75
76	61·49	44·67	61·29	44·94	61·09	45·21	60·90	45·47	76
77	62·29	45·26	62·10	45·53	61·90	45·80	61·70	46·07	77
78	63·10	45·85	62·90	46·12	62·70	46·40	62·50	46·67	78
79	63·91	46·43	63·71	46·71	63·50	46·99	63·30	47·27	79
80	64·72	47·02	64·52	47·30	64·31	47·59	64·10	47·87	80
81	65·53	47·61	65·32	47·90	65·11	48·18	64·90	48·46	81
82	66·34	48·20	66·13	48·49	65·92	48·78	65·70	49·06	82
83	67·15	48·79	66·93	49·08	66·72	49·37	66·50	49·66	83
84	67·96	49·37	67·74	49·67	67·52	49 97	67·31	50·26	84
85	68·77	49·96	68·55	50·26	68·33	50·56	68·11	50·86	85
86	69·58	50·55	69·35	50·85	69·13	51·15	68·91	51·46	86
87	70·38	51·14	70·16	51·44	69·94	51·75	69·71	52·05	87
88	71·19	51·73	70·97	52·04	70·74	52·34	70·51	52·65	88
89	72·00	52·31	71·77	52·63	71·54	52·94	71·31	53·25	89
90	72·81	52·90	72·58	53·22	72·35	53·53	72·11	53·85	90
91	73·62	53·49	73·39	53·81	73·15	54·13	72·91	54·45	91
92	74·43	54·08	74·19	54·40	73·95	54·72	73·72	55·05	92
93	75·24	54·66	75·00	54.99	74·76	55·32	74·52	55·64	93
94	76·05	55·25	75·81	55.58	75·56	55·91	75·32	56·24	94
95	76·86	55·84	76·61	56.17	76·37	56·51	76·12	56·84	95
96	77·67	56·43	77·42	56.77	77·17	57·10	76·92	57·44	96
97	78·47	57·02	78·23	57.36	77·97	57·70	77·72	58·04	97
98	79.28	57·60	79·03	57·95	78·78	58·29	78·52	58·64	98
99	80·09	58·19	79·84	58·54	79·58	58·89	79·32	59·23	99
100	80·90	58·78	80·64	59·13	80·39	59·48	80·13	59·83	100
Distance.	Dep.	Lat.	Dep.	Lat.	Dep.	Lat.	Dep.	Lat.	Distance.
	54 Deg.		53¾ Deg.		53½ Deg.		53¼ Deg.		

Distance.	37 Deg.		37¼ Deg.		37½ Deg.		37¾ Deg.		Distance.
	Lat.	Dep.	Lat.	Dep.	Lat.	Dep.	Lat.	Dep.	
1	0·80	0·60	0·80	0·61	0·79	0·61	0·79	0·61	1
2	1·60	1·20	1·59	1·21	1·59	1·22	1·58	1·22	2
3	2·40	1·81	2·39	1·82	2·38	1·83	2·37	1·84	3
4	3·19	2·41	3·18	2·42	3·17	2·43	3·16	2·45	4
5	3·99	3·01	3·98	3·03	3·97	3·04	3·95	3·06	5
6	4·79	3·61	4·78	3·63	4·76	3·65	4·74	3·67	6
7	5·59	4·21	5·57	4·24	5·55	4·26	5·53	4·29	7
8	6·39	4·81	6·37	4·84	6·35	4·87	6·33	4·90	8
9	7·19	5·42	7·16	5·45	7·14	5·48	7·12	5·51	9
10	7·99	6·02	7·96	6·05	7·93	6·09	7·91	6·12	10
11	8·78	6·62	8·76	6·66	8·73	6·70	8·70	6·73	11
12	9·58	7·22	9·55	7·26	9·52	7·31	9·49	7·35	12
13	10·38	7·82	10·35	7·87	10·31	7·91	10·28	7·96	13
14	11·18	8·43	11·14	8·47	11·11	8·52	11·07	8·57	14
15	11·98	9·03	11·94	9·08	11·90	9·13	11·86	9·18	15
16	12·78	9·63	12·74	9·68	12·69	9·74	12·65	9·80	16
17	13·58	10·23	13·53	10·29	13·49	10·35	13·44	10·41	17
18	14·38	10·83	14·33	10·90	14·28	10·96	14·23	11·02	18
19	15·17	11·43	15·12	11·50	15·07	11·57	15·02	11·63	19
20	15·97	12·04	15·92	12·11	15·87	12·18	15·81	12·24	20
21	16·77	12·64	16·72	12·71	16·66	12·78	16·60	12·86	21
22	17·57	13·24	17·51	13·32	17·45	13·39	17·40	13·47	22
23	18·37	13·84	18·31	13·92	18·25	14·00	18·19	14·08	23
24	19·17	14·44	19·10	14·53	19·04	14·61	18·98	14·69	24
25	19·97	15·05	19·90	15·13	19·83	15·22	19·77	15·31	25
26	20·76	15·65	20·70	15·74	20·63	15·83	20·56	15·92	26
27	21·56	16·25	21·49	16·34	21·42	16·44	21·35	16·53	27
28	22·36	16·85	22·29	16·95	22·21	17·05	22·14	17·14	28
29	23·16	17·45	23·08	17·55	23·01	17·65	22·93	17·75	29
30	23·96	18·05	23·88	18·16	23·80	18·26	23·72	18·37	30
31	24·76	18·66	24·68	18·76	24·59	18·87	24·51	18·98	31
32	25·56	19·26	25·47	19·37	25·39	19·48	25·30	19·59	32
33	26·35	19·86	26·27	19·97	26·18	20·09	26·09	20·20	33
34	27·15	20·46	27·06	20·58	26·97	20·70	26·88	20·82	34
35	27·95	21·06	27·86	21·19	27·77	21·31	27·67	21·43	35
36	28·75	21·67	28·66	21·79	28·56	21·92	28·46	22·04	36
37	29·55	22·27	29·45	22·40	29·35	22·52	29·26	22·65	37
38	30·35	22·87	30·25	23·00	30·15	23·13	30·05	23·26	38
39	31·15	23·47	31·04	23·61	30·94	23·74	30·84	23·88	39
40	31·95	24·07	31·84	24·21	31·73	24·35	31·63	24·49	40
41	32·74	24·67	32·64	24·82	32·53	24·96	32·42	25·10	41
42	33·54	25·28	33·43	25·42	33·32	25·57	33·21	25·71	42
43	34·34	25·88	34·23	26·03	34·11	26·18	34·00	26·33	43
44	35·14	26·48	35·02	26·63	34·91	26·79	34·79	26·94	44
45	35·94	27·08	35·82	27·24	35·70	27·39	35·58	27·55	45
46	36·74	27·68	36·62	27·84	36·49	28·00	36·37	28·16	46
47	37·54	28·29	37·41	28·45	37·29	28·61	37·16	28·77	47
48	38·33	28·89	38·21	29·05	38·08	29·22	37·95	29·39	48
49	39·13	29·49	39·00	29·66	38·87	29·83	38·74	30·00	49
50	39·93	30·09	39·80	30·26	39·67	30·44	39·53	30·61	50
Distance.	Dep.	Lat.	Dep.	Lat.	Dep.	Lat.	Dep.	Lat.	Distance.
	53 Deg.		52¾ Deg.		52½ Deg.		52¼ Deg.		

Distance.	37 Deg.		37¼ Deg.		37½ Deg.		37¾ Deg.		Distance.
	Lat.	Dep.	Lat.	Dep.	Lat.	Dep.	Lat.	Dep.	
51	40·73	30·69	40·60	30·87	40·46	31·05	40·33	31·22	51
52	41·53	31·29	41·39	31·48	41·25	31·66	41·12	31·84	52
53	42·33	31·90	42·19	32·08	42·05	32·26	41·91	32·45	53
54	43·13	32·50	42·98	32·69	42·84	32·87	42·70	33·06	54
55	43·92	33·10	43·78	33·29	43·63	33·48	43·49	33·67	55
56	44·72	33·70	44·58	33·90	44·43	34·09	44·28	34·28	56
57	45·52	34·30	45·37	34·50	45·22	34·70	45·07	34·90	57
58	46·32	34·91	46·17	35·11	46·01	35·31	45·86	35·51	58
59	47·12	35.51	46·96	35·71	46·81	35·92	46·65	36·12	59
60	47·92	36·11	47·76	36·32	47·60	36·53	47·44	36·73	60
61	48·72	36·71	48·56	36·92	48·39	37·13	48·23	37·35	61
62	49·52	37·31	49·35	37·53	49·19	37·74	49·02	37·96	62
63	50·31	37·91	50·15	38·13	49·98	38·35	49·81	38·57	63
64	51·11	38·52	50·94	38·74	50·77	38·96	50·60	39·18	64
65	51·91	39·12	51·74	39·34	51·57	39·57	51·39	39·79	65
66	52·71	39·72	52·54	39·95	52·36	40·18	52·19	40·41	66
67	53·51	40·32	53·33	40·55	53·15	40·79	52·98	41·02	67
68	54·31	40·92	54·13	41·16	53·95	41·40	53·77	41·63	68
69	55·11	41·53	54·92	41·77	54·74	42·00	54·56	42·24	69
70	55·90	42·13	55·72	42·37	55·53	42·61	55·35	42·86	70
71	56·70	42·73	56·52	42·98	56·33	43·22	56·14	43·47	71
72	57·50	43·33	57·31	43·58	57·12	43·83	56·93	44·08	72
73	58·30	43·93	58·11	44·19	57·91	44·44	57·72	44·69	73
74	59·10	44·53	58·90	44·79	58·71	45·05	58·51	45·30	74
75	59·90	45·14	59·70	45·40	59·50	45·66	59·30	45·92	75
76	60·70	45·74	60·50	46·00	60·29	46·27	60·09	46·53	76
77	61·49	46·34	61·29	46·61	61·09	46·87	60·88	47·14	77
78	62·29	46·94	62·09	47·21	61·88	47·48	61·67	47·75	78
79	63·09	47·54	62·88	47·82	62·67	48·09	62·46	48·37	79
80	63·89	48·15	63·68	48·42	63·47	48·70	63·26	48·98	80
81	64·69	48·75	64·48	49·03	64·26	49·31	64·05	49·59	81
82	65·49	49·35	65·27	49·63	65·05	49·92	64·84	50·20	82
83	66·29	49·95	66·07	50·24	65·85	50·53	65·63	50·81	83
84	67·09	50·55	66·86	50·84	66·64	51·14	66·42	51·43	84
85	67·88	51·15	67·66	51·45	67·43	51·74	67·21	52·04	85
86	68·68	51·76	68·46	52·06	68·23	52·35	68·00	52·65	86
87	69·48	52·36	69·25	52·66	69·02	52·96	68·79	53·26	87
88	70·28	52·96	70·05	53·27	69·82	53·57	69·58	53·88	88
89	71·08	53·56	70·84	53·87	70·61	54·18	70·37	54·49	89
90	71·88	54·16	71·64	54·48	71·40	54·79	71·16	55·10	90
91	72·68	54·77	72·44	55·08	72·20	55·40	71·95	55·71	91
92	73·47	55·37	73·23	55·69	72·99	56·01	72·74	56·32	92
93	74·27	55·97	74·03	56·29	73·78	56·61	73·53	56·94	93
94	75·07	56·57	74·82	56·90	74·58	57·22	74·32	57·55	94
95	75·87	57·17	75·62	57·50	75·37	57·83	75·12	58·16	95
96	76·67	57·77	76·42	58·11	76·16	58·44	75·91	58·77	96
97	77·47	58·38	77·21	58·71	76·96	59·05	76·70	59·39	97
98	78·27	58·98	78·01	59·32	77·75	59·66	77·49	60·00	98
99	79·06	59·58	78·80	59·92	78·54	60·27	78·28	60·61	99
100	79 86	60·18	79·60	60·53	79·34	60·88	79·07	61·22	100
Distance.	Dep.	Lat.	Dep.	Lat.	Dep.	Lat.	Dep.	Lat.	Distance.
	53 Deg.		52¾ Deg.		52½ Deg.		52¼ Deg.		

Distance.	38 Deg.		38¼ Deg.		38½ Deg.		38¾ Deg.		Distance.
	Lat.	Dep.	Lat.	Dep.	Lat.	Dep.	Lat.	Dep.	
1	0·79	0·62	0·79	0·62	0·78	0·62	0·78	0·63	1
2	1·58	1·23	1·57	1·24	1·57	1·24	1·56	1·25	2
3	2·36	1·85	2·36	1·86	2·35	1·87	2·34	1·88	3
4	3·15	2·46	3·14	2·48	3·13	2·49	3·12	2·50	4
5	3·94	3·08	3·93	3·10	3·91	3·11	3·90	3·13	5
6	4·73	3·69	4·71	3·71	4·70	3·74	4·68	3·76	6
7	5·52	4·31	5·50	4·33	5·48	4·36	5·46	4·38	7
8	6·30	4·93	6·28	4·95	6·26	4·98	6·24	5·01	8
9	7·09	5·54	7·07	5·57	7·04	5·60	7·02	5·63	9
10	7·88	6·16	7·85	6·19	7·83	6·23	7·80	6·26	10
11	8·67	6·77	8·64	6·81	8·61	6·85	8·58	6·89	11
12	9·46	7·39	9·42	7·43	9·39	7·47	9·36	7·51	12
13	10·24	8·00	10·21	8·05	10·17	8·09	10·14	8·14	13
14	11·03	8·62	10·99	8·67	10·96	8·72	10·92	8·76	14
15	11·82	9·23	11·78	9·29	11·74	9·34	11·70	9·39	15
16	12·61	9·85	12·57	9·91	12·52	9·96	12·48	10·01	16
17	13·40	10·47	13·35	10·52	13·30	10·58	13·26	10·64	17
18	14·18	11·08	14·14	11·14	14·09	11·21	14·04	11·27	18
19	14·97	11·70	14·92	11·76	14·87	11·83	14·82	11·89	19
20	15·76	12.31	15·71	12·38	15·65	12·45	15·60	12·52	20
21	16·55	12·93	16·49	13·00	16·43	13·07	16·38	13·14	21
22	17·34	13·54	17·28	13·62	17·22	13·70	17·16	13·77	22
23	18·12	14·16	18·06	14·24	18·00	14·32	17·94	14·40	23
24	18·91	14·78	18·85	14·86	18·78	14·94	18·72	15·02	24
25	19·70	15·39	19·63	15·48	19·57	15·56	19·50	15·65	25
26	20·49	16·01	20·42	16·10	20·35	16·19	20·28	16·27	26
27	21·28	16·62	21·20	16·72	21·13	16·81	21·06	16·90	27
28	22·06	17·24	21·99	17·33	21·91	17·43	21·84	17·53	28
29	22·85	17·85	22·77	17·95	22·70	18·05	22·62	18·15	29
30	23·64	18·47	23·56	18·57	23·48	18·68	23·40	18·78	30
31	24·43	19·09	24·34	19·19	24·26	19·30	24·18	19·40	31
32	25·22	19·70	25·13	19·81	25·04	19·92	24·96	20·03	32
33	26·00	20·32	25·92	20·43	25·83	20·54	25·74	20·66	33
34	26·79	20·93	26·70	21·05	26·61	21·17	26·52	21·28	34
35	27·58	21·55	27·49	21·67	27·39	21·79	27·30	21·91	35
36	28·37	22·16	28·27	22·29	28·17	22·41	28·08	22·53	36
37	29·16	22·78	29·06	22·91	28·96	23·03	28·86	23·16	37
38	29·94	23·40	29·84	23·53	29·74	23·66	29·64	23·79	38
39	30·73	24·01	30·63	24·14	30·52	24·28	30·42	24·41	39
40	31·52	24·63	31·41	24·76	31·30	24·90	31·20	25·04	40
41	32·31	25·24	32·20	25·38	32·09	25·52	31·98	25·66	41
42	33·10	25·86	32·98	26·00	32·87	26·15	32·76	26·29	42
43	33·88	26·47	33·77	26·62	33·65	26·77	33·53	26·91	43
44	34·67	27·09	34·55	27·24	34·43	27·39	34·31	27·54	44
45	35·46	27·70	35·34	27·86	35·22	28·01	35·09	28·17	45
46	36·25	28·32	36·12	28·48	36·00	28·64	35·87	28·79	46
47	37·04	28·94	36·91	29·10	36·78	29·26	36·65	29·42	47
48	37·82	29·55	37·79	29·72	37·57	29·88	37·43	30·04	48
49	38·61	30·17	38·48	30·34	38·35	30·50	38·21	30·67	49
50	39·40	30·78	39·27	30·95	39·13	31·13	38·99	31·30	50
Distance.	Dep.	Lat.	Dep.	Lat.	Dep.	Lat.	Dep.	Lat.	Distance.
	52 Deg.		51¾ Deg.		51½ Deg.		51¼ Deg.		

Distance.	38 Deg.		38¼ Deg.		38½ Deg.		38¾ Deg.		Distance.
	Lat.	Dep.	Lat.	Dep.	Lat.	Dep.	Lat.	Dep.	
51	40·19	31·40	40·05	31·57	39·91	31·75	39·77	31·92	51
52	40·98	32·01	40·84	32·19	40·70	32·37	40·55	32·55	52
53	41·76	32·63	41·62	32·81	41·48	32·99	41·33	33·17	53
54	42·55	33·25	42·41	33·43	42·26	33·62	42·11	33·80	54
55	43·34	33·86	43·19	34·05	43·04	34·24	42·89	34·43	55
56	44·13	34·48	43·98	34·67	43·83	34·86	43·67	35·05	56
57	44·92	35·09	44·76	35·29	44·61	35·48	44·45	35·68	57
58	45·70	35·71	45·55	35·91	45·39	36·11	45·23	36·30	58
59	46·49	36·32	46·33	36·53	46·17	36·73	46·01	36·93	59
60	47·28	36·94	47·12	37·15	46·96	37·35	46·79	37·56	60
61	48·07	37·56	47·90	37·76	47·74	37·97	47·57	38·18	61
62	48·86	38·17	48·69	38·38	48·52	38·60	48·35	38·81	62
63	49·64	38·79	49·47	39·00	49·30	39·22	49·13	39·43	63
64	50·43	39·40	50·26	39·62	50·09	39·84	49·91	40·06	64
65	51·22	40·02	51·05	40·24	50·87	40·46	50·69	40·68	65
66	52·01	40·63	51·83	40·86	51·65	41·09	51·47	41·31	66
67	52·80	41·25	52·62	41·48	52·43	41·71	52·25	41·94	67
68	53·58	41·86	53·40	42·10	53·22	42·33	53·03	42·56	68
69	54·37	42·48	54·19	42·72	54·00	42·95	53·81	43·19	69
70	55·16	43·10	54·97	43·34	54·78	43·58	54·59	43·81	70
71	55·95	43·71	55·76	43·96	55·57	44·20	55·37	44·44	71
72	56·74	44·33	56·54	44·57	56·35	44·82	56·15	45·07	72
73	57·52	44·94	57·33	45·19	57·13	45·44	56·93	45·69	73
74	58·31	45·56	58·11	45·81	57·91	46·07	57·71	46·32	74
75	59·10	46·17	58·90	46·43	58·70	46·69	58·49	46·94	75
76	59·89	46·79	59·68	47·05	59·48	47·31	59·27	47·57	76
77	60·68	47·41	60·47	47·67	60·26	47·93	60·05	48·20	77
78	61·46	48·02	61·25	48·29	61·04	48·56	60·83	48·82	78
79	62·25	48·64	62·04	48·91	61·83	49·18	61·61	49·45	79
80	63·04	49·25	62·83	49·53	62·61	49·80	62·39	50·07	80
81	63·83	49·87	63·61	50·15	63·39	50·42	63·17	50·70	81
82	64·62	50·48	64·40	50·77	64·17	51·05	63·95	51·33	82
83	65·40	51·10	65·18	51·38	64·96	51·67	64·73	51·95	83
84	66·19	51·72	65·97	52·00	65·74	52·29	65·51	52·58	84
85	66·98	52·33	66·75	52·62	66·52	52·91	66·29	53·20	85
86	67·77	52·95	67·54	53·24	67·30	53·54	67·07	53·83	86
87	68·56	53·56	68·32	53·86	68·09	54·16	67·85	54·46	87
88	69·34	54·18	69·11	54·48	68·87	54·78	68·63	55·08	88
89	70·13	54·79	69·89	55·10	69·65	55·40	69·41	55·71	89
90	70·92	55·41	70·68	55·72	70·43	56·03	70·19	56·33	90
91	71·71	56·03	71·46	56·34	71·22	56·65	70·97	56·96	91
92	72·50	56·64	72·25	56·96	72·00	57·27	71·75	57·58	92
93	73·28	57·26	73·03	57·58	72·78	57·89	72·53	58·21	93
94	74·07	57·87	73·82	58·19	73·57	58·52	73·31	53·84	94
95	74·86	58·49	74·61	58·81	74·35	59·14	74·09	59·46	95
96	75·65	59·10	75·39	59·43	75·13	59·76	74·87	60·09	96
97	76·44	59·72	76·18	60·05	75·91	60·38	75·65	60·71	97
98	77·22	60·33	76·96	60·67	76·70	61·01	76·43	61·34	98
99	78·01	60·95	77·75	61·29	77·48	61·63	77·21	61·97	99
100	78·80	61·57	78·53	61·91	78·26	62·25	77·99	62·59	100
Distance.	Dep.	Lat.	Dep.	Lat.	Dep.	Lat.	Dep.	Lat.	Distance.
	52 Deg.		51¾ Deg.		51½ Deg.		51¼ Deg.		

Distance.	39 Deg.		39¼ Deg.		39½ Deg.		39¾ Deg.		Distance.
	Lat.	Dep.	Lat.	Dep.	Lat.	Dep.	Lat.	Dep.	
1	0·78	0·63	0·77	0·63	0·77	0·64	0·77	0·64	1
2	1·55	1·26	1·55	1·27	1·54	1·27	1·54	1·28	2
3	2·33	1·89	2·32	1·90	2·31	1·91	2·31	1·92	3
4	3·11	2·52	3·10	2·53	3·09	2·54	3·08	2·56	4
5	3·89	3·15	3·87	3·16	3·86	3·18	3·84	3·20	5
6	4·66	3·78	4·65	3·80	4·63	3·82	4·61	3·84	6
7	5·44	4·41	5·42	4·43	5·40	4·45	5·38	4·48	7
8	6·22	5·03	6·20	5·06	6·17	5·09	6·15	5·12	8
9	6·99	5·66	6·97	5·69	6·94	5·72	6·92	5·75	9
10	7·77	6·29	7·74	6·33	7·72	6·36	7·69	6·39	10
11	8·55	6·92	8·52	6·96	8·49	7·00	8·46	7·03	11
12	9·33	7·55	9·29	7·59	9·26	7·63	9·23	7·67	12
13	10·10	8·18	10·07	8·23	10·03	8·27	9·99	8·31	13
14	10·88	8·81	10·84	8·86	10·80	8·91	10·76	8·95	14
15	11·66	9·44	11·62	9·49	11·57	9·54	11·53	9·59	15
16	12·43	10·07	12·39	10·12	12·35	10·18	12·30	10·23	16
17	13·21	10·70	13·16	10·76	13·12	10·81	13·07	10·87	17
18	13·99	11·33	13·94	11·39	13·89	11·45	13·84	11·51	18
19	14·77	11·96	14·71	12·02	14·66	12·09	14·61	12·15	19
20	15·54	12·59	15·49	12·65	15·43	12·72	15·38	12·79	20
21	16·32	13·22	16·26	13·29	16·20	13·36	16·15	13·43	21
22	17·10	13·84	17·04	13·92	16·98	13·99	16·91	14·07	22
23	17·87	14·47	17·81	14·55	17·75	14·63	17·68	14·71	23
24	18·65	15·10	18·59	15·18	18·52	15·27	18·45	15·35	24
25	19·43	15·73	19·36	15·82	19·29	15·90	19·22	15·99	25
26	20·21	16·36	20·13	16·45	20·06	16·54	19·99	16·63	26
27	20·98	16·99	20·91	17·08	20·83	17·17	20·76	17·26	27
28	21·76	17·62	21·68	17·72	21·61	17·81	21·53	17·90	28
29	22·54	18·25	22·46	18·35	22·38	18·45	22·30	18·54	29
30	23·31	18·88	23·23	18·98	23·15	19·08	23·07	19·18	30
31	24·09	19·51	24·01	19·61	23·92	19·72	23·83	19·82	31
32	24·87	20·14	24·78	20·25	24·69	20·35	24·60	20·46	32
33	25·65	20·77	25·55	20·88	25·46	20·99	25·37	21·10	33
34	26·42	21·40	26·33	21·51	26·24	21·63	26·14	21·74	34
35	27·20	22·03	27·10	22·14	27·01	22·26	26·91	22·38	35
36	27·98	22·66	27·88	22·78	27·78	22·90	27·68	23·02	36
37	28·75	23·28	28·65	23·41	28·55	23·53	28·45	23·66	37
38	29·53	23·91	29·43	24·04	29·32	24·17	29·22	24·30	38
39	30·31	24·54	30·20	24·68	30·09	24·81	29·98	24·94	39
40	31·09	25·17	30·98	25·31	30·86	25·44	30·75	25·58	40
41	31·86	25·80	31·75	25·94	31·64	26·08	31·52	26·22	41
42	32·64	26·43	32·52	26·57	32·41	26·72	32·29	26·86	42
43	33·42	27·06	33·30	27·21	33·18	27·35	33·06	27·50	43
44	34·19	27·69	34·07	27·84	33·95	27·99	33·83	28·14	44
45	34·97	28·32	34·85	28·47	34·72	28·62	34·60	28·77	45
46	35·75	28·95	35·62	29·10	35·49	29·26	35·37	29·41	46
47	36·53	29·58	36·40	29·74	36·27	29·90	36·14	30·05	47
48	37·30	30·21	37·17	30·37	37·04	30·53	36·90	30·69	48
49	38·08	30·84	37·95	31·00	37·81	31·17	37·67	31·33	49
50	38·86	31·47	38·72	31·64	38·58	31·80	38·44	31·97	50
	Dep.	Lat.	Dep.	Lat.	Dep.	Lat.	Dep.	Lat.	
Distance.	51 Deg.		50¾ Deg.		50½ Deg.		50¼ Deg.		Distance.

Distance.	39 Deg.		39¼ Deg.		39½ Deg.		39¾ Deg.		Distance.
	Lat.	Dep.	Lat.	Dep.	Lat.	Dep.	Lat.	Dep.	
51	39.63	32·10	39·49	32·27	39·35	32·44	39·21	32·61	51
52	40.41	32·72	40·27	32·90	40·12	33·08	39·98	33·25	52
53	41·19	33·35	41·04	33·53	40·90	33·71	40·75	33·89	53
54	41·97	33·98	41·82	34·17	41·67	34·35	41·52	34·53	54
55	42·74	34·61	42·59	34·80	42·44	34·98	42·29	35·17	55
56	43·52	35·24	43·37	35·43	43·21	35·62	43·06	35·81	56
57	44·30	35·87	44·14	36·06	43·98	36·26	43·82	36·45	57
58	45·07	36·50	44·91	36·70	44·75	36·89	44·59	37·09	58
59	45·85	37·13	45·69	37·33	45·53	37·53	45·36	37·73	59
60	46·63	37·76	46·46	37·96	46·30	38·16	46·13	38·37	60
61	47·41	38·39	47·24	38·60	47·07	38·80	46·90	39·01	61
62	48·18	39·02	48·01	39·23	47·84	39·44	47·67	39·65	62
63	48·96	39·65	48·79	39·86	48·61	40·07	48·44	40·28	63
64	49·74	40·28	49·56	40·49	49·38	40·71	49·21	40·92	64
65	50·51	40·91	50·34	41·13	50·16	41·35	49·97	41·56	65
66	51·29	41·54	51·11	41·76	50·93	41·98	50·74	42·20	66
67	52·07	42·16	51·88	42·39	51·70	42·62	51·51	42·84	67
68	52·85	42·79	52·66	43·02	52·47	43·25	52·28	43·48	68
69	53·52	43·42	53·43	43·66	53·24	43·89	53·05	44·12	69
70	54·40	44·05	54·21	44·29	54·01	44·53	53·82	44·76	70
71	55·18	44·68	54·98	44·92	54·79	45·16	54·59	45·40	71
72	55·95	45·31	55·76	45·55	55·56	45·80	55·36	46·04	72
73	56·73	45·94	56·53	46·19	56·33	46·43	56·13	46·68	73
74	57·51	46·57	57·31	46·82	57·10	47·07	56·89	47·32	74
75	58·29	47·20	58·08	47·45	57·87	47·71	57·66	47·96	75
76	59·06	47·83	58·85	48·09	58·64	48·34	58·43	48·60	76
77	59·84	48·46	59·63	48·72	59·42	48·98	59·20	49·24	77
78	60·62	49·09	60·40	49·35	60·19	49·61	59·97	49·88	78
79	61·39	49·72	61·18	49·98	60·96	50·25	60·74	50·52	79
80	62·17	50·35	61·95	50·62	61·73	50·89	61·51	51·16	80
81	62·95	50·97	62·73	51·25	62·50	51·52	62·28	51·79	81
82	63·73	51·60	63·50	51·88	63·27	52·16	63·04	52·43	82
83	64·50	52·23	64·27	52·51	64·04	52·79	63·81	53·07	83
84	65·28	52·86	65·05	53·15	64·82	53·43	64·58	53·71	84
85	66·06	53·49	65·82	53·78	65·59	54·07	65·35	54·35	85
86	66·83	54·12	66·60	54·41	66·36	54·70	66·12	54·99	86
87	67·61	54·75	67·37	55·05	67·13	55·34	66·89	55·63	87
88	68·39	55·38	68·15	55·68	67·90	55·97	67·66	56·27	88
89	69·17	56·01	68·92	56·32	68·67	56·61	68·43	56·91	89
90	69·94	56·64	69·70	56·94	69·45	57·25	69·20	57·55	90
91	70·72	57·27	70·47	57·58	70·22	57·88	69·96	58·19	91
92	71·50	57·90	71·24	58·21	70·99	58·52	70·73	58·83	92
93	72·27	58·53	72·02	58.84	71·76	59·16	71·50	59·47	93
94	73·05	59·16	72·79	59.47	72·53	59·79	72·27	60·11	94
95	73·83	59·79	73·57	60.11	73·30	60·43	73·04	60·75	95
96	74·61	60·41	74·34	60.74	74·08	61·06	73·81	61·39	96
97	75·38	61·04	75·12	61.37	74·85	61·70	74·58	62·03	97
98	76·16	61·67	75·89	62·01	75·62	62·34	75·35	62·66	98
99	76·94	62·30	76·66	62·64	76·39	62·97	76·12	63·30	99
100	77·71	62·93	77·44	63·27	77·16	63·61	76·88	63·94	100
	Dep.	Lat.	Dep.	Lat.	Dep.	Lat.	Dep.	Lat.	
Distance.	51 Deg.		50¾ Deg.		50½ Deg.		50¼ Deg.		Distance.

Distance.	40 Deg.		40¼ Deg.		40½ Deg.		40¾ Deg.		Distance.
	Lat.	Dep.	Lat.	Dep.	Lat.	Dep.	Lat.	Dep.	
1	0·77	0·64	0·76	0·65	0·76	0·65	0·76	0·65	1
2	1·53	1·29	1·53	1·29	1·52	1·30	1·52	1·31	2
3	2·30	1·93	2·29	1·94	2·28	1·95	2·27	1·96	3
4	3·06	2·57	3·05	2·58	3·04	2·60	3·03	2·61	4
5	3·83	3·21	3·82	3·23	3·80	3·25	3·79	3·26	5
6	4·60	3·86	4·58	3·88	4·56	3·90	4·55	3·92	6
7	5·36	4·50	5·34	4·52	5·32	4·55	5·30	4·57	7
8	6·13	5·14	6·11	5·17	6·08	5·20	6·06	5·22	8
9	6·89	5·79	6·87	5·82	6·84	5·84	6·82	5·87	9
10	7·66	6·43	7·63	6·46	7·60	6·49	7·58	6·53	10
11	8·43	7·07	8·40	7·11	8·36	7·14	8·33	7·18	11
12	9·19	7·71	9·16	7·75	9·12	7·79	9·09	7·83	12
13	9·96	8·36	9·92	8·40	9·89	8·44	9·85	8·49	13
14	10·72	9·00	10·69	9·05	10·65	9·09	10·61	9·14	14
15	11·49	9·64	11·45	9·69	11·41	9·74	11·36	9·79	15
16	12·26	10·28	12·21	10·34	12·17	10·39	12·12	10·44	16
17	13·02	10·93	12·97	10·98	12·93	11·04	12·88	11·10	17
18	13·79	11·57	13·74	11·63	13·69	11·69	13·64	11·75	18
19	14·55	12·21	14·50	12·28	14·45	12·34	14·39	12·40	19
20	15·32	12·86	15·26	12·92	15·21	12·99	15·15	13·06	20
21	16·09	13·50	16·03	13·57	15·97	13·64	15·91	13·71	21
22	16·85	14·14	16·79	14·21	16·73	14·29	16·67	14·36	22
23	17·62	14·78	17·55	14·86	17·49	14·94	17·42	15·01	23
24	18·39	15·43	18·32	15·51	18·25	15·59	18·18	15·67	24
25	19·15	16·07	19·08	16·15	19·01	16·24	18·94	16·32	25
26	19·92	16·71	19·84	16·80	19·77	16·89	19·70	16·97	26
27	20·68	17·36	20·61	17·45	20·53	17·54	20·45	17·62	27
28	21·45	18·00	21·37	18·09	21·29	18·18	21·21	18·28	28
29	22·22	18·64	22·13	18·74	22·05	18·83	21·97	18·93	29
30	22·98	19·28	22·90	19·38	22·81	19·48	22·73	19·58	30
31	23·75	19·93	23·66	20·03	23·57	20·13	23·48	20·24	31
32	24·51	20·57	24·42	20·68	24·33	20·78	24·24	20·89	32
33	25·28	21·21	25·19	21·32	25·09	21·43	25·00	21·54	33
34	26·05	21·85	25·95	21·97	25·85	22·08	25·76	22·19	34
35	26·81	22·50	26·71	22·61	26·61	22·73	26·51	22·85	35
36	27·58	23·14	27·48	23·26	27·37	23·38	27·27	23·50	36
37	28·34	23·78	28·24	23·91	28·13	24·03	28·03	24·15	37
38	29·11	24·43	29·00	24·55	28·90	24·68	28·79	24·80	38
39	29·88	25·07	29·77	25·20	29·66	25·33	29·54	25·46	39
40	30·64	25·71	30·53	25·84	30·42	25·98	30·30	26·11	40
41	31·41	26·35	31·29	26·49	31·18	26·63	31·06	26·76	41
42	32·17	27·00	32·06	27·14	31·94	27·28	31·82	27·42	42
43	32·94	27·64	32·82	27·78	32·70	27·93	32·58	28·07	43
44	33·71	28·28	33·58	28·43	33·46	28·58	33·33	28·72	44
45	34·47	28·93	34·35	29·08	34·22	29·23	34·09	29·37	45
46	35·24	29·57	35·11	29·72	34·98	29·87	34·85	30·03	46
47	36·00	30·21	35·87	30·37	35·74	30·52	35·61	30·68	47
48	36·77	30·85	36·64	31·01	36·50	31·17	36·36	31·33	48
49	37·54	31·50	37·40	31·66	37·26	31·82	37·12	31·99	49
50	38·30	32·14	38·16	32·31	38·02	32·47	37·88	32·64	50
Distance.	Dep.	Lat.	Dep.	Lat.	Dep.	Lat.	Dep.	Lat.	Distance.
	50 Deg.		49¾ Deg.		49½ Deg.		49¼ Deg.		

Distance.	40 Deg.		40¼ Deg.		40½ Deg.		40¾ Deg.		Distance.
	Lat.	Dep.	Lat.	Dep.	Lat.	Dep.	Lat.	Dep.	
51	39·07	32·78	38·92	32·95	38·78	33·12	38·64	33·29	51
52	39·83	33·42	39·69	33·60	39·54	33·77	39·39	33·94	52
53	40·60	34·07	40·45	34·24	40·30	34·42	40·15	34·60	53
54	41·37	34·71	41·21	34·89	41·06	35·07	40·91	35·25	54
55	42·13	35·35	41·98	35·54	41·82	35·72	41·67	35·90	55
56	42·90	36·00	42·74	36·18	42·58	36·37	42·42	36·55	56
57	43·66	36·64	43·50	36·83	43·34	37·02	43·18	37·21	57
58	44·43	37·28	44·27	37·48	44·10	37·67	43·94	37·86	58
59	45·20	37·92	45·03	38·12	44·86	38·32	44·70	38·51	59
60	45·96	38·57	45·79	38·77	45·62	38·97	45·45	39·17	60
61	46·73	39·21	46·56	39·41	46·38	39·62	46·21	39·82	61
62	47·49	39·85	47·32	40·06	47·15	40·27	46·97	40·47	62
63	48·26	40·50	48·08	40·71	47·91	40·92	47·73	41·12	63
64	49·03	41·14	48·85	41·35	48·67	41·56	48·48	41·78	64
65	49·79	41·78	49·61	42·00	49·43	42·21	49·24	42·43	65
66	50·56	42·42	50·37	42·64	50·19	42·86	50·00	43·08	66
67	51·32	43·07	51·14	43·29	50·95	43·51	50·76	43·73	67
68	52·09	43·71	51·90	43·94	51·71	44·16	51·51	44·39	68
69	52·86	44·35	52·66	44·58	52·47	44·81	52·27	45·04	69
70	53·62	45·00	53·43	45·23	53·23	45·46	53·03	45·69	70
71	54·39	45·64	54·19	45·87	53·99	46·11	53·79	46·35	71
72	55·16	46·28	54·95	46·52	54·75	46·76	54·54	47·00	72
73	55·92	46·92	55·72	47·17	55·51	47·41	55·30	47·65	73
74	56·69	47·57	56·48	47·81	56·27	48·06	56·06	48·30	74
75	57·45	48·21	57·24	48·46	57·03	48·71	56·82	48·96	75
76	58·22	48·85	58·01	49·11	57·79	49·36	57·57	49·61	76
77	58·99	49·49	58·77	49·75	58·55	50·01	58·33	50·26	77
78	59·75	50·14	59·53	50·40	59·31	50·66	59·09	50·92	78
79	60·52	50·78	60·30	51·04	60·07	51·31	59·85	51·57	79
80	61·28	51·42	61·06	51·69	60·83	51·96	60·61	52·22	80
81	62·05	52·07	61·82	52·34	61·59	52·61	61·36	52·87	81
82	62·82	52·71	62·59	52·98	62·35	53.25	62·12	53·53	82
83	63·58	53·35	63·35	53·63	63·11	53·90	62·88	54·18	83
84	64·35	53·99	64·11	54·27	63·87	54·55	63·64	54·83	84
85	65·11	54·64	64·87	54·92	64·63	55·20	64·39	55·48	85
86	65·88	55·28	65·64	55·57	65·39	55·85	65·15	56·14	86
87	66·65	55·92	66·40	56·21	66·16	56·50	65·91	56·79	87
88	67·41	56·57	67·16	56·86	66·92	57·15	66·67	57·44	88
89	68·18	57·21	67·93	57·50	67·68	57·80	67·42	58·10	89
90	68·94	57·85	68·69	58·15	68·44	58·45	68·18	58·75	90
91	69·71	58·49	69·45	58·80	69·20	59·10	68·94	59·40	91
92	70·48	59·14	70·22	59·44	69·96	59·75	69·70	60·05	92
93	71·24	59·78	70·98	60·09	70·72	60·40	70·45	60·71	93
94	72·01	60·42	71·74	60·74	71·48	61·05	71·21	61·36	94
95	72·77	61·06	72·51	61·38	72·24	61·70	71·97	62·01	95
96	73·54	61·71	73·27	62·03	73·00	62·35	72·73	62·66	96
97	74·31	62·35	74·03	62 67	73·76	63·00	73·48	63.32	97
98	75·07	62·99	74·80	63·32	74·52	63·65	74·24	63·97	98
99	75·84	63·64	75·56	63·97	75·28	64·30	75·00	64·62	99
100	76·60	64·28	76·32	64·61	76·04	64·94	75·76	65·28	100
Distance.	Dep.	Lat.	Dep.	Lat.	Dep.	Lat.	Dep.	Lat.	Distance.
	50 Deg.		49¾ Deg.		49½ Deg.		49¼ Deg.		

Distance.	41 Deg.		41¼ Deg.		41½ Deg.		41¾ Deg.		Distance.
	Lat.	Dep.	Lat.	Dep.	Lat.	Dep.	Lat.	Dep.	
1	0·75	0·66	0·75	0·66	0·75	0·66	0·75	0·67	1
2	1·51	1·31	1·50	1·32	1·50	1·33	1·49	1·33	2
3	2·26	1·97	2·26	1·98	2·25	1·99	2·24	2·00	3
4	3·02	2·62	3·01	2·64	3·00	2·65	2·98	2·66	4
5	3·77	3·28	3·76	3·30	3·74	3·31	3·73	3·33	5
6	4·53	3·94	4·51	3·96	4·49	3·98	4·48	4·00	6
7	5·28	4·59	5·26	4·62	5·24	4·64	5·22	4·66	7
8	6·04	5·25	6·01	5·27	5·99	5·30	5·97	5·33	8
9	6·79	5·90	6·77	5·93	6·74	5·96	6·71	5·99	9
10	7·55	6·56	7·52	6·59	7·49	6·63	7·46	6·66	10
11	8·30	7·22	8·27	7·25	8·24	7·29	8·21	7·32	11
12	9·06	7·87	9·02	7·91	8·99	7·95	8·95	7·99	12
13	9·81	8·53	9·77	8·57	9·74	8·61	9·70	8·66	13
14	10·57	9·18	10·53	9·23	10·49	9·28	10·44	9·32	14
15	11·32	9·84	11·28	9·89	11·23	9·94	11·19	9·99	15
16	12·08	10·50	12·03	10·55	11·98	10·60	11·94	10·65	16
17	12·83	11·15	12·78	11·21	12·73	11·26	12·68	11·32	17
18	13·58	11·81	13·53	11·87	13·48	11·93	13·43	11·99	18
19	14·34	12·47	14·28	12·53	14·23	12·59	14·18	12·65	19
20	15·09	13·12	15·04	13·19	14·98	13·25	14·92	13·32	20
21	15·85	13·78	15·79	13·85	15·73	13·91	15·67	13·98	21
22	16·60	14·43	16·54	14·51	16·48	14·58	16·41	14·65	22
23	17·36	15·09	17·29	15·16	17·23	15·24	17·16	15·32	23
24	18·11	15·75	18·04	15·82	17·97	15·90	17·91	15·98	24
25	18·87	16·40	18·80	16·48	18·72	16·57	18·65	16·65	25
26	19·62	17·06	19·55	17·14	19·47	17·23	19·40	17·31	26
27	20·38	17·71	20·30	17·80	20·22	17·89	20·14	17·98	27
28	21·13	18·37	21·05	18·46	20·97	18·55	20·89	18·64	28
29	21·89	19·03	21·80	19·12	21·72	19·22	21·64	19·31	29
30	22·64	19·68	22·56	19·78	22·47	19·88	22·38	19·98	30
31	23·40	20·34	23·31	20·44	23·22	20·54	23·13	20·64	31
32	24·15	20·99	24·06	21·10	23·97	21·20	23·87	21·31	32
33	24·91	21·65	24·81	21·76	24·72	21·87	24·62	21·97	33
34	25·66	22·31	25·56	22·42	25·46	22·53	25·37	22·64	34
35	26·41	22·96	26·31	23·08	26·21	23·19	26·11	23·31	35
36	27·17	23·62	27·07	23·74	26·96	23·85	26·86	23·97	36
37	27·92	24·27	27·82	24·40	27·71	24·52	27·60	24·64	37
38	28·68	24·93	28·57	25·06	28·46	25·18	28·35	25·30	38
39	29·43	25·59	29·32	25·71	29·21	25·84	29·10	25·97	39
40	30·19	26·24	30·07	26·37	29·96	26·50	29·84	26·64	40
41	30·94	26·90	30·83	27·03	30·71	27·17	30·59	27·30	41
42	31·70	27·55	31·58	27·69	31·46	27·83	31·33	27·97	42
43	32·45	28·21	32·33	28·35	32·21	28·49	32·08	28·63	43
44	33·21	28·87	33·08	29·01	32·95	29·16	32·83	29·30	44
45	33·96	29·52	33·83	29·67	33·70	29·82	33·57	29·97	45
46	34·72	30·18	34·58	30·33	34·45	30·48	34·32	30·63	46
47	35·47	30·83	35·34	30·99	35·20	31·14	35·06	31·30	47
48	36·23	31·49	36·09	31·65	35·95	31·81	35·81	31·96	48
49	36·98	32·15	36·84	32·31	36·70	32·47	36·56	32·63	49
50	37·74	32·80	37·59	32·97	37·45	33·13	37·30	33·29	50
Distance.	Dep.	Lat.	Dep.	Lat.	Dep.	Lat.	Dep.	Lat.	Distance.
	49 Deg.		48¾ Deg.		48½ Deg.		48¼ Deg.		

Distance.	41 Deg.		41¼ Deg.		41½ Deg.		41¾ Deg.		Distance.
	Lat.	Dep.	Lat.	Dep.	Lat.	Dep.	Lat.	Dep.	
51	38·49	33·46	38·34	33·63	38·20	33·79	38·05	33·96	51
52	39·24	34·12	39·10	34·29	38·95	34·46	38·79	34·63	52
53	40·00	34·77	39·85	34·95	39·69	35·12	39·54	35·29	53
54	40·75	35·43	40·60	35·60	40·44	35·78	40·29	35·96	54
55	41·51	36·08	41·35	36·26	41·19	36·44	41·03	36·62	55
56	42·26	36·74	42·10	36·92	41·94	37·11	41·78	37·29	56
57	43·02	37·40	42·85	37·58	42·69	37·77	42·53	37·96	57
58	43·77	38·05	43·61	38·24	43·44	38·43	43·27	38·62	58
59	44·53	38·71	44·36	38·90	44·19	39·09	44·02	39·29	59
60	45·28	39·36	45·11	39·56	44·94	39·76	44·76	39·95	60
61	46·04	40·02	45·86	40·22	45·69	40·42	45·51	40·62	61
62	46·79	40·68	46·61	40·88	46·44	41·08	46·26	41·28	62
63	47·55	41·33	47·37	41·54	47·18	41·75	47·00	41·95	63
64	48·30	41·99	48·12	42·20	47·93	42·41	47·75	42·62	64
65	49·06	42·64	48·87	42·86	48·68	43·07	48·49	43·28	65
66	49·81	43·30	49·62	43·52	49·43	43·73	49·24	43·95	66
67	50·57	43·96	50·37	44·18	50·18	44·40	49·99	44·61	67
68	51·32	44·61	51·13	44·84	50·93	45·06	50·73	45·28	68
69	52·07	45·27	51·88	45·49	51·68	45·72	51·48	45·95	69
70	52·83	45·92	52·63	46·15	52·43	46·38	52·22	46·61	70
71	53·58	46·58	53·38	46·81	53·18	47·05	52·97	47·28	71
72	54·34	47·24	54·13	47·47	53·92	47·71	53·72	47·94	72
73	55·09	47·89	54·88	48·13	54·67	48·37	54·46	48·61	73
74	55·85	48·55	55·64	48·79	55·42	49·03	55·21	49·28	74
75	56·60	49·20	56·39	49·45	56·17	49·70	55·95	49·94	75
76	57·36	49·86	57·14	50·11	56·92	50·36	56·70	50·61	76
77	58·11	50·52	57·89	50·77	57·67	51·02	57·45	51·27	77
78	58·87	51·17	58·64	51·43	58·42	51·68	58·19	51·94	78
79	59·62	51·83	59·40	52·09	59·17	52·35	58·94	52·60	79
80	60·38	52·48	60·15	52·75	59·92	53·01	59·68	53·27	80
81	61·13	53·14	60·90	53·41	60·67	53·67	60·43	53·94	81
82	61·89	53·80	61·65	54·07	61·41	54·33	61·18	54·60	82
83	62·64	54·45	62·40	54·73	62·16	55·00	61·92	55·27	83
84	63·40	55·11	63·15	55·38	62·91	55·66	62·67	55·93	84
85	64·15	55·76	63·91	56·04	63·66	56·32	63·41	56·60	85
86	64·90	56·42	64·66	56·70	64·41	56·99	64·16	57·27	86
87	65·66	57·08	65·41	57·36	65·16	57·65	64·91	57·93	87
88	66·41	57·73	66·16	58·02	65·91	58·31	65·65	58·60	88
89	67·17	58·39	66·91	58·68	66·66	58·97	66·40	59·26	89
90	67·92	59·05	67·67	59·34	67·41	59·64	67·15	59·93	90
91	68·68	59·70	68·42	60·00	68·15	60·30	67·89	60·60	91
92	69·43	60·36	69·17	60·66	68·90	60·96	68·64	61·26	92
93	70·19	61·01	69·92	61·32	69·65	91·62	69·38	61·93	93
94	70·94	61·67	70·67	61·98	70·40	62·29	70·13	62·59	94
95	71·70	62·33	71·43	62·64	71·15	62·95	70·88	63·26	95
96	72·45	62·98	72·18	63·30	71·90	63·61	71·62	63·92	96
97	73·21	63·64	72·93	63·96	72·65	64·27	72·37	64·59	97
98	73·96	64·29	73·68	64·62	73·40	64·94	73·11	65·26	98
99	74·72	64·95	74·43	65·28	74·15	65·60	73·86	65·92	99
100	75·47	65·61	75·18	65·93	74·90	66·26	74·61	66·59	100
Distance.	Dep.	Lat.	Dep.	Lat.	Dep.	Lat.	Dep.	Lat.	Distance.
	49 Deg.		48¾ Deg.		48½ Deg.		48¼ Deg.		

Distance.	42 Deg.		42¼ Deg.		42½ Deg.		42¾ Deg.		Distance.
	Lat.	Dep.	Lat.	Dep.	Lat.	Dep.	Lat.	Dep.	
1	0·74	0·67	0·74	0·67	0·74	0·68	0·73	0·68	1
2	1·49	1·34	1·48	1·34	1·47	1·35	1·47	1·36	2
3	2·23	2·01	2·22	2·02	2·21	2·03	2·20	2·04	3
4	2·97	2·68	2·96	2·69	2·95	2·70	2·94	2·72	4
5	3·72	3·35	3·70	3·36	3·69	3·38	3·67	3·39	5
6	4·46	4·01	4·44	4·03	4·42	4·05	4·41	4·07	6
7	5·20	4·68	5·18	4·71	5·16	4·73	5·14	4·75	7
8	5·95	5·35	5·92	5·38	5·90	5·40	5·87	5·43	8
9	6·69	6·02	6·66	6·05	6·64	6·08	6·61	6·11	9
10	7·43	6·69	7·40	6·72	7·37	6·76	7·34	6·79	10
11	8·17	7·36	8·14	7·40	8·11	7·43	8·08	7·47	11
12	8·92	8·03	8·88	8·07	8·85	8·11	8·81	8·15	12
13	9·66	8·70	9·62	8·74	9·58	8·78	9·55	8·82	13
14	10·40	9·37	10·36	9·41	10·32	9·46	10·28	9·50	14
15	11·15	10·04	11·10	10·09	11·06	10·13	11·01	10·18	15
16	11·89	10·71	11·84	10·76	11·80	10·81	11·75	10·86	16
17	12·63	11·38	12·58	11·43	12·53	11·48	12·48	11·54	17
18	13·38	12·04	13·32	12·10	13·27	12·16	13·22	12·22	18
19	14·12	12·71	14·06	12·77	14·01	12·84	13·95	12·90	19
20	14·86	13.38	14·80	13·45	14·75	13·51	14·69	13·58	20
21	15·61	14·05	15·54	14·12	15·48	14·19	15·42	14·25	21
22	16·35	14·72	16·28	14·79	16·22	14·86	16·16	14·93	22
23	17·09	15·39	17·02	15·46	16·96	15·54	16·89	15·61	23
24	17·84	16·06	17·77	16·14	17·69	16·21	17·62	16·29	24
25	18·58	16·73	18·51	16·81	18·43	16·89	18·36	16·97	25
26	19·32	17·40	19·25	17·48	19·17	17·57	19·09	17·65	26
27	20·06	18·07	19·99	18·15	19·91	18·24	19·83	18·33	27
28	20·81	18·74	20·73	18·83	20·64	18·92	20·56	19·01	28
29	21·55	19·40	21·47	19·50	21·38	19·59	21·30	19·69	29
30	22·29	20·07	22·21	20·17	22·12	20·27	22·03	20·36	30
31	23·04	20·74	22·95	20·84	22·86	20·94	22·76	21·04	31
32	23·78	21·41	23·69	21·52	23·59	21·62	23·50	21·72	32
33	24·52	22·08	24·43	22·19	24·33	22·29	24·23	22·40	33
34	25·27	22·75	25·17	22·86	25·07	22·97	24·97	23·08	34
35	26·01	23·42	25·91	23·53	25·80	23·65	25·70	23·76	35
36	26·75	24·09	26·65	24·21	26·54	24·32	26·44	24·44	36
37	27·50	24·76	27·39	24·88	27·28	25·00	27·17	25·12	37
38	28·24	25·43	28·13	25·55	28·02	25·67	27·90	25·79	38
39	28·98	26·10	28·87	26·22	28·75	26·35	28·64	26·47	39
40	29·73	26·77	29·61	26·89	29·49	27·02	29·37	27·15	40
41	30·47	27·43	30·35	27·57	30·23	27·70	30·11	27·83	41
42	31·21	28·10	31·09	28·24	30·97	28·37	30·84	28·51	42
43	31·96	28·77	31·83	28·91	31·70	29·05	31·58	29·19	43
44	32·70	29·44	32·57	29·58	32·44	29·73	32·31	29·87	44
45	33·44	30·11	33·31	30·26	33·18	30·40	33·04	30·55	45
46	34·18	30·78	34·05	30·93	33·91	31·08	33·78	31·22	46
47	34·93	31·45	34·79	31·60	34·65	31·75	34·51	31·90	47
48	35·67	32·12	35·53	32·27	35·39	32·43	35·25	32·58	48
49	36·41	32·79	36·27	32·95	36·13	33·10	35·98	33·26	49
50	37·16	33·46	37·01	33·62	36·86	33·78	36·72	33·94	50
	Dep.	Lat.	Dep.	Lat.	Dep.	Lat.	Dep.	Lat.	
Distance.	48 Deg.		47¾ Deg.		47½ Deg.		47¼ Deg.		Distance.

Distance.	42 Deg.		42¼ Deg.		42½ Deg.		42¾ Deg.		Distance.
	Lat.	Dep.	Lat.	Dep.	Lat.	Dep.	Lat.	Dep.	
51	37·90	34·13	37·75	34·29	37·60	34·46	37·45	34·62	51
52	38·64	34·79	38·49	34·96	38·34	35·13	38·18	35·30	52
53	39·39	35·46	39·23	35·64	39·08	35·81	38·92	35·98	53
54	40·13	36·13	39·97	36·31	39·81	36·48	39·65	36·66	54
55	40·87	36·80	40·71	36·98	40·55	37·16	40·39	37·33	55
56	41·62	37·47	41·45	37·65	41·29	37·83	41·12	38·01	56
57	42·36	38·14	42·19	38·32	42·02	38·51	41·86	38·69	57
58	43·10	38·81	42·93	39·00	42·76	39·18	42·59	39·37	58
59	43·85	39·48	43·67	39·67	43·50	39·86	43·32	40·05	59
60	44·59	40·15	44·41	40·34	44·24	40·54	44·06	40·73	60
61	45·33	40·82	45·15	41·01	44·97	41·21	44·79	41·41	61
62	46·07	41·49	45·89	41·69	45·71	41·89	45·53	42·09	62
63	46·82	42·16	46·63	42·36	46·45	42·56	46·26	42·76	63
64	47·56	42·82	47·37	43·03	47·19	43·24	47·00	43·44	64
65	48·30	43·49	48·11	43·70	47·92	43·91	47·73	44·12	65
66	49·05	44·16	48·85	44·38	48·66	44·59	48·47	44·80	66
67	49·79	44·83	49·59	45·05	49·40	45·26	49·20	45·48	67
68	50·53	45·50	50·33	45·72	50·13	45·94	49·93	46·16	68
69	51·28	46·17	51·07	46·39	50·87	46·62	50·67	46·84	69
70	52·02	46·84	51·82	47·07	51·61	47·29	51·40	47·52	70
71	52.76	47·51	52·56	47·74	52·35	47·97	52·14	48·19	71
72	53·51	48·18	53·30	48·41	53·08	48·64	52·87	48·87	72
73	54·25	48·85	54·04	49·08	53·82	49·32	53·61	49·55	73
74	54·99	49·52	54·78	49·76	54·56	49·99	54·34	50·23	74
75	55·74	50·18	55·52	50·43	55·30	50·67	55·07	50·91	75
76	56·48	50·85	56·26	51·10	56·03	51·34	55·81	51·59	76
77	57·22	51·52	57·00	51·77	56·77	52·02	56·54	52·27	77
78	57·97	52·19	57·74	52·44	57·51	52·70	57·28	52·95	78
79	58·71	52·86	58·48	53·12	58·24	53·37	58·01	53·63	79
80	59·45	53·53	59·22	53·79	58·98	54·05	58·75	54·30	80
81	60·19	54·20	59·96	54·46	59·72	54·72	59·48	54·98	81
82	60·94	54·87	60·70	55·13	60·46	55·40	60·21	55·66	82
83	61·68	55·54	61·44	55·81	61·19	56·07	60·95	56·34	83
84	62·42	56·21	62·18	56·48	61·93	56·75	61·68	57·02	84
85	63·17	56·88	62·92	57·15	62·67	57·43	62·42	57·70	85
86	63·91	57·55	63·66	57·82	63·41	58·10	63·15	58·38	86
87	64·65	58·21	64·40	58·50	64·14	58·78	63·89	59·06	87
88	65·40	58·88	65·14	59·17	64·88	59·45	64·62	59·73	88
89	66·14	59·55	65·88	59·84	65·62	60·13	65·35	60·41	89
90	66·88	60·22	66·62	60·51	66·35	60·80	66·09	61·09	90
91	67·63	60·89	67·36	61·19	67·09	61·48	66·82	61·77	91
92	68·37	61·56	68·10	61·86	67·83	62·15	67·56	62·45	92
93	69·11	62·23	68·84	62·53	68·57	62·83	68·29	63·13	93
94	69·86	62·90	69·58	63·20	69·30	63·51	69·03	63·81	94
95	70·60	63·57	70·32	63·87	70·04	64·18	69·76	64·49	95
96	71·34	64·24	71·06	64·55	70·78	64·86	70·49	65·16	96
97	72·08	64·91	71·80	65·22	71·52	65·53	71·23	65·84	97
98	72·83	65·57	72·54	65·89	72·25	66·21	71·96	66·52	98
99	73·57	66·24	73·28	66·56	72·99	66·88	72·70	67·20	99
100	74·31	66·91	74·02	67·24	73·73	67·56	73·43	67·88	100
	Dep.	Lat.	Dep.	Lat.	Dep.	Lat.	Dep.	Lat.	
Distance.	48 Deg.		47¾ Deg.		47½ Deg.		47¼ Deg.		Distance.

Distance.	43 Deg.		43¼ Deg.		43½ Deg.		43¾ Deg.		Distance.
	Lat.	Dep.	Lat.	Dep.	Lat.	Dep.	Lat.	Dep.	
1	0·73	0·68	0·73	0·69	0·73	0·69	0·72	0·69	1
2	1·46	1·36	1·46	1·37	1·45	1·38	1·44	1·38	2
3	2·19	2·05	2·19	2·06	2·18	2·07	2·17	2·07	3
4	2·93	2·73	2·91	2·74	2·90	2·75	2·89	2·77	4
5	3·66	3·41	3·64	3·43	3·63	3·44	3·61	3·46	5
6	4·39	4·09	4·37	4·11	4·35	4·13	4·33	4·15	6
7	5·12	4·77	5·10	4·80	5·08	4·82	5·06	4·84	7
8	5·85	5·46	5·83	5·48	5·80	5·51	5·78	5·53	8
9	6·58	6·14	6·56	6·17	6·53	6·20	6·50	6·22	9
10	7·31	6·82	7·28	6·85	7·25	6·88	7·22	6·92	10
11	8·04	7·50	8·01	7·54	7·98	7·57	7·95	7·61	11
12	8·78	8·18	8·74	8·22	8·70	8·26	8·67	8·30	12
13	9·51	8·87	9·47	8·91	9·43	8·95	9·39	8·99	13
14	10·24	9·55	10·20	9·59	10·16	9·64	10·11	9·68	14
15	10·97	10·23	10·93	10·28	10·88	10·33	10·84	10·37	15
16	11·70	10·91	11·65	10·96	11·61	11·01	11·56	11·06	16
17	12·43	11·59	12·38	11·65	12·33	11·70	12·28	11·76	17
18	13·16	12·28	13·11	12·33	13·06	12·39	13·00	12·45	18
19	13·90	12·96	13·84	13·02	13·78	13·08	13·72	13·14	19
20	14·63	13·64	14·57	13·70	14·51	13·77	14·45	13·83	20
21	15·36	14·32	15·30	14·39	15·23	14·46	15·17	14·52	21
22	16·09	15·00	16·02	15·07	15·96	15·14	15·89	15·21	22
23	16·82	15·69	16·75	15·76	16·68	15·83	16·51	15·90	23
24	17·55	16·37	17·48	16·44	17·41	16·52	17·34	16·60	24
25	18·28	17·05	18·21	17·13	18·13	17·21	18·06	17·29	25
26	19·02	17·73	18·94	17·81	18·86	17·90	18·78	17·98	26
27	19·75	18·41	19·67	18·50	19·59	18·59	19·50	18·67	27
28	20·48	19·10	20·39	19·19	20·31	19·27	20·23	19·36	28
29	21·21	19·78	21·12	19·87	21·04	19·96	20·95	20·05	29
30	21·94	20·46	21·85	20·56	21·76	20·65	21·67	20·75	30
31	22·67	21·14	22·58	21·24	22·49	21·34	22·39	21·44	31
32	23·40	21·82	23·31	21·93	23·21	22·03	23·12	22·13	32
33	24·13	22·51	24·04	22·61	23·94	22·72	23·84	22·82	33
34	24·87	23·19	24·76	23·30	24·66	23·40	24·56	23·51	34
35	25·60	23·87	25·49	23·98	25·39	24·09	25·28	24·20	35
36	26·33	24·55	26·22	24·67	26·11	24·78	26·01	24·89	36
37	27·06	25·23	26·95	25·35	26·84	25·47	26·73	25·59	37
38	27·79	25·92	27·68	26·04	27·56	26·16	27·45	26·28	38
39	28·52	26·60	28·41	26·72	28·29	26·85	28·17	26·97	39
40	29·25	27·28	29·13	27·41	29·01	27·53	28·89	27·66	40
41	29·99	27·96	29·86	28·09	29·74	28·22	29·62	28·35	41
42	30·72	28·64	30·59	28·78	30·47	28·91	30·34	29·04	42
43	31·45	29·33	31·32	29·46	31·19	29·60	31·06	29·74	43
44	32·18	30·01	32·05	30·15	31·92	30·29	31·78	30·43	44
45	32·91	30·69	32·78	30·83	32·64	30·98	32·51	31·12	45
46	33·64	31·37	33·51	31·52	33·37	31·66	33·23	31·81	46
47	34·37	32·05	34·23	32·20	34·09	32·35	33·95	32·50	47
48	35·10	32·74	34·96	32·89	34·82	33·04	34·67	33·19	48
49	35·84	33·42	35·69	33·57	35·54	33·73	35·40	33·88	49
50	36·57	34·10	36·42	34·26	36·27	34·42	36·12	34·58	50
	Dep.	Lat.	Dep.	Lat.	Dep.	Lat.	Dep.	Lat.	
Distance.	47 Deg.		46¾ Deg.		46½ Deg.		46¼ Deg.		Distance.

Distance.	43 Deg.		43¼ Deg.		43½ Deg.		43¾ Deg.		Distance.
	Lat.	Dep.	Lat.	Dep.	Lat.	Dep.	Lat.	Dep.	
51	37·30	34·78	37·15	34·94	36·99	35·11	36·84	35·27	51
52	38·03	35·46	37·88	35·63	37·72	35·79	37·56	35·96	52
53	38·76	36·15	38·60	36·31	38·44	36·48	38·29	36·65	53
54	39·49	36·83	39·33	37·00	39·17	37·17	39·01	37·34	54
55	40·22	37·51	40·06	37·69	39·90	37·86	39·73	38·03	55
56	40·96	38·19	40·79	38·37	40·62	38·55	40·45	38·72	56
57	41·69	38·87	41·52	39·06	41·35	39·24	41·17	39·42	57
58	42·42	39·56	42·25	39·74	42·07	39·92	41·90	40·11	58
59	43·15	40·24	42·97	40·43	42·80	40·61	42·62	40·80	59
60	43·88	40·92	43·70	41·11	43·52	41·30	43·34	41·49	60
61	44·61	41·60	44·43	41·80	44·25	41·99	44·06	42·18	61
62	45·34	42·28	45·16	42·48	44·97	42·68	44·79	42·87	62
63	46·08	42·97	45·89	43·17	45·70	43·37	45·51	43·57	63
64	46·81	43·65	46·62	43·85	46·42	44·05	46·23	44·26	64
65	47·54	44·33	47·34	44·54	47·15	44·74	46·95	44·95	65
66	48·27	45·01	48·07	45·22	47·87	45·43	47·68	45·64	66
67	49·00	45·69	48·80	45·91	48·60	46·12	48·40	46·33	67
68	49·73	46·38	49·53	46·59	49·33	46·81	49·12	47·02	68
69	50·46	47·06	50·26	47·28	50·05	47·50	49·84	47·71	69
70	51·19	47·74	50·99	47·96	50·78	48·18	50·57	48·41	70
71	51·93	48·42	51·71	48·65	51·50	48·87	51·29	49·10	71
72	52·66	49·10	52·44	49·33	52·23	49·56	52·01	49·79	72
73	53·39	49·79	53·17	50·02	52·95	50·25	52·73	50·48	73
74	54·12	50·47	53·90	50·70	53·68	50·94	53·45	51·17	74
75	54·85	51·15	54·63	51·39	54·40	51·63	54·18	51·86	75
76	55·58	51·83	55·36	52·07	55·13	52·31	54·90	52·55	76
77	56·31	52·51	56·08	52·76	55·85	53·00	55·62	53·25	77
78	57·05	53·20	56·81	53·44	56·58	53·69	56·34	53·94	78
79	57·78	53·88	57·54	54·13	57·30	54·38	57·07	54·63	79
80	58·51	54·56	58·27	54·81	58·03	55·07	57·79	55·32	80
81	59·24	55·24	59·00	55·50	58·76	55·76	58·51	56·01	81
82	59·97	55·92	59·73	56·18	59·48	56·45	59·23	56·70	82
83	60·70	56·61	60·45	56·87	60·21	57·13	59·96	57·40	83
84	61·43	57·29	61·18	57·56	60·93	57·82	60·68	58·09	84
85	62·17	57·97	61·91	58·24	61·66	58·51	61·40	58·78	85
86	62·90	58·65	62·64	58·93	62·38	59·20	62·12	59·47	86
87	63·63	59·33	63·37	59·61	63·11	59·89	62·85	60·16	87
88	64·36	60·02	64·10	60·30	63·83	60·58	63·57	60·85	88
89	65·09	60·70	64·82	60·98	64·56	61·26	64·29	61·54	89
90	65·82	61·38	65·55	61·67	65·28	61·95	65·01	62·24	90
91	66·55	62·06	66·28	62·35	66·01	62·64	65·74	62·93	91
92	67·28	62·74	67·01	63·04	66·73	63·33	66·46	63·62	92
93	68·02	63·43	67·74	63·72	67·46	64·02	67·18	64·31	93
94	68·75	64·11	68·47	64·41	68·19	64·71	67·90	65·00	94
95	69·48	64·79	69·20	65·09	68·91	65·39	68·62	65·69	95
96	70·21	65·47	69·92	65·78	69·64	66·08	69·35	66·39	96
97	70·94	66·15	70·65	66·46	70·36	66·77	70·07	67·08	97
98	71·67	66·84	71·37	67·15	71·09	67·46	70·79	67·77	98
99	72·40	67·52	72·11	67·83	71·81	68·15	71·51	68·46	99
100	73·14	68·20	72·84	68·52	72·54	68·84	72·24	69·15	100
Distance.	Dep.	Lat.	Dep.	Lat.	Dep.	Lat.	Dep.	Lat.	Distance.
	47 Deg.		46¾ Deg.		46½ Deg.		46¼ Deg.		

Distance.	44 Deg.		44¼ Deg.		44½ Deg.		44¾ Deg.		45 Deg.		Distance.
	Lat.	Dep.	Lat.	Dep.	Lat.	Dep.	Lat.	Dep.	Lat.	Dep.	
1	0·72	0·69	0·72	0·70	0·71	0·70	0·71	0·71	0·71	0·71	1
2	1·44	1·39	1·43	1·40	1·43	1·40	1·42	1·41	1·41	1·41	2
3	2·16	2·08	2·15	2·09	2·14	2·10	2·13	2·11	2·12	2·12	3
4	2·88	2·78	2·87	2·79	2·85	2·80	2·84	2·82	2·83	2·83	4
5	3·60	3·47	3·58	3·49	3·57	3·50	3·55	3·52	3·54	3·54	5
6	4·32	4·17	4·30	4·19	4·28	4·21	4·26	4·22	4·24	4·24	6
7	5·04	4·86	5·01	4·88	4·99	4·91	4·97	4·93	4·95	4·95	7
8	5·75	5·56	5·73	5·58	5·71	5·61	5·68	5·63	5·66	5·66	8
9	6·47	6·25	6·45	6·28	6·42	6·31	6·39	6·34	6·36	6·36	9
10	7·19	6·95	7·16	6·98	7·13	7·01	7·10	7·04	7·07	7·07	10
11	7·91	7·64	7·88	7·68	7·85	7·71	7·81	7·74	7·78	7·78	11
12	8·63	8·34	8·60	8·37	8·56	8·41	8·52	8·45	8·49	8·49	12
13	9·35	9·03	9·31	9·07	9·27	9·11	9·23	9·15	9·19	9·19	13
14	10·07	9·73	10·03	9·77	9·99	9·81	9·94	9·86	9·90	9·90	14
15	10·79	10·42	10·74	10·47	10·70	10·51	10·65	10·56	10·61	10·61	15
16	11·51	11·11	11·46	11·16	11·41	11·21	11·36	11·26	11·31	11·31	16
17	12·23	11·81	12·18	11·86	12·13	11·92	12·07	11·97	12·02	12·02	17
18	12·95	12·50	12·89	12·56	12·84	12·62	12·78	12·67	12·73	12·73	18
19	13·67	13·20	13·61	13·26	13·55	13·32	13·49	13·38	13·43	13.43	19
20	14·39	13·89	14·33	13·96	14·26	14·02	14·20	14·08	14·14	14·14	20
21	15·11	14·59	15·04	14·65	14·98	14·72	14·91	14·78	14·85	14·85	21
22	15·83	15·28	15·76	15·35	15·69	15·42	15·62	15·49	15·56	15·56	22
23	16·54	15·98	16·47	16·05	16·40	16·12	16·33	16·19	16·26	16·26	23
24	17·26	16·67	17·19	16·75	17·12	16·82	17·04	16·90	16·97	16·97	24
25	17·98	17·37	17·91	17·44	17·83	17·52	17·75	17·60	17·68	17·68	25
26	18·70	18·06	18·62	18·14	18·54	18·22	18·46	18·30	18·38	18·38	26
27	19·42	18·76	19·34	18·84	19·26	18·92	19·17	19·01	19·09	19·09	27
28	20·14	19·45	20·06	19·54	19·97	19·63	19·89	19·71	19·80	19·80	28
29	20·86	20·15	20·77	20·24	20·68	20·33	20·60	20·42	20·51	20·51	29
30	21·58	20·84	21·49	20·93	21·40	21·03	21·31	21·12	21·21	21·21	30
31	22·30	21·53	22·21	21·63	22·11	21·73	22·02	21·82	21·92	21·92	31
32	23·02	22·23	22·92	22·33	22·82	22·43	22·73	22·53	22·63	22·63	32
33	23·74	22·92	23·64	23·03	23·54	23·13	23·44	23·23	23·33	23·33	33
34	24·46	23·62	24·35	23·72	24·25	23·83	24·15	23·94	24·04	24·04	34
35	25·18	24·31	25·07	24·42	24·96	24·53	24·86	24·64	24·75	24·75	35
36	25·90	25·01	25·79	25·12	25·68	25·23	25·57	25·34	25·46	25·46	36
37	26·62	25·70	26·50	25·82	26·39	25·93	26·28	26·05	26·16	26·16	37
38	27·33	26·40	27·22	26·52	27·10	26·63	26·99	26·75	26·87	26·87	38
39	28·05	27·09	27·94	27·21	27·82	27·34	27·70	27·46	27·58	27·58	39
40	28·77	27·79	28·65	27·91	28·53	28·04	28·41	28·16	28·28	28·28	40
41	29·49	28·48	29·37	28·61	29·24	28·74	29·12	28·86	28·99	28·99	41
42	30·21	29·18	30·08	29·31	29·96	29·44	29·83	29·57	29·70	29·70	42
43	30·93	29·87	30·80	30·00	30·67	30·14	30·54	30·27	30·41	30·41	43
44	31·65	30·56	31·52	30·70	31·38	30·84	31·25	30·98	31·11	31·11	44
45	32·37	31·26	32·23	31·40	32·10	31·54	31·96	31·68	31·82	31·82	45
46	33·09	31·95	32·95	32·10	32·81	32·24	32·67	32·38	32·53	32·53	46
47	33·81	32·65	33·67	32·80	33·52	32·94	33·38	33·09	33·23	33·23	47
48	34·53	33·34	34·38	33·49	34·24	33·64	34·09	33·79	33·94	33·94	48
49	35·25	34·04	35·10	34·19	34·95	34·34	34·80	34·50	34·65	34·65	49
50	35·97	34·73	35·82	34·89	35·66	35·05	35·51	35·20	35·36	35·36	50
Distance.	Dep.	Lat.	Dep.	Lat.	Dep.	Lat.	Dep.	Lat.	Dep.	Lat.	Distance.
	46 Deg.		45¾ Deg.		45½ Deg.		45¼ Deg.		45 Deg.		

Distance.	44 Deg.		44¼ Deg.		44½ Deg.		44¾ Deg.		45 Deg.		Distance.
	Lat.	Dep.	Lat.	Dep.	Lat.	Dep.	Lat.	Dep.	Lat.	Dep.	
51	36·69	35·43	36·53	35·59	36·38	35·75	36·22	35·90	36·06	36·06	51
52	37·41	36·12	37·25	36·29	37·09	36·45	36·93	36·61	36·77	36·77	52
53	38·12	36·82	37·96	36·98	37·80	37·15	37·64	37·31	37·48	37·48	53
54	38·84	37·51	38·68	37·68	38·52	37.85	38·35	38·02	38·18	38·18	54
55	39·56	38·21	39·40	38·38	39·23	38·55	39·06	38·72	38·89	38·89	55
56	40·28	38·90	40·11	39·08	39·94	39·25	39·77	39·42	39·60	39·60	56
57	41·00	39·60	40·83	39·77	40·66	39·95	40·48	40·13	40·31	40·31	57
58	41·72	40·29	41·55	40·47	41·37	40·65	41·19	40·83	41·01	41·01	58
59	42·44	40·98	42·26	41·17	42·08	41·35	41·90	41·54	41·72	41·72	59
60	43·16	41·68	42·98	41·87	42·79	42·05	42·61	42·24	42·43	42·43	60
61	43·88	42·37	43·69	42·57	43·51	42·76	43·32	42·94	43·13	43·13	61
62	44·60	43·07	44·41	43·26	44·22	43·46	44·03	43·65	43·84	43·84	62
63	45·32	43·76	45·13	43·96	44·93	44·16	44·74	44·35	44·55	44·55	63
64	46·04	44·46	45·84	44·66	45·65	44·86	45·45	45·06	45·25	45·25	64
65	46·76	45·15	46·56	45·36	46·36	45·56	46·16	45·76	45·96	45·96	65
66	47·48	45·85	47·28	46·05	47·07	46·26	46·87	46·46	46·67	46·67	66
67	48·20	46·54	47·99	46·75	47·79	46·96	47·58	47·17	47·38	47·38	67
68	48·92	47·24	48·71	47·45	48·50	47·66	48·29	47·87	48·08	48·08	68
69	49·63	47·93	49·42	48·15	49·21	48·36	49·00	48·58	48·79	48·79	69
70	50·35	48·63	50·14	48·85	49·93	49·06	49·71	49·28	49·50	49·50	70
71	51·07	49·32	50·86	49·54	50·64	49·76	50·42	49·98	50·20	50·20	71
72	51·79	50·02	51·57	50·24	51·35	50·47	51·13	50·69	50·91	50·91	72
73	52·51	50·71	52·29	50·94	52·07	51·17	51·84	51·39	51·62	51·62	73
74	53·23	51·40	53·01	51·64	52·78	51·87	52·55	52·10	52·33	52·33	74
75	53·95	52·10	53·72	52·33	53·49	52·57	53·26	52·80	53·03	53·03	75
76	54·67	52·79	54·44	53·03	54·21	53·27	53·97	53·51	53·74	53·74	76
77	55·39	53·49	55·16	53·73	54·92	53·97	54·68	54·21	54·45	54·45	77
78	56·11	54·18	55·87	54·43	55·63	54·67	55·39	54·91	55·15	55·15	78
79	56·83	54·88	56·59	55·13	56·35	55·37	56·10	55·62	55·86	55·86	79
80	57·55	55·57	57·30	55·82	57·06	56·07	56·81	56·32	56·57	56·57	80
81	58·27	56·27	58·02	56·52	57·77	56·77	57·52	57·03	57·28	57·28	81
82	58·99	56·96	58·74	57·22	58·49	57·47	58·24	57·73	57·98	57·98	82
83	59·71	57·56	59·45	57·92	59·20	58·18	58·95	58·43	58·69	58·69	83
84	60·42	58·35	60·17	58·61	59·01	58·88	59·66	59·14	59·40	59·40	84
85	61·14	59·05	60·89	59·31	60·63	59·58	60·37	59·84	60·10	60·10	85
86	61·86	59·74	61·60	60·01	61·34	60·28	61·08	60·55	60·81	60·81	86
87	62·58	60·44	62·32	60·71	62·05	60·98	61·79	61·25	61·52	61·52	87
88	63·30	61·13	63·03	61·41	62·77	61·68	62·50	61·95	62·23	62·23	88
89	64·02	61·82	63·75	62·10	63·48	62·38	63·21	62·66	62·93	62·93	89
90	64·74	62·52	64·47	62·80	64·19	63·08	63·92	63·36	63·64	63·64	90
91	65·46	63·21	65·18	63·50	64·91	63·78	64·63	64·07	64·35	64·35	91
92	66·18	63·91	65·90	64·20	65·62	64·48	65·34	64·77	65·05	65·05	92
93	66·90	64·60	66·62	64·89	66·33	65·18	66·05	65·47	65·76	65·76	93
94	67·62	65·30	67·33	65·59	67·05	65·89	66·76	66·18	66·47	66·47	94
95	68·34	65·99	68·05	66·29	67·76	66·59	67·47	66·88	67·18	67·18	95
96	69·06	66·69	68·76	66·99	68·47	67·29	68·18	67·59	67·88	67·88	96
97	69·78	67·38	69·48	67·69	69·19	67·99	68·89	68·29	68·59	68·59	97
98	70·50	68·08	70·20	68·38	69·90	68·69	69·60	68·99	69·30	69·30	98
99	71·21	68·77	70·91	69·08	70·61	69·39	70·31	69·70	70·00	70·00	99
100	71·93	69·47	71·63	69·78	71·33	70·09	71·02	70·40	70·71	70·71	100
	Dep.	Lat.	Dep.	Lat.	Dep.	Lat.	Dep.	Lat.	Dep.	Lat.	
Distance.	46 Deg.		45¾ Deg.		45½ Deg.		45¼ Deg.		45 Deg.		Distance.

′	0°	1°	2°	3°	4°	5°	6°	7°	′
0	·000 0000	·017 4524	·034 8995	·052 3360	·069 7565	·087 1557	·104 5285	·121 8693	60
1	2909	7432	·035 1902	6264	·070 0467	4455	8178	·122 1581	59
2	5818	·018 0341	4809	9169	3368	7353	·105 1070	4468	58
3	8727	3249	7716	·053 2074	6270	·088 0251	3963	7355	57
4	·001 1636	6158	·036 0623	4979	9171	3148	6856	·123 0241	56
5	4544	9066	3530	7883	·071 2073	6046	9748	3128	55
6	7453	·019 1974	6437	·054 0788	4974	8943	·106 2641	6015	54
7	·002 0362	4883	9344	3693	7876	·089 1840	5533	8901	53
8	3271	7791	·037 2251	6597	·072 0777	4738	8425	·124 1788	52
9	6180	·020 0699	5158	9502	3678	7635	·107 1318	4674	51
10	9089	3608	8065	·055 2406	6580	·090 0532	4210	7560	50
11	·003 1998	6516	·038 0971	5311	9481	3429	7102	·125 0446	49
12	4907	9424	3878	8215	·073 2382	6326	9994	3332	48
13	7815	·021 2332	6785	·056 1119	5283	9223	·108 2885	6218	47
14	·004 0724	5241	9692	4024	8184	·091 2119	5777	9104	46
15	3633	8149	·039 2598	6928	·074 1085	5016	8669	·126 1990	45
16	6542	·022 1057	5505	9832	3986	7913	·109 1560	4875	44
17	9451	3965	8411	·057 2736	6887	·092 0809	4452	7761	43
18	·005 2360	6873	·040 1318	5640	9787	3706	7343	·127 0646	42
19	5268	9781	4224	8544	·075 2688	6602	·110 0234	3531	41
20	8177	·023 2690	7131	·058 1448	5589	9499	3126	6416	40
21	·006 1086	5598	·041 0037	4352	8489	·093 2395	6017	9302	39
22	3995	8506	2944	7256	·076 1390	5291	8908	·128 2186	38
23	6904	·024 1414	5850	·059 0160	4290	8187	·111 1799	5071	37
24	9813	4322	8757	3064	7190	·094 1083	4689	7956	36
25	·007 2721	7230	·042 1663	5967	·077 0091	3979	7580	·129 0841	35
26	5630	·025 0138	4569	8871	2991	6875	·112 0471	3725	34
27	8539	3046	7475	·060 1775	5891	9771	3361	6609	33
28	·008 1448	5954	·043 0382	4678	8791	·095 2666	6252	9494	32
29	4357	8862	3288	7582	·078 1691	5562	9142	·130 2378	31
30	7265	·026 1769	6194	·061 0485	4591	8458	·113 2032	5262	30
31	·009 0174	4677	9100	3389	7491	·096 1353	4922	8146	29
32	3083	7585	·044 2006	6292	·079 0391	4248	7812	·131 1030	28
33	5992	·027 0493	4912	9196	3290	7144	·114 0702	3913	27
34	8900	3401	7818	·062 2099	6190	·097 0039	3592	6797	26
35	·010 1809	6309	·045 0724	5002	9090	2934	6482	9681	25
36	4718	9216	3630	7905	·080 1989	5829	9372	·132 2564	24
37	7627	·028 2124	6536	·063 0808	4889	8724	·115 2261	5447	23
38	·011 0535	5032	9442	3711	7788	·098 1619	5151	8330	22
39	3444	7940	·046 2347	6614	·081 0687	4514	8040	·133 1213	21
40	6353	·029 0847	5253	9517	3587	7408	·116 0929	4096	20
41	9261	3755	8159	·064 2420	6486	·099 0303	3818	6979	19
42	·012 2170	6662	·047 1065	5323	9385	3197	6707	9862	18
43	5079	9570	3970	8226	·082 2284	6092	9596	·134 2744	17
44	7987	·030 2478	6876	·065 1129	5183	8986	·117 2485	5627	16
45	·013 0896	5385	9781	4031	8082	·100 1881	5374	8509	15
46	3805	8293	·048 2687	6934	·083 0981	4775	8263	·135 1392	14
47	6713	·031 1200	5592	9836	3880	7669	·118 1151	4274	13
48	9622	4108	8498	·066 2739	6778	·101 0563	4040	7156	12
49	·014 2530	7015	·049 1403	5641	9677	3457	6928	·136 0038	11
50	5439	9922	4308	8544	·084 2576	6351	9816	2919	10
51	8348	·032 2830	7214	·067 1446	5474	9245	·119 2704	5801	9
52	·015 1256	5737	·050 0119	4349	8373	·102 2138	5593	8683	8
53	4165	8644	3024	7251	·085 1271	5032	8481	·137 1564	7
54	7073	·033 1552	5929	·068 0153	4169	7925	·120 1368	4445	6
55	9982	4459	8835	3055	7067	·103 0819	4256	7327	5
56	·016 2890	7366	·051 1740	5957	9966	3712	7144	·138 0208	4
57	5799	·034 0274	4645	8859	·086 2864	6605	·121 0031	3089	3
58	8707	3181	7550	·069 1761	5762	9499	2919	5970	2
59	·017 1616	6088	·052 0455	4663	8660	·104 2392	5806	8850	1
60	4524	8995	3360	7565	·087 1557	5285	8693	·139 1731	0
′	89°	88°	87°	86°	85°	84°	83°	82°	′

NAT. COSINE.

′	8°	9°	10°	11°	12°	13°	14°	15°	′
0	·139 1731	·156 4345	·173 6482	·190 8090	·207 9117	·224 9511	·241 9219	·258 8190	60
1	4612	7218	9346	·191 0945	·208 1962	·225 2345	·242 2041	·259 1000	59
2	7492	·157 0091	·174 2211	3801	4807	5179	4863	3810	58
3	·140 0372	2962	5075	6656	7652	8013	7685	6619	57
4	3252	5830	7939	9510	·209 0497	·226 0846	·243 0507	9428	56
5	6132	8708	·175 0803	·192 2365	3341	3680	3329	·260 2237	55
6	9012	·158 1581	3667	5220	6186	6513	6150	5045	54
7	·141 1892	4458	6531	8074	9030	9346	8971	7852	53
8	4772	7325	9395	·193 0928	·210 1874	·227 2179	·244 1792	·261 0662	52
9	7651	·159 0197	·176 2258	3782	4718	5012	4613	3469	51
10	·142 0531	3069	5121	6636	7561	7844	7433	6277	50
11	3410	5940	7984	9490	·211 0405	·228 0677	·245 0254	9085	49
12	6289	8812	·177 0847	·194 2344	3248	3509	3074	·262 1892	48
13	9168	·160 1683	3710	5197	6091	6341	5894	4699	47
14	·143 2047	4555	6573	8050	8934	9172	8713	7506	46
15	4926	7426	9435	·195 0903	·212 1777	·229 2004	·246 1533	·263 0312	45
16	7805	·161 0297	·178 2298	3756	4619	4835	4352	3118	44
17	·144 0684	3167	5160	6609	7462	7666	7171	5925	43
18	3562	6038	8022	9461	·213 0304	·230 0497	9990	8730	42
19	6440	8909	·179 0884	·196 2314	3146	3328	·247 2809	·264 1530	41
20	9319	·162 1779	3746	5166	5988	6159	5027	4342	40
21	·145 2197	4650	6607	8018	8829	8989	8445	7147	39
22	5075	7520	9469	·197 0870	·214 1671	·231 1819	·248 1263	9952	38
23	7953	·163 0390	·180 2330	3722	4512	4649	4081	·265 2757	37
24	·146 0830	3260	5191	6573	7353	7479	6899	5561	36
25	3708	6129	8052	9425	·215 0194	·232 0309	9716	8366	35
26	6585	8999	·181 0913	·198 2276	3035	3138	·249 2533	·266 1170	34
27	9463	·164 1868	3774	5127	5876	5967	5350	3973	33
28	·147 2340	4738	6635	7978	8716	8796	8167	6777	32
29	5217	7607	9495	·199 0829	·216 1556	·233 1625	·250 0984	9581	31
30	8094	·165 0470	·182 2355	3679	4396	4454	3800	·267 2384	30
31	·148 0971	3345	5215	6530	7236	7282	6616	5187	29
32	3848	6214	8075	9380	·217 0076	·234 0110	9432	7989	28
33	6724	9082	·183 0935	·200 2230	2915	2938	·251 2248	·268 0792	27
34	9601	·166 1951	3795	5080	5754	5766	5063	3594	26
35	·149 2477	4819	6654	7930	8593	8594	7879	6396	25
36	5353	7687	9514	·201 0779	·218 1432	·235 1421	·252 0694	9198	24
37	8230	·167 0556	·184 2373	3629	4271	4248	3508	·269 2000	23
38	·150 1106	3423	5232	6478	7110	7075	6323	4801	22
39	3981	6291	8091	9327	9948	9902	9137	7602	21
40	6857	9159	·185 0949	·202 2176	·219 2780	·236 2729	·253 1952	·270 0403	20
41	9733	·168 2026	3808	5024	5624	5555	4766	3204	19
42	151 2608	4894	6666	7873	8462	8381	7579	6004	18
43	5484	7761	9524	·203 0721	·220 1300	·237 1207	·254 0393	8805	17
44	8359	·169 0628	·186 2382	3569	4137	4033	3206	·271 1605	16
45	·152 1234	3495	5240	6418	6974	6859	6019	4404	15
46	4109	6362	8098	9265	9811	9684	8832	7204	14
47	6984	9228	·187 0956	·204 2113	·221 2648	·238 2510	·255 1645	·272 0003	13
48	9858	·170 2095	3813	4961	5485	5335	4458	2802	12
49	·153 2733	4961	6670	7808	8321	8159	7270	5601	11
50	5607	7828	9528	·205 0655	·222 1158	·239 0984	·256 0082	8400	10
51	8482	·171 0694	·188 2385	3502	3994	3808	2894	·273 1198	9
52	·154 1356	3560	5241	6349	6830	6633	5705	3997	8
53	4230	6425	8098	9195	9666	9457	8517	6794	7
54	7104	9291	·189 0954	·206 2042	·223 2501	·240 2280	·257 1328	9592	6
55	9978	·172 2156	3811	4888	5337	5104	4139	·274 2390	5
56	·155 2851	5022	6667	7734	8172	7927	6950	5187	4
57	5725	7887	9523	·207 0580	·224 1007	·241 0751	9760	7984	3
58	8598	·173 0752	·190 2379	3426	3842	3574	·258 2570	·275 0781	2
59	·156 1472	3617	5234	6272	6676	6396	5381	3577	1
60	4345	6482	8090	9117	9511	9219	8190	6374	0
′	81°	80°	79°	78°	77°	76°	75°	74°	′

NAT. COSINE.

′	16°	17°	18°	19°	20°	21°	22°	23°	′
0	.275 6374	·292 3717	·309 0170	·325 5682	·342 0201	·358 3679	·374 6066	·390 7311	60
1	9170	6499	2936	8432	2935	6395	8763	9989	59
2	·276 1965	9280	5702	·326 1182	5668	9110	·375 1459	.391 2666	58
3	4761	·293 2061	8468	3932	8400	·359 1825	4156	5343	57
4	7556	4842	·310 1234	6681	·343 1133	4540	6852	8019	56
5	·277 0352	7623	3999	9430	3865	7254	9547	·392 0695	55
6	3147	·294 0403	6764	·327 2179	6597	9968	·376 2243	3371	54
7	5941	3183	9529	4928	9329	·360 2682	4938	6047	53
8	8736	5963	·311 2294	7676	·344 2060	5395	7632	8722	52
9	·278 1530	8743	5058	·328 0424	4791	8108	·377 0327	·393 1397	51
10	4324	·295 1522	7822	3172	7521	·361 0821	3021	4071	50
11	7118	4302	·312 0586	5919	·345 0252	3534	5714	6745	49
12	9911	7081	3349	8666	2982	6246	8408	9419	48
13	·279 2704	9859	6112	·329 1413	5712	8958	·378 1101	·394 2093	47
14	5497	·296 2638	8875	4160	8441	·362 1669	3794	4766	46
15	8290	5416	·313 1638	6906	·346 1171	4380	6486	7439	45
16	·280 1083	8194	4400	9653	3900	7091	9178	·395 0111	44
17	3875	·297 0971	7163	·330 2398	6628	9802	·379 1870	2783	43
18	6667	3749	9925	5144	9357	·363 2512	4562	5455	42
19	9459	6526	·314 2680	7889	·347 2085	5222	7253	8127	41
20	·281 2251	9303	5448	·331 0634	4812	7932	9944	·396 0798	40
21	5042	·298 2079	8209	3379	7540	·364 0641	·380 2634	3468	39
22	7833	4856	·315 0969	6123	·348 0267	3351	5324	6139	38
23	·282 0624	7632	3730	8867	2994	6059	8014	8809	37
24	3415	·299 0408	6490	·332 1611	5720	8768	·381 0704	·397 1479	36
25	6205	3184	9250	4355	8447	·365 1476	3393	4148	35
26	8995	5959	·316 2010	7098	·349 1173	4184	6082	6818	34
27	·283 1785	8734	4770	9841	3898	6891	8770	9486	33
28	4575	·300 1509	7529	·333 2584	6624	9599	·382 1459	·398 2155	32
29	7364	4284	·317 0288	5326	9349	·366 2306	4147	4823	31
30	·284 0153	7058	3047	8069	·350 2074	5012	6834	7491	30
31	2942	9832	5805	·334 0810	4798	7719	9522	.399 0158	29
32	5731	·301 2606	8563	3552	7523	·367 0425	·383 2209	2825	28
33	8520	5380	·318 1321	6293	·351 0246	3130	4895	5492	27
34	·285 1308	8153	4079	9034	2970	5836	7582	8158	26
35	4096	·302 0926	6836	·335 1775	5693	8541	·384 0268	·400 0825	25
36	6884	3699	9593	4516	8416	·368 1246	2953	3490	24
37	9671	6471	·319 2350	7256	·352 1139	3950	5639	6156	23
38	·286 2458	9244	5106	9996	3862	6654	8324	8821	22
39	5246	·303 2016	7863	·336 2735	6584	9358	·385 1008	·401 1486	21
40	8032	4788	·320 0619	5475	9306	·369 2061	3693	4150	20
41	·287 0819	7559	3374	8214	·353 2027	4765	6377	6814	19
42	3605	·304 0331	6130	·337 0953	4748	7468	9060	9478	18
43	6391	3102	8885	3691	7469	·370 0170	·386 1744	·402 2141	17
44	9177	5872	·321 1640	6429	·354 0190	2872	4427	4804	16
45	·288 1963	8643	4395	9167	2910	5574	7110	7467	15
46	4748	·305 1413	7149	·338 1905	5630	8276	9792	·403 0129	14
47	7533	4183	9903	4642	8350	·371 0977	·387 2474	2791	13
48	·289 0318	6953	·322 2657	7379	·355 1070	3678	5156	5453	12
49	3103	9723	5411	·339 0116	3789	6379	7837	8114	11
50	5887	·306 2492	8164	2852	6508	9079	·388 0518	·404 0775	10
51	8671	5261	·323 0917	5589	9226	·372 1780	3199	3436	9
52	·290 1455	8030	3670	8325	·356 1944	4479	5880	6096	8
53	4239	·307 0798	6422	·340 1060	4662	7179	8560	8756	7
54	7022	3566	9174	3796	7380	9878	·389 1240	·405 1416	6
55	9805	6334	·324 1926	6531	·357 0097	·373 2577	3919	4075	5
56	·291 2588	9102	4678	9265	2814	5275	6598	6734	4
57	5371	·308 1869	7420	·341 2000	5531	7973	9277	9393	3
58	8153	4636	·325 0180	4734	8248	·374 0671	·390 1955	·406 2051	2
59	·292 0935	7403	2931	7468	·358 0964	3369	4633	4709	1
60	3717	·309 0170	5682	·342 0201	3679	6066	7311	7366	0
′	73°	72°	71°	70°	69°	68°	67°	66°	′

NAT. COSINE.

′	24°	25°	26°	27°	28°	29°	30°	31°	′
0	·406 7366	·422 6183	·438 3711	·453 9905	·469 4716	·484 8096	·500 0000	·515 0381	60
1	·407 0024	8819	6326	·454 2497	7284	·485 0640	2519	2874	59
2	2681	·423 1455	8940	5088	9852	3184	5037	5367	58
3	5337	4090	·439 1553	7679	·470 2419	5727	7556	7859	57
4	7993	6725	4166	·455 0269	4986	8270	·501 0073	·516 0351	56
5	·408 0649	9360	6779	2859	7553	·486 0812	2591	2842	55
6	3305	·424 1994	9392	5449	·471 0119	3354	5107	5333	54
7	5960	4628	·440 2004	8038	2685	5895	7624	7824	53
8	8615	7262	4615	·456 0627	5250	8436	·502 0140	·517 0314	52
9	·409 1269	9895	7227	3216	7815	·487 0977	2655	2804	51
10	3923	·425 2528	9838	5804	·472 0380	3517	5170	5293	50
11	6577	5161	·441 2448	8392	2944	6057	7685	7782	49
12	9230	7793	5059	·457 0979	5508	8597	·503 0199	·518 0270	48
13	·410 1883	·426 0425	7668	3566	8071	·488 1136	2713	2758	47
14	4536	3056	·442 0278	6153	·473 0634	3674	5227	5246	46
15	7189	5687	2887	8739	3197	6212	7740	7733	45
16	9841	8318	5496	·458 1325	5759	8750	·504 0252	519 0219	44
17	·411 2492	·427 0949	8104	3910	8321	·489 1288	2765	2705	43
18	5144	3579	·443 0712	6496	·474 0882	3325	5276	5191	42
19	7795	6208	3319	9080	3443	6361	7788	7676	41
20	·412 0445	8838	5927	·459 1665	6004	8897	·505 0298	·520 0161	40
21	3096	·428 1467	8534	4248	8564	·490 1433	2809	2646	39
22	5745	4095	·444 1140	6832	·475 1124	3968	5319	5130	38
23	8395	6723	3746	9415	3683	6503	7828	7613	37
24	·413 1044	9351	6352	·460 1998	6242	9038	·506 0338	521 0096	36
25	3693	·429 1979	8957	4580	8801	·491 1572	2846	2579	35
26	6342	4606	·445 1562	7162	·476 1359	4105	5355	5061	34
27	8990	7233	4167	9744	3917	6638	7863	7543	33
28	·414 1638	9859	6771	·461 2325	6474	9171	·507 0370	·522 0024	32
29	4285	·430 2485	9375	4906	9031	·492 1704	2877	2505	31
30	6932	5111	·446 1978	7486	·477 1588	4236	5384	4986	30
31	9579	7736	4581	·462 0066	4144	6767	7890	7466	29
32	·415 2226	·431 0361	7184	2646	6700	9298	·508 0396	9945	28
33	4872	2986	9786	5225	9255	·493 1829	2901	·523 2424	27
34	7517	5610	·447 2388	7804	·478 1810	4359	5406	4903	26
35	·416 0163	8234	4990	·463 0382	4364	6889	7910	7381	25
36	2808	·432 0857	7591	2960	6919	9419	·509 0414	9859	24
37	5453	3481	·448 0192	5538	9472	·494 1948	2918	·524 2336	23
38	8097	6103	2792	8115	·479 2026	4476	5421	4813	22
39	·417 0741	8726	5392	·464 0692	4579	7005	7924	7290	21
40	3385	·433 1348	7992	3269	7131	9532	·510 0426	9766	20
41	6028	3970	·449 0591	5845	9683	·495 2060	2928	·525 2241	19
42	8671	6591	3190	8420	·480 2235	4587	5429	4717	18
43	·418 1313	9212	5789	·465 0996	4786	7113	7930	7191	17
44	3956	·434 1832	8387	3571	7337	9639	·511 0431	9665	16
45	6597	4453	·450 0984	6145	9888	·496 2165	2931	·526 2139	15
46	9239	7072	3582	8719	·481 2438	4690	5431	4613	14
47	·419 1880	9692	6179	·466 1293	4987	7215	7930	7085	13
48	4521	·435 2311	8775	3866	7537	9740	·512 0429	9558	12
49	7161	4930	·451 1372	6439	·482 0086	·497 2264	2927	·527 2030	11
50	9801	7548	3967	9012	2634	4787	5425	4502	10
51	·420 2441	·436 0166	6563	·467 1584	5182	7310	7923	6973	9
52	5080	2784	9158	4156	7730	9833	·513 0420	9443	8
53	7719	5401	·452 1753	6727	·483 0277	·498 2355	2916	·528 1914	7
54	·421 0358	8018	4347	9298	2824	4877	5413	4383	6
55	2996	·437 0634	6941	·468 1869	5370	7399	7908	6852	5
56	5634	3251	9535	4439	7916	9920	·514 0404	9322	4
57	8272	5866	·453 2128	7009	·484 0462	·499 2441	2899	·529 1790	3
58	·422 0909	8482	4721	9578	3007	4961	5393	4258	2
59	3546	·438 1097	7313	·469 2147	5552	7481	7887	6726	1
60	6183	3711	9905	4716	8096	·500 0000	·515 0381	9193	0
′	65°	64°	63°	62°	61°	60°	59°	58°	′

′	32°	33°	34°	35°	36°	37°	38°	39°	′
0	·529 9193	·544 6390	·559 1929	·573 5764	·587 7853	·601 8150	·615 6615	·629 3204	60
1	·530 1659	8830	4340	8147	·588 0206	·602.0473	8907	5464	59
2	4125	·545 1269	6751	·574 0529	2558	2795	·616 1198	7724	58
3	6591	3707	9162	2911	4910	5117	3489	9983	57
4	9057	6145	·560 1572	5292	7262	7439	5780	·630 2242	56
5	·531 1521	8583	3981	7672	9613	9760	8069	4500	55
6	3986	·546 1020	6390	·575 0053	·589 1964	·603 2080	·617 0359	6758	54
7	6450	3456	8798	2432	4314	4400	2648	9015	53
8	8913	5892	·561 1206	4811	6663	6719	4936	·631 1272	52
9	·532 1376	8328	3614	7190	9012	9038	7224	3528	51
10	3839	·547 0763	6021	9568	·590 1361	.604 1356	9511	5784	50
11	6301	3198	8428	·576.1946	3709	3674	·618 1798	8039	49
12	8763	5632	·562 0834	4323	6057	5991	4084	·632 0293	48
13	·533 1224	8066	3239	6700	8404	8308	6370	2547	47
14	3685	·548 0499	5645	9076	·591 0750	·605 0624	8655	4800	46
15	6145	2932	8049	·577 1452	3096	2940	·619 0939	7053	45
16	8605	5365	·563 0453	3827	5442	5255	3224	9306	44
17	·534 1065	7797	2857	6202	7787	7570	5507	·633 1557	43
18	3523	·549 0228	5260	8576	·592 0132	9884	7790	3809	42
19	5982	2659	7663	·578 0950	2476	·606 2198	·620 0073	6059	41
20	8440	5090	·564 0066	3323	4819	4511	2355	8310	40
21	·535 0898	7520	2467	5696	7163	6824	4636	·634 0559	39
22	3355	9950	4869	8069	9505	9136	6917	2808	38
23	5812	·550 2379	7270	·579 0440	·593 1847	·607 1447	9198	5057	37
24	8268	4807	9670	2812	4189	3758	·621 1478	7305	36
25	·536 0724	7236	·565 2070	5183	6530	6069	3757	9553	35
26	3179	9663	4469	7553	8871	8379	6036	·635 1800	34
27	5634	·551 2091	6868	9923	·594 1211	·608 0689	8314	4046	33
28	8089	4518	9267	·580 2292	3550	2908	·622 0592	6292	32
29	·537 0543	6944	·566.1665	4661	5889	5306	2870	8537	31
30	2996	9370	4062	7030	8228	7614	5146	·636 0782	30
31	5449	·552 1795	6459	9397	·595 0566	9922	7423	3026	29
32	7902	4220	8856	·581 1765	2904	·609 2229	9698	5270	28
33	·538 0354	6645	·567 1252	4132	5241	4535	·623 1974	7513	27
34	2806	9069	3648	6498	7577	6841	4248	9756	26
35	5257	·553 1492	·6043	8864	9913	9147	6522	·637 1998	25
36	7708	3915	8437	·582 1230	·596 2249	·610 1452	8796	4240	24
37	·539 0158	6338	·568 0832	3595	4584	3756	·624 1069	6481	23
38	2608	8760	3225	5959	6918	6060	3342	8721	22
39	5058	·554 1182	5619	8323	9252	8363	5614	·638 0961	21
40	7507	3603	8011	·583 0687	·597 1586	·611 0666	7885	3201	20
41	9955	6024	·569 0403	3050	3919	2969	·625 0156	5440	19
42	·540 240[illegible]	8444	2795	5412	6251	5270	2427	7678	18
43	4851	·555 0364	5187	7774	8583	7572	4696	9916	17
44	7298	3283	7577	·584 0136	·598 0915	9873	6966	·639 2153	16
45	9745	5702	9968	2497	3246	·612 2173	9235	4390	15
46	·541 2191	8121	·570 2357	4857	5577	4473	·626 1503	6620	14
47	4637	·556 0539	4747	7217	7906	6772	3771	8862	13
48	7082	2956	7136	9577	·599 0236	9071	6038	·640 1097	12
49	9527	5373	9524	·585 1936	2565	·613 1369	8305	3332	11
50	·542 1971	7790	·571 1912	4294	4893	3666	·627 0571	5566	10
51	4415	·557 0206	4299	6652	7221	5964	2837	7799	9
52	6859	2621	6686	9010	9549	8260	5102	·641 0032	8
53	9302	5036	9073	·586 1367	·600 1876	·614 0556	7366	2264	7
54	·543 1744	7451	·572 1459	3724	4202	2852	9631	4496	6
55	4187	9865	3844	6080	6528	5147	·628 1894	6728	5
56	6628	·558 2279	6229	8435	8854	7442	4157	8958	4
57	9069	4692	8614	·587 0790	·601 1179	9736	6420	·642 1189	3
58	·544 1510	7105	·573 0998	3145	3503	·615 2029	8682	3418	2
59	3951	9517	3381	5499	5827	4322	0943	5647	1
60	6390	·559 1929	5764	7853	8150	6615	·629 3204	7876	0
′	57°	56°	55°	54°	53°	52°	51°	50°	′

′	40°	41°	42°	43°	44°	45°	46°	47°	′
0	·642 7876	·656 0590	·669 1306	·681 9984	·694 6584	·707 1068	·719 3398	·731 3537	60
1	·643 0104	2785	3468	·682 2111	8676	3124	5418	5521	59
2	2332	4980	5628	4237	·695 0767	5180	7438	7503	58
3	4559	7174	7789	6363	2858	7236	9457	9486	57
4	6785	9367	9948	8489	4949	9291	·720 1476	·782 1467	56
5	9011	·657 1560	·670 2108	·683 0613	7039	·708 1345	3494	3449	55
6	·644 1236	3752	4266	2738	9128	3398	5511	5429	54
7	3461	5944	6424	4861	·696 1217	5451	7528	7409	53
8	5685	8135	8582	6984	3305	7504	9544	9388	52
9	7909	·658 0326	·671 0739	9107	5392	9556	·721 1559	.733 1367	51
10	·645 0132	2516	2895	·684 1229	7479	·709 1607	3574	3345	50
11	2355	4706	5051	3350	9565	3657	5589	5322	49
12	4577	6895	7206	5471	·697 1651	5707	7602	7299	48
13	6798	9083	9361	7591	3736	7757	9615	9275	47
14	9019	·659 1271	·672 1515	9711	5821	9806	·722 1628	.734 1250	46
15	·646 1240	3458	3668	·685 1830	7905	·710 1854	3640	3225	45
16	3460	5645	5821	3948	9988	3901	5651	5199	44
17	5679	7831	7973	6066	·698 2071	5948	7661	7173	43
18	7898	·660 0017	·673 0125	8184	4153	7995	9671	9146	42
19	·647 0116	2202	2276	·686 0300	6234	·711 0041	·723 1681	·735 1118	41
20	2334	4386	4427	2416	8315	2086	3690	3090	40
21	4551	6570	6577	4532	·699 0396	4130	5698	5061	39
22	6767	8754	8727	6647	2476	6174	7705	7032	38
23	8984	·661 0936	·674 0876	8761	4555	8218	9712	9002	37
24	·648 1199	3119	3024	·687 0875	6633	·712 0260	·724 1719	·736 0971	36
25	3414	5300	5172	2988	8711	2303	3724	2940	35
26	5628	7482	7319	5101	·700 0789	4344	5729	4908	34
27	7842	9662	9466	7213	2866	6385	7734	6875	33
28	·649 0056	·662 1842	·675 1612	9325	4942	8426	9738	8842	32
29	2268	4022	3757	·688 1435	7018	·713 0465	·725 1741	·737 0808	31
30	4480	6200	5902	3546	9093	2504	3744	2773	30
31	6692	8379	8046	5655	·701 1167	4543	5746	4738	29
32	8903	·663 0557	·676 0190	7765	3241	6581	7747	6703	28
33	·650 1114	2734	2333	9873	5314	8618	9748	8666	27
34	3324	4910	4476	·689 1981	7387	·714 0655	·726 1748	·738 0629	26
35	5533	7087	6618	4089	9459	2691	3748	2592	25
36	7742	9262	8760	6195	·702 1531	4727	5747	4553	24
37	9951	·664 1437	·677 0901	8302	3601	6762	7745	6515	23
38	·651 2158	3612	3041	·690 0407	5672	8796	9743	8475	22
39	4366	5785	5181	2512	7741	·715 0830	·727 1740	·739 0435	21
40	6572	7959	7320	4617	9811	2863	3736	2394	20
41	8778	·665 0131	9459	6721	·703 1879	4895	5732	4353	19
42	·652 0984	2304	·678 1597	8824	3947	6927	7728	6311	18
43	3189	4475	3734	·691 0927	6014	8959	9722	8268	17
44	5394	6646	5871	3029	8081	·716 0989	·728 1716	·740 0225	16
45	7598	8817	8007	5131	·704 0147	3019	3710	2181	15
46	9801	·666 0987	·679 0143	7232	2213	5049	5703	4137	14
47	·653 2004	3156	2278	9332	4278	7078	7695	6092	13
48	4206	5325	4413	·692 1432	6342	9106	9686	8046	12
49	6408	7493	6547	3531	8406	·717·1134	·729 1677	·741 0000	11
50	8609	9661	8681	5630	·705 0469	3161	3668	1953	10
51	·654 0810	·667 1828	·680 0813	7728	2532	5187	5657	3905	9
52	3010	3994	2946	9825	4594	7213	7646	5857	8
53	5209	6160	5078	·693 1922	6655	9238	9635	7808	7
54	7408	8326	7209	4018	8716	·718 1263	·730 1623	9758	6
55	9607	·668 0490	9339	6114	·706 0776	3287	3610	·742 1708	5
56	·655 1804	2655	·681 1469	8209	2835	5310	5597	3658	4
57	4002	4818	3599	·694 0304	4894	7333	7583	5606	3
58	6198	6981	5728	2398	6953	9355	9568	7554	2
59	8395	9144	7856	4491	9011	·719 1377	·731 1553	9502	1
60	·656 0590	·669 1306	9984	6584	·707 1068	3398	3537	·743 1448	0
′	49°	48°	47°	46°	45°	44°	43°	42°	′

′	48°	49°	50°	51°	52°	53°	54°	′
0	·743 1448	·754 7096	·766 0444	·777 1460	·788 0108	·798 6355	·809 0170	60
1	3394	9004	2314	3290	1898	8105	1879	59
2	5340	·755 0911	4183	5120	3688	9855	3588	58
3	7285	2818	6051	6949	5477	·799 1604	5296	57
4	9229	4724	7918	8777	7266	3352	7004	56
5	·744 1173	6630	9785	·778 0604	9054	5100	8710	55
6	3115	8535	·767 1652	2431	·789 0841	6847	·810 0416	54
7	5058	·756 0439	3517	4258	2627	8593	2122	53
8	6999	2342	5382	6084	4413	·800 0388	3826	52
9	8941	4246	7246	7909	6198	2083	5530	51
10	·745 0881	6148	9110	9733	7983	3827	7234	50
11	2821	8050	·768 0973	·779 1557	9767	5571	8936	49
12	4760	9951	2835	3380	·790 1550	7314	·811 0638	48
13	6699	·757 1851	4697	5202	3333	9056	2339	47
14	8636	3751	6558	7024	5115	·801 0797	4040	46
15	·746 0574	5650	8418	8845	6896	2538	5740	45
16	2510	7548	·769 0278	·780 0665	8676	4278	7439	44
17	4446	9446	2137	2485	·791 0456	6018	9137	43
18	6382	·758 1343	3996	4304	2235	7756	·812 0835	42
19	8317	3240	5853	6123	4014	9495	2532	41
20	·747 0251	5136	7710	7940	5792	·802 1232	4229	40
21	2184	7031	9567	9757	7569	2969	5925	39
22	4117	8926	·770 1423	·781 1574	9345	4705	7620	38
23	6049	·759 0820	3278	3390	·792 1121	6440	9314	37
24	7981	2713	5132	5205	2896	8175	·813 1008	36
25	9912	4606	6986	7019	4671	9909	2701	35
26	·748 1842	6498	8840	8833	6445	·803 1642	4393	34
27	3772	8389	·771 0692	·782 0646	8218	3375	6084	33
28	5701	·760 0280	2544	2459	9990	5107	7775	32
29	7629	2170	4395	4270	·793 1762	6838	9466	31
30	9557	4060	6246	6082	3533	8569	·814 1155	30
31	·749 1484	5949	8096	7892	5304	·804 0299	2844	29
32	3411	7837	9945	9702	7074	2028	4532	28
33	5337	9724	·772 1794	·783 1511	8843	3756	6220	27
34	7262	·761 1611	3642	3320	·794 0611	5484	7906	26
35	9187	3497	5489	5127	2379	7211	9593	25
36	·750 1111	5383	7336	6935	4146	8938	·815 1278	24
37	3034	7268	9182	8741	5913	·805 0664	2963	23
38	4957	9152	·773 1027	·784 0547	7678	2389	4647	22
39	6879	·762 1036	2872	2352	9444	4113	6330	21
40	8800	2919	4716	4157	·795 1208	5837	8013	20
41	·751 0721	4802	6559	5961	2972	7560	9695	19
42	2641	6683	8402	7764	4735	9283	·816 1376	18
43	4561	8564	·774 0244	9566	6497	·806 1005	3056	17
44	6480	·763 0445	2086	·785 1368	8259	2726	4736	16
45	8398	2325	3926	3169	·796 0020	4446	6416	15
46	·752 0316	4204	5767	4970	1780	6166	8094	14
47	2233	6082	7606	6770	3540	7885	9772	13
48	4149	7960	9445	8569	5299	9603	·817 1449	12
49	6065	9838	·775 1283	·786 0367	7058	·807 1321	3125	11
50	7980	·764 1714	3121	2165	8815	3038	4801	10
51	9894	3590	4957	3963	·797 0572	4754	6476	9
52	·753 1808	5465	6794	5759	2329	6470	8151	8
53	3721	7340	8629	7555	4084	8185	9824	7
54	5634	9214	·776 0464	9350	5839	9899	·818 1497	6
55	7546	·765 1087	2298	·787 1145	7594	·808 1612	3169	5
56	9457	2960	4132	2939	9347	3325	4841	4
57	·754 1368	4832	5965	4732	·798 1100	5037	6512	3
58	3278	6704	7797	6524	2853	6749	8182	2
59	5187	8574	9629	8316	4604	8460	9852	1
60	7096	·766 0444	·777 1460	·788 0108	6355	·809 0170	·819 1520	0
′	41°	40°	39°	38°	37°	36°	35°	′

NAT. COSINE.

′	55°	56°	57°	58°	59°	60°	61°	′
0	·819 1520	·829 0376	·838 6706	·848 0481	·857 1673	·866 0254	·874 6197	60
1	3189	2002	8290	2022	3171	1708	7607	59
2	4856	3628	9873	3562	4668	3161	9016	58
3	6523	5252	·839 1455	5102	6164	4614	·875 0425	57
4	8189	6877	3037	6641	7660	6066	1832	56
5	9854	8500	4618	8179	9155	7517	3239	55
6	·820 1519	·830 0123	6199	9717	·858 0649	8967	4645	54
7	3183	1745	7778	·849 1254	2143	·867 0417	6051	53
8	4846	3366	9357	2790	3635	1866	7455	52
9	6509	4987	·840 0936	4325	5127	3314	8859	51
10	8170	6607	2513	5860	6619	4762	·876 0263	50
11	9832	8226	4090	7394	8109	6209	1665	49
12	·821 1492	9845	5666	8927	9599	7655	3067	48
13	3152	·831 1463	7241	·850 0459	·859 1088	9100	4468	47
14	4811	3080	8816	1991	2576	·868 0544	5868	46
15	6469	4696	·841 0390	3522	4064	1988	7268	45
16	8127	6312	1963	5053	5551	3431	8666	44
17	9784	7927	3536	6582	7037	4874	·877 0064	43
18	·822 1440	9541	5108	8111	8523	6315	1462	42
19	3096	·832 1155	6679	9639	·860 0007	7756	2858	41
20	4751	2768	8249	·851 1167	1491	9196	4254	40
21	6405	4380	9819	2693	2975	·869 0636	5649	39
22	8059	5991	·842 1388	4219	4457	2074	7043	38
23	9712	7602	2956	5745	5939	3512	8437	37
24	·823 1364	9212	4524	7269	7420	4949	9830	36
25	3015	·833 0822	6091	8793	8901	6386	·878 1222	35
26	4666	2430	7657	·852 0316	·861 0380	7821	2613	34
27	6316	4038	9222	1839	1859	9256	4004	33
28	7965	5646	·843 0787	3360	3337	·870 0691	5394	32
29	9614	7252	2351	4881	4815	2124	6783	31
30	·824 1262	8858	3914	6402	6292	3557	8171	30
31	2909	·834 0463	5477	7921	7768	4989	9559	29
32	4556	2068	7039	9440	9243	6420	·879 0946	28
33	6202	3672	8600	·853 0958	·862 0717	7851	2332	27
34	7847	5275	·844 0161	2475	2191	9281	3717	26
35	9491	6877	1720	3992	3664	·871 0710	5102	25
36	·825 1135	8479	3279	5508	5137	2138	6486	24
37	2778	·835 0080	4838	7023	6608	3566	7869	23
38	4420	1680	6395	8538	8079	4993	9251	22
39	6062	3279	7952	·854 0051	9549	6419	·880 0633	21
40	7703	4878	9508	1564	·863 1019	7844	2014	20
41	9343	6476	·845 1064	3077	2488	9269	3394	19
42	·826 0983	8074	2618	4588	3956	·872 0693	4774	18
43	2622	9670	4172	6099	5423	2116	6152	17
44	4260	·836 1266	5726	7609	6889	3538	7530	16
45	5897	2862	7278	9119	8355	4960	8907	15
46	7534	4456	8830	·855 0627	9820	6381	·881 0284	14
47	9170	6050	·846 0381	2135	·864 1284	7801	1660	13
48	·827 0806	7643	1932	3643	2748	9221	3035	12
49	2440	9236	3481	5149	4211	·873 0640	4409	11
50	4074	·837 0827	5030	6655	5673	2058	5782	10
51	5708	2418	6579	8160	7134	3475	7155	9
52	7340	4009	8126	9664	8595	4891	8527	8
53	8972	5598	9673	·856 1168	·865 0055	6307	9898	7
54	·828 0603	7187	·847 1219	2671	1514	7722	·882 1269	6
55	2234	8775	2765	4173	2973	9137	2638	5
56	3864	·838 0363	4309	5674	4430	·874 0550	4007	4
57	5493	1950	5853	7175	5887	1963	5376	3
58	7121	3536	7397	8675	7344	3375	6743	2
59	8749	5121	8939	·857 0174	8799	4786	8110	1
60	·829 0376	6706	·848 0481	1673	·866 0254	6197	9476	0
′	34°	33°	32°	31°	30°	29°	28°	′

NAT. COSINE.

′	62°	63°	64°	65°	66°	67°	68°	′
0	·882 9476	.891 0065	·898 7940	·906 3078	·913 5455	·920 5049	·927 1839	60
1	·883 0841	1385	9215	4307	6637	6185	2928	59
2	2206	2705	·899 0489	5535	7819	7320	4016	58
3	3569	4024	1763	6762	9001	8455	5104	57
4	4933	5342	3035	7989	·914 0181	9589	6191	56
5	6295	6659	4307	9215	1361	.921 0722	7277	55
6	7656	7975	5578	·907 0440	2540	1854	8363	54
7	9017	9291	6848	1665	3718	2986	9447	53
8	·884 0377	·892 0606	8117	2888	4895	4116	·928 0531	52
9	1736	1920	9386	4111	6072	5246	1614	51
10	3095	3234	·900 0654	5333	7247	6375	2696	50
11	4453	4546	1921	6554	8422	7504	3778	49
12	5810	5858	3188	7775	9597	8632	4858	48
13	7166	7169	4453	8995	·915 0770	9758	5938	47
14	8522	8480	5718	·908 0214	1943	·922 0884	7017	46
15	9876	9789	6982	1432	3115	2010	8096	45
16	·885 1230	·893 1098	8246	2649	4286	3134	9173	44
17	2584	2406	9508	3866	5456	4258	·929 0250	43
18	3936	3714	·901 0770	5082	6626	5381	1326	42
19	5288	5021	2031	6297	7795	6503	2401	41
20	6639	6326	3292	7511	8963	7624	3475	40
21	7989	7632	4551	8725	·916 0130	8745	4549	39
22	9339	8936	5810	9938	1297	9865	5622	38
23	·886 0688	·894 0240	7068	·909 1150	2462	·923 0984	6694	37
24	2036	1542	8325	2361	3627	2102	7765	36
25	3383	2844	9582	3572	4791	3220	8835	35
26	4730	4146	·902 0838	4781	5955	4336	9905	34
27	6075	5446	2092	5990	7118	5452	·930 0974	33
28	7420	6746	3347	7199	8279	6567	2042	32
29	8765	8045	4600	8406	9440	7682	3109	31
30	·887 0108	9344	5853	9613	·917 0601	8795	4176	30
31	1451	·895 0641	7105	·910 0819	1760	9908	5241	29
32	2793	1938	8356	2024	2919	·924 1020	6306	28
33	4134	3234	9606	3228	4077	2131	7370	27
34	5475	4529	·903 0856	4432	5234	3242	8434	26
35	6815	5824	2105	5635	6391	4351	9496	25
36	8154	7118	3353	6837	7546	5460	·931 0553	24
37	9492	8411	4600	8038	8701	6568	1619	23
38	·888 0830	9703	5847	9238	9855	7676	2679	22
39	2166	.896 0994	7093	·911 0438	·918 1009	8782	3739	21
40	3503	2285	8338	1637	2161	9888	4797	20
41	4838	3575	9582	2835	3313	·925 0993	5855	19
42	6172	4864	·904 0825	4033	4464	2097	6912	18
43	7506	6153	2068	5229	5614	3201	7969	17
44	8839	7440	3310	6425	6763	4303	9024	16
45	·889 0171	8727	4551	7620	7912	5405	·932 0079	15
46	1503	·897 0014	5792	8815	9060	6506	1133	14
47	2834	1299	7032	·912 0008	·919 0207	7606	2186	13
48	4164	2584	8271	1201	1353	8706	3238	12
49	5493	3868	9509	2393	2499	9805	4290	11
50	6822	5151	·905 0746	3584	3644	·926 0902	5340	10
51	8149	6433	1983	4775	4788	2000	6390	9
52	9476	7715	3219	5965	5931	3096	7439	8
53	·890 0803	8996	4454	7154	7073	4192	8488	7
54	2128	·898 0276	5688	8342	8215	5286	9535	6
55	3453	1555	6922	9529	9356	6380	·933 0582	5
56	4777	2834	8154	·913 0716	·920 0496	7474	1628	4
57	6100	4112	9386	1902	1635	8566	2673	3
58	7423	5389	·906 0618	3087	2774	9658	3718	2
59	8744	6665	1848	4271	3912	·927 0748	4761	1
60	·891 0065	7940	3078	5455	5049	1839	5804	0
′	27°	26°	25°	24°	23°	22°	21°	′

′	69°	70°	71°	72°	73°	74°	75°	′
0	·933 5804	·939 6926	·945 5186	·951 0565	·956 3048	·961 2617	·965 9258	60
1	6846	7921	6132	1464	3898	3418	·966 0011	59
2	7388	8914	7078	2361	4747	4219	0762	58
3	8928	9907	8023	3258	5595	5019	1513	57
4	9968	·940 0899	8968	4154	6443	5818	2263	56
5	·934 1007	1891	9911	5050	7290	6616	3012	55
6	2045	2881	·946 0854	5944	8136	7413	3761	54
7	3082	3871	1795	6838	8981	8210	4508	53
8	4119	4860	2736	7731	9825	9005	5255	52
9	5154	5848	3677	8623	·957 0669	9800	6001	51
10	6189	6835	4616	9514	1512	·962 0594	6746	50
11	7223	7822	5555	·952 0404	2354	1387	7490	49
12	8257	8808	6493	1294	3195	2180	8234	48
13	9289	9793	7430	2183	4035	2972	8977	47
14	·935 0321	·941 0777	8366	3071	4875	3762	9718	46
15	1352	1760	9301	3958	5714	4552	·967 0459	45
16	2382	2743	·947 0236	4844	6552	5342	1200	44
17	3412	3721	1170	5730	7389	6130	1939	43
18	4440	4705	2103	6615	8225	6917	2678	42
19	5468	5686	3035	7499	9060	7704	3415	41
20	6495	6665	3966	8382	9895	8490	4152	40
21	7521	7644	4897	9264	·958 0729	9275	4888	39
22	8547	8621	5827	·953 0146	1562	·963 0060	5624	38
23	9571	9598	6756	1027	2394	0843	6358	37
24	·936 0595	·942 0575	7684	1907	3226	1626	7092	36
25	1618	1550	8612	2786	4056	2408	7825	35
26	2641	2525	9538	3664	4886	3189	8557	34
27	3662	3498	·948 0464	4542	5715	3969	9288	33
28	4683	4471	1389	5418	6543	4748	·968 0018	32
29	5703	5444	2313	6294	7371	5527	0748	31
30	6722	6415	3237	7170	8197	6305	1476	30
31	7740	7386	4159	8044	9023	7081	2204	29
32	8758	8355	5081	8917	9848	7858	2931	28
33	9774	9324	6002	9790	·959 0672	8633	3658	27
34	·937 0790	·943 0293	6922	·954 0662	1496	9407	4383	26
35	1806	1260	7842	1533	2318	·964 0181	5108	25
36	2820	2227	8760	2403	3140	0954	5832	24
37	3833	3192	9678	3273	3961	1726	6555	23
38	4846	4157	·949 0595	4141	4781	2497	7277	22
39	5858	5122	1511	5009	5600	3268	7998	21
40	6869	6085	2426	5876	6418	4037	8719	20
41	7880	7048	3341	6743	7236	4806	9438	19
42	8889	8010	4255	7608	8053	5574	·969 0157	18
43	9898	8971	5168	8473	8869	6341	0875	17
44	·938 0906	9931	6080	9336	9684	7108	1593	16
45	1913	·944 0890	6991	·955 0199	0499	7873	2309	15
46	2920	1849	7902	1062	·960 1312	8638	3025	14
47	3925	2807	8812	1923	2125	9402	3740	13
48	4930	3764	9721	2784	2937	·965 0165	4453	12
49	5934	4720	·950 0629	3643	3748	0927	5167	11
50	6938	5675	1536	4502	4558	1689	5879	10
51	7940	6630	2443	5361	5368	2449	6591	9
52	8942	7584	3348	6218	6177	3209	7301	8
53	9943	8537	4253	7074	6984	3968	8011	7
54	·939 0943	9489	5157	7930	7792	4726	8720	6
55	1942	·945 0441	6061	8785	8598	5484	9428	5
56	2940	1391	6963	9639	9403	6240	·970 0135	4
57	3938	2341	7865	·956 0492	·961 0208	6996	0842	3
58	4935	3290	8766	1345	1012	7751	1548	2
59	5931	4238	9666	2197	1815	8505	2253	1
60	6926	5186	·951 0565	3048	2617	9258	2957	0
′	20°	19°	18°	17°	16°	15°	14°	′

NAT. COSINE.

′	76°	77°	78°	79°	80°	81°	82°	′
0	·970 2957	·974 3701	·978 1476	·981 6272	·9848 078	·9876 883	·9902 681	60
1	3660	4355	2080	6826	582	·9877 338	·9903 085	59
2	4363	5008	2684	7380	·9849 086	792	489	58
3	5065	5660	3287	7933	589	·9878 245	891	57
4	5766	6311	3889	8485	·9850 091	697	·9904 293	56
5	6466	6962	4490	9037	593	·9879 148	694	55
6	7165	7612	5090	9587	·9851 093	599	·9905 095	54
7	7863	8261	5689	·982 0137	593	·9880 048	494	53
8	8561	8909	6288	0686	·9852 092	497	893	52
9	9258	9556	6886	1234	590	945	·9906 290	51
10	9953	·975 0203	7483	1781	·9853 087	·9881 392	687	50
11	·971 0649	0849	8079	2327	583	838	·9907 083	49
12	1343	1494	8674	2873	·9854 079	·9882 284	478	48
13	2036	2138	9268	3417	574	728	873	47
14	2729	2781	9862	3961	·9855 068	·9883 172	·9908 266	46
15	3421	3423	·979 0455	4504	561	615	659	45
16	4112	4065	1047	5046	·9856 053	·9884 057	·9909 051	44
17	4802	4706	1638	5587	544	498	442	43
18	5491	5345	2228	6128	·9857 035	939	832	42
19	6180	5985	2818	6668	524	·9885 378	·9910 221	41
20	6867	6623	3406	7206	·9858 013	817	610	40
21	7554	7260	3994	7744	501	·9886 255	997	39
22	8240	7897	4581	8282	988	692	·9911 384	38
23	8926	8533	5167	8818	·9859 475	·9887 128	770	37
24	9610	9168	5752	9353	960	564	·9912 155	36
25	·972 0294	9802	6337	9888	·9860 445	998	540	35
26	0976	·976 0435	6921	·983 0422	929	·9888 432	923	34
27	1658	1067	7504	0955	·9861 412	865	·9913 306	33
28	2339	1699	8086	1487	894	·9889 297	688	32
29	3020	2330	8667	2019	·9862 375	728	·9914 069	31
30	3699	2960	9247	2549	856	·9890 159	449	30
31	4378	3589	9827	3079	·9863 336	588	828	29
32	5056	4217	·980 0405	3608	815	·9891 017	·9915 206	28
33	5733	4845	0983	4136	·9864 293	445	584	27
34	6409	5472	1560	4663	770	872	961	26
35	7084	6098	2136	5189	·9865 246	·9892 298	·9916 337	25
36	7759	6723	2712	5715	722	723	712	24
37	8432	7347	3286	6239	·9866 196	·9893 148	·9917 086	23
38	9105	7970	3860	6763	670	572	459	22
39	9777	8593	4433	7286	·9867 143	994	832	21
40	·973 0449	9215	5005	7808	615	·9894 416	·9918 204	20
41	1119	9836	5576	8330	·9868 087	838	574	19
42	1789	·977 0456	6147	8850	557	·9895 258	944	18
43	2458	1075	6716	9370	·9869 027	677	·9919 314	17
44	3125	1693	7285	9889	496	·9896 096	682	16
45	3793	2311	7853	·984 0407	964	514	·9920 049	15
46	4458	2928	8420	0924	·9870 431	931	416	14
47	5124	3544	8986	1441	897	·9897 347	782	13
48	5789	4159	9552	1956	·9871 363	762	·9921 147	12
49	6453	4773	·981 0116	2471	827	·9898 177	511	11
50	7116	5386	0680	2985	·9872 291	590	874	10
51	7778	5999	1243	3498	754	·9899 003	·9922 237	9
52	8439	6611	1805	4010	·9873 216	415	599	8
53	9100	7222	2366	4521	678	826	959	7
54	9760	7832	2927	5032	·9874 138	·9900 237	·9923 319	6
55	·974 0419	8441	3486	5542	598	646	679	5
56	1077	9050	4045	6050	·9875 057	·9901 055	·9924 037	4
57	1734	9658	4603	6558	514	462	394	3
58	2390	·978 0265	5160	7066	972	869	751	2
59	3046	0871	5716	7572	·9876 428	·9902 275	·9925 107	1
60	3701	1476	6272	8078	883	681	462	0
′	13°	12°	11°	10°	9°	8°	7°	′

NAT. COSINE.

′	83°	84°	85°	86°	87°	88°	89°	′
0	·9925 462	·9945 219	·9961 947	·9975 641	·9986 295	·9993 908	·9998 477	60
1	816	523	·9962 200	843	447	·9994 009	527	59
2	·9926 169	825	452	·9976 045	598	110	577	58
3	521	·9946 127	704	245	748	209	625	57
4	873	428	954	445	898	308	673	56
5	·9927 224	729	·9963 204	645	·9987 046	405	720	55
6	573	·9947 028	453	843	194	502	766	54
7	922	327	701	·9977 040	340	598	812	53
8	·9928 271	625	948	237	486	693	856	52
9	618	921	·9964 195	433	631	788	900	51
10	965	·9948 217	440	627	775	881	942	50
11	·9929 310	513	685	821	919	974	984	49
12	655	807	929	·9978 015	·9988 061	·9995 066	·9999 025	48
13	999	·9949 101	·9965 172	207	203	157	065	47
14	·9930 342	393	414	399	344	247	105	46
15	685	685	655	589	484	336	143	45
16	·9931 026	976	895	779	623	424	181	44
17	367	·9950 266	·9966 135	968	761	512	218	43
18	706	556	374	·9979 156	899	599	254	42
19	·9932 045	844	612	343	·9989 035	684	289	41
20	384	·9951 132	849	530	171	770	323	40
21	721	419	·9967 085	716	306	854	357	39
22	·9933 057	705	321	900	440	937	389	38
23	393	990	555	·9980 084	573	·9996 020	421	37
24	728	·9952 274	789	267	706	101	452	36
25	·9934 062	557	·9968 022	450	837	182	482	35
26	395	840	254	631	968	262	511	34
27	727	·9953 122	485	811	·9990 098	341	539	33
28	·9935 058	403	715	991	227	419	567	32
29	389	683	945	·9981 170	355	497	593	31
30	719	962	·9969 173	348	482	573	619	30
31	·9936 047	·9954 240	401	525	609	649	644	29
32	375	517	628	701	734	724	668	28
33	703	794	854	877	859	798	692	27
34	·9937 029	·9955 070	·9970 080	·9982 052	983	871	714	26
35	355	345	304	225	·9991 106	943	736	25
36	679	620	528	398	228	·9997 015	756	24
37	·9938 003	893	750	570	350	086	776	23
38	326	·9956 165	972	742	470	156	795	22
39	648	437	·9971 193	912	590	224	813	21
40	969	708	413	·9983 082	709	292	831	20
41	·9939 290	978	633	250	827	360	847	19
42	610	·9957 247	851	418	944	426	863	18
43	928	515	·9972 069	585	·9992 060	492	878	17
44	·9940 246	783	286	751	176	556	892	16
45	563	·9958 049	502	917	290	620	905	15
46	880	315	717	·9984 081	404	683	917	14
47	·9941 195	580	931	245	517	745	928	13
48	510	844	·9973 145	408	629	807	939	12
49	823	·9959 107	357	570	740	867	949	11
50	·9942 136	370	569	731	851	927	958	10
51	448	631	780	891	960	986	966	9
52	760	892	990	·9985 050	·9993 069	·9998 044	973	8
53	·9943 070	·9960 152	·9974 199	209	177	101	979	7
54	379	411	408	367	284	157	985	6
55	688	669	615	524	390	213	989	5
56	996	926	822	680	495	267	993	4
57	·9944 303	·9961 183	·9975 028	835	600	321	996	3
58	609	438	233	989	704	374	998	2
59	914	693	437	·9986 143	806	426	1·0000 000	1
60	·9945 219	947	641	295	908	477	000	0
′	6°	5°	4°	3°	2°	1°	0°	′

′	0°	1°	2°	3°	4°	5°	6°	7°	′
0	·000 0000	·017 4551	·034 9208	·052 4078	·069 9268	·087 4887	·105 1042	·122 7846	60
1	2909	7460	·035 2120	6995	·070 2191	7818	3983	·123 0798	59
2	5818	·018 0370	5033	9912	5115	·088 0749	6925	3752	58
3	8727	3280	7945	·053 2829	8038	3681	9866	6705	57
4	·001 1636	6190	·036 0858	5746	·071 0961	6612	·106 2808	9658	56
5	4544	9100	3771	8663	3885	9544	5750	·124 2612	55
6	7453	·019 2010	6683	·054 1581	6809	·089 2476	8692	5566	54
7	·002 0362	4920	9596	4498	9733	5408	·107 1634	8520	53
8	3271	7830	·037 2500	7416	·072 2657	8341	4576	125 1474	52
9	6180	·020 0740	5422	·055 0333	5581	·090 1273	7519	4429	51
10	9089	3650	8335	3251	8505	4200	·108 0462	7384	50
11	·003 1998	6560	·038 1248	6169	·073 1430	7138	3405	·126 0339	49
12	4907	9470	4161	9087	4354	·091 0071	6348	3294	48
13	7816	·021 2380	7074	·056 2005	7279	3004	9291	6249	47
14	·004 0725	5291	9988	4923	·074 0203	5938	·109 2234	9205	46
15	3634	8201	·039 2901	7841	3128	8871	5178	·127 2161	45
16	6542	·022 1111	5814	·057 0759	6053	·092 1804	8122	5117	44
17	9451	4021	8728	3678	8979	4738	·110 1060	8073	43
18	·005 2360	6932	·040 1641	6596	·075 1904	7672	4010	·128 1030	42
19	5269	9842	4555	9515	4829	·093 0606	6955	3986	41
20	8178	023. 2753	7469	·058 2434	7755	3540	9899	6943	40
21	·006 1087	5663	·041 0383	5352	·076 0680	6474	·111 2844	9900	39
22	3996	8574	3296	8271	3606	9409	5789	·129 2858	38
23	6905	·024 1484	6210	·059 1190	6532	·094 2344	8734	5815	37
24	9814	4395	9124	4109	9458	5278	·112 1680	8773	36
25	·007 2723	7305	·042 2038	7029	·077 2384	8213	4625	·130 1731	35
26	5632	·025 0216	4952	9948	5311	·095 1148	7571	4690	34
27	8541	3127	7866	·060 2867	8237	4084	·113 0517	7648	33
28	·008 1450	6038	·043 0781	5787	·078 1164	7019	3463	·131 0607	32
29	4360	8948	3695	8706	4090	9955	6410	3566	31
30	7269	·026 1859	6609	·061 1626	7017	·096 2890	9356	6525	30
31	·009 0178	4770	9524	4546	9944	5826	·114 2303	9484	29
32	3087	7681	·044 2438	7466	·079 2871	8763	5250	·132 2444	28
33	5996	·027 0592	5353	·062 0386	5798	·097 1699	8197	5404	27
34	8905	3503	8268	3306	8726	4635	·115 1144	8364	26
35	·010 1814	6414	·045 1183	6226	·080 1653	7572	4092	·133 1324	25
36	4724	9325	4097	9147	4581	·098 0509	7039	4285	24
37	7633	·028 2236	7012	·063 2067	7509	3446	9987	7246	23
38	·011 0542	5148	9927	4988	·081 0437	6383	·116 2936	·134 0207	22
39	3451	8059	·046 2842	7908	3365	9320	5884	3168	21
40	6361	·029 0970	5757	·064 0829	6293	·099 2257	8832	6129	20
41	9270	3882	8673	3750	9221	5194	·117 1781	9091	19
42	·012 2179	6793	·047 1588	6671	·082 2150	8133	4730	·135 2053	18
43	5088	9705	4503	9592	5078	·100 1071	7679	5015	17
44	7998	·030 2616	7419	·065 2513	8007	4009	·118 0628	7978	16
45	·013 0907	5528	·048 0334	5435	·083 0936	6947	3578	·136 0940	15
46	3817	8439	3250	8356	3865	9886	6528	3903	14
47	6726	·031 1351	6166	·066 1278	6794	·101 2824	9478	6866	13
48	9635	4263	9082	4199	9723	5763	·119 2428	9830	12
49	·014 2545	7174	·049 1997	7121	·084 2653	8702	5378	·137 2793	11
50	5454	·032 0086	4913	·067 0043	5583	·102 1641	8329	5757	10
51	8364	2998	7829	2965	8512	4580	·120 1279	8721	9
52	·015 1273	5910	·050 0746	5887	·085 1442	7520	4230	·138 1685	8
53	4183	8822	3662	8809	4372	·103 0460	7182	4650	7
54	7093	·033 1734	6578	·068 1732	7302	3399	·121 0133	7615	6
55	·016 0002	4646	9495	4654	·086 0233	6340	3085	·139 0580	5
56	2912	7558	·051 2411	7577	3163	9280	6036	3545	4
57	5821	·034 0471	5328	·069 0499	6094	·104 2220	8988	6510	3
58	8731	3383	8244	3422	9025	5161	·122 1941	9476	2
59	·017 1641	6295	·052 1161	6345	·087 1956	8101	4893	·140 2442	1
60	4551	9208	4078	9268	4887	·105 1042	7846	5408	0
′	89°	88°	87°	86°	85°	84°	83°	82°	′

NAT. COTAN.

′	8°	9°	10°	11°	12°	13°	14°	15°	′
0	·140 5408	·158 3844	.176 3270	·194 3803	·212 5566	·230 8682	·249 3280	·267 9492	60
1	8375	6826	6269	6822	8606	·231 1746	6370	·268 2610	59
2	·141 1342	9809	9269	9841	·213 1647	4811	9460	5728	58
3	4308	·159 2791	·177 2269	·195 2861	4688	7876	·250 2551	8847	57
4	7276	5774	5270	5881	7730	·232 0941	5642	·269 1967	56
5	·142 0243	8757	8270	8901	·214 0772	4007	8734	.5087	55
6	3211	·160 1740	·178 1271	·196 1922	3814	7073	·251 1826	8207	54
7	6179	4724	4273	4943	6857	·233 0140	4919	·270 1328	53
8	9147	7708	7274	7964	9900	3207	8012	4449	52
9	·143 2115	·161 0692	·179 0276	·197 0986	·215 2944	6274	·252 1106	7571	51
10	5084	3677	3279	4008	5988	9342	4200	·271 0694	50
11	8053	6662	6281	7031	9032	·234 2410	7294	3817	49
12	·144 1022	9647	9284	·198 0053	·216 2077	5479	·253 0389	6940	48
13	3991	·162 2632	·180 2287	3076	5122	8548	3484	·272 0064	47
14	6961	5618	5291	6100	8167	·235 1617	6580	3188	46
15	9931	8603	8295	9124	·217 1213	4687	9676	6313	45
16	·145 2901	·163 1590	·181 1299	·199 2148	4259	7758	·254 2773	9438	44
17	5872	4576	4303	5172	7306	·236 0829	5870	·273 2564	43
18	8842	7563	7308	8197	·218 0353	3900	8968	5690	42
19	·146 1813	·164 0550	·182 0313	·200 1222	3400	6971	·255 2066	8817	41
20	4784	3537	3319	4248	6448	.237 0044	5165	·274 1945	40
21	7756	6525	6324	7274	9496	3116	8264	5072	39
22	·147 0727	9513	9330	·201 0300	·219 2544	6189	·256 1363	8201	38
23	3699	·165 2501	·183 2337	3327	5593	9262	4463	·275 1330	37
24	6672	5489	5343	6354	8643	·238 2336	7564	4459	36
25	9644	8478	8350	9381	·220 1692	5410	·257 0664	7589	35
26	·148 2617	·166 1467	·184 1358	·202 2409	4742	8485	3766	·276 0719	34
27	5590	4456	4365	5437	7793	·239 1560	6868	3850	33
28	8563	7446	7373	8465	·221 0844	4635	9970	6981	32
29	·149 1536	·167 0436	·185 0382	·203 1494	3895	7711	·258 3073	·277 0113	31
30	4510	3426	3390	4523	6947	·240 0788	6176	3245	30
31	7484	6417	6399	7552	9999	3864	9280	6378	29
32	·150 0458	9407	9409	·204 0582	·222 3051	6942	·259 2384	9512	28
33	3433	·168 2398	·186 2418	3612	6104	·241 0019	5488	·278 2646	27
34	6408	5390	5428	6643	9157	3097	8593	5780	26
35	9383	8381	8439	9674	·223 2211	6176	·260 1699	8915	25
36	·151 2358	·169 1373	·187 1449	·205 2705	5265	9255	4805	·279 2050	24
37	5333	4366	4460	5737	8319	·242 2334	7911	5186	23
38	8309	7358	7471	8769	·224 1374	5414	·261 1018	8322	22
39	·152 1285	·170 0351	·188 0483	·206 1801	4429	8494	4126	·280 1459	21
40	4262	3344	3495	4834	7485	·243 1575	7234	4597	20
41	7238	6338	6507	7867	·225 0541	4656	·262 0342	7735	19
42	·153 0215	9331	9520	·207 0900	3597	7737	3451	·281 0873	18
43	3192	·171 2325	·189 2533	3934	6654	·244 0819	6560	4012	17
44	6170	5320	5546	6968	9711	3902	9670	7152	16
45	9147	8314	8559	·208 0003	·226 2769	6934	·263 2780	·282 0292	15
46	·154 2125	·172 1309	·190 1573	3038	5827	·245 0068	5891	3432	14
47	5103	4304	4587	6073	8885	3151	9002	6573	13
48	8082	7300	7602	9109	·227 1944	6236	·264 2114	9715	12
49	·155 1061	·173 0296	·191 0617	·209 2145	5003	9320	5226	·283 2857	11
50	4040	3292	3632	5181	8063	·246 2405	8339	5999	10
51	7019	6288	6648	8218	·228 1123	5491	·265 1452	9143	9
52	9998	9285	9664	·210 1255	4184	8577	4566	·284 2286	8
53	·156 2978	·174 2282	·192 2680	4293	7244	·247 1663	7680	5430	7
54	5958	5279	5696	7331	·229 0306	4750	·266 0794	8575	6
55	8939	8277	8713	·211 0369	3367	7837	3909	·285 1720	5
56	·157 1919	·175 1275	·193 1731	3407	6429	·248 0925	7025	4866	4
57	4900	4273	4748	6446	9492	4013	·267 0141	8012	3
58	7881	7272	7766	9486	·230 2555	7102	3257	·286 1159	2
59	·158 0863	·176 0271	·194 0784	·212 2525	5618	·249 0191	6374	4306	1
60	3844	3270	3803	5566	8682	3230	9492	7454	-0
′	81°	80°	79°	78°	77°	76°	75°	74°	′

NAT. COTAN.

′	16°	17°	18°	19°	20°	21°	22°	23°	′
0	·286 7454	·305 7307	·324 9197	·344 3276	·363 9702	·383 8640	·404 0262	·424 4748	60
1	·287 0602	·306 0488	·325 2413	6530	·364 2997	·384 1978	3646	8182	59
2	3751	3670	5630	9785	6292	5317	7031	·425 1610	58
3	6900	6852	8848	·345 3040	9588	8656	·405 0417	5051	57
4	·288 0050	·307 0034	·326 2060	6296	·365 2885	·385 1996	3804	8487	56
5	3201	3218	5284	9553	6182	5337	7191	·426 1924	55
6	6352	6402	8504	·346 2810	9480	8679	·406 0579	5361	54
7	9503	9586	·327 1724	6068	·366 2779	·386 2021	3968	8800	53
8	·289 2655	·308 2771	4944	9327	6079	5364	7358	·427 2239	52
9	5808	5957	8165	·347 2586	9379	8708	·407 0748	5680	51
10	8961	9143	·328 1387	5846	·367 2680	387 2053	4139	9121	50
11	·290 2114	·309 2330	4610	9107	5981	5398	7531	·428 2563	49
12	5269	5517	7833	·348 2368	9284	8744	·408 0924	6005	48
13	8423	8705	·329 1056	5630	·368 2587	·388 2091	4318	9449	47
14	·291 1578	·310 1893	4281	8893	5890	5439	7713	·429 2894	46
15	4734	5083	7505	·349 2156	9195	8787	·409 1108	6339	45
16	7890	8272	·330 0731	5420	·369 2500	·389 2136	4504	9785	44
17	·292 1047	·311 1462	3957	8685	5800	5486	7901	·430 3232	43
18	4205	4653	7184	·350 1950	9112	8837	·410 1299	6680	42
19	7363	7845	·331 0411	5216	·370 2420	·390 2189	4697	·431 0129	41
20	·293 0521	·312 1036	3639	8483	5728	5541	8097	3579	40
21	3680	4229	6868	·351 1750	9030	8894	·411 1497	7030	39
22	6839	7422	·332 0097	5018	·371 2346	·391 2247	4898	·432 0481	38
23	9999	·313 0616	3327	8287	5656	5602	8300	3933	37
24	·294 3160	3810	6557	·352 1556	8967	8957	·412 1703	7386	36
25	6321	7005	9788	4826	·372 2278	·392 2313	5106	·433 0840	35
26	9483	·314 0200	·333 3020	8096	5596	5670	8510	4295	34
27	·295 2645	3396	6252	·353 1368	8902	9027	·413 1915	7751	33
28	5808	6593	9485	4640	.373 2217	·393 2386	5321	·434 1208	32
29	8971	9790	·334 2719	7912	5532	5745	8728	4665	31
30	·296 2135	·315 2988	5953	·354 1186	8847	9105	·414 2136	8124	30
31	5299	6186	9188	4460	·374 2163	·394 2465	5544	·435 1583	29
32	8464	9385	·335 2424	7734	5479	5827	8953	5043	28
33	·297 1630	·316 2585	5660	·355 1010	8797	9189	·415 2363	8504	27
34	4796	5785	8896	4286	·375 2115	·395 2552	5774	·436 1966	26
35	7962	8986	·336 2134	7562	5433	5916	9186	5429	25
36	·298 1129	·317 2187	5372	·356 0840	8753	9280	·416 2598	8893	24
37	4297	5389	8610	4118	·376 2073	·396 2645	6012	·437 2357	23
38	7465	8591	·337 1850	7397	5394	6011	9426	5823	22
39	·299 0634	·318 1794	5090	·357 0676	8716	9378	·417 2841	9289	21
40	3803	4998	8330	3956	·377 2038	·397 2746	6257	·438 2756	20
41	6973	8202	·338 1571	7237	5361	6114	9673	6224	19
42	·300 0144	·319 1407	4813	·358 0518	8685	9483	·418 3091	9693	18
43	3315	4613	8056	3801	·378 2010	·398 2853	6509	·439 3163	17
44	6486	7819	·339 1299	7083	5335	6224	9928	6634	16
45	9658	·320 1025	4543	·359 0367	8661	9595	·419 3348	·440 0105	15
46	·301 2831	4232	7787	3651	·379 1988	·399 2968	6769	3578	14
47	6004	7440	·340 1032	6936	5315	6341	·420 0190	7051	13
48	9178	·321 0649	4278	·360 0222	8644	9715	3613	·441 0526	12
49	·302 2352	3858	7524	3508	·380 1973	·400 3089	7036	4001	11
50	5527	7067	.341 0771	6795	5302	6465	·421 0460	7477	10
51	8703	·322 0278	4019	·361 0082	8633	9841	3885	·442 0954	9
52	·303 1879	3489	7267	3371	·381 1964	.401 3218	7311	4432	8
53	5055	6700	·342 0516	6660	5296	6590	·422 0738	7910	7
54	8232	9912	3765	9949	8629	9974	4165	·443 1390	6
55	·304 1410	·323 3125	7015	·362 3240	·382 1962	·402 3354	7594	4871	5
56	4588	6338	·343 0266	6531	5296	6734	·423 1023	8352	4
57	7767	9552	3518	9823	8631	·403 0115	4453	·444 1834	3
58	·305 0946	·324 2766	6770	·363 3115	·383 1967	3496	7884	5318	2
59	4126	5981	·344 0023	6408	5303	6879	·424 1316	8802	1
60	7307	9197	3276	9702	8640	·404 0262	4748	·445 2287	0
′	73°	72°	71°	70°	69°	68°	67°	66°	′

′	24°	25°	26°	27°	28°	29°	30°	31°	′
0	·445 2287	·466 3077	·487 7326	·509 5254	·531 7094	·554 3091	·577 3503	·600 8606	60
1	5773	6618	·488 0927	8919	·532 0826	6894	7382	·601 2566	59
2	9260	·467 0161	4530	·510 2585	4559	·555 0698	·578 1262	6527	58
3	·446 2747	3705	8133	6252	8293	4504	5144	·602 0490	57
4	6236	7250	·489 1737	9919	·533 2029	8311	9027	4454	56
5	9726	·468 0796	5343	·511 3588	5765	·556 2119	·579 2912	8419	55
6	·447 3216	4342	8949	7259	9503	5929	6797	·603 2386	54
7	6708	7890	·490 2557	·512 0930	·534 3242	9739	·580 0684	6354	53
8	·448 0200	·469 1439	6166	4602	6981	·557 3551	4573	·604 0323	52
9	3693	4988	9775	8275	·535 0723	7364	8462	4294	51
10	7187	8539	·491 3386	·513 1950	4465	·558 1179	·581 2353	8266	50
11	·449 0682	·470 2090	6997	5625	8208	4994	6245	·605 2240	49
12	4178	5643	·492 0610	9302	·536 1953	8811	·582 0139	6215	48
13	7675	9196	4224	·514 2980	5699	·559 2629	4034	·606 0192	47
14	·450 1173	·471 2751	7838	6658	9446	6449	7930	4170	46
15	4672	6306	·493 1454	·515 0338	·537 3194	·560 0269	·583 1828	8149	45
16	8171	9863	5071	4019	6943	4091	5726	·607 2130	44
17	·451 1672	·472 3420	8689	7702	·538 0694	7914	9627	6112	43
18	5173	6978	·494 2308	·516 1385	4445	·561 1738	·584 3528	·608 0095	42
19	8676	·473 0538	5928	5069	8198	5564	7431	4080	41
20	·452 2179	4098	9549	8755	·539 1952	9391	·585 1335	8067	40
21	5683	7659	·495 3171	·517 2441	5707	·562 3219	5241	·609 2054	39
22	9188	·474 1222	6794	6129	9464	7048	9148	6043	38
23	·453 2694	4785	·496 0418	9818	·540 3221	·563 0879	·586 3056	·610 0034	37
24	6201	8349	4043	·518 3508	6980	4710	6965	4026	36
25	9709	·475 1914	7669	7199	·541 0740	8543	·587 0876	8019	35
26	·454 3218	5481	·497 1297	·519 0891	4501	·564 2378	4788	·611 2014	34
27	6728	9048	4925	4584	8263	6213	8702	6011	33
28	·455 0238	·476 2616	8554	8278	·542 2027	·565 0050	·588 2616	·612 0008	32
29	3750	6185	·498 2185	·520 1974	5791	3888	6533	4007	31
30	7263	9755	5816	5671	9557	7728	·589 0450	8008	30
31	·456 0776	·477 3326	9449	9368	·543 3324	·566 1568	4369	·613 2010	29
32	4290	6899	·499 3082	·521 3067	7092	5410	8289	6013	28
33	7806	·478 0472	6717	6767	·544 0862	9254	·590 2211	·614 0018	27
34	·457 1322	4046	·500 0352	·522 0468	4632	·567 3098	6134	4024	26
35	4839	7621	3989	4170	8404	6944	·591 0058	8032	25
36	8357	·479 1197	7627	7874	·545 2177	·568 0791	3984	·615 2041	24
37	·458 1877	4774	·501 1266	·523 1578	5951	4639	7910	6052	23
38	5397	8352	4906	5284	9727	8488	·592 1839	·616 0064	22
39	8918	·480 1932	8547	8990	·546 3503	·569 2339	5768	4077	21
40	·459 2439	5512	·502 2189	·524 2698	7281	6191	9699	8092	20
41	5962	9093	5832	6407	·547 1060	·570 0045	·593 3632	·617 2108	19
42	9486	·481 2675	9476	·525 0117	4840	3899	7565	6126	18
43	·460 3011	6258	·503 3121	3829	8621	7755	·594 1501	·618 0145	17
44	6537	9842	6768	7541	·548 2404	·571 1612	5437	4166	16
45	·461 0063	·482 3427	·504 0415	·526 1255	6188	5471	9375	8188	15
46	3591	7014	4063	4969	9973	9331	·595 3314	·619 2211	14
47	7119	·483 0601	7713	8685	·549 3759	·572 3192	7255	6236	13
48	·462 0649	4189	·505 1363	·527 2402	7547	7054	·596 1196	·620 0263	12
49	4179	7778	5015	6120	·550 1335	·573 0918	5140	4291	11
50	7710	·484 1368	8668	9839	5125	4783	9084	8320	10
51	·463 1243	4959	·506 2322	·528 3560	8916	8649	·597 3030	·621 2351	9
52	4776	8552	5977	7281	·551 2708	·574 2516	6978	6383	8
53	8310	·485 2145	9633	·529 1004	6502	6385	·598 0926	·622 0417	7
54	·464 1845	5739	·507 3290	4727	·552 0297	·575 0255	4877	4452	6
55	5382	9334	6948	8452	4093	4126	8828	8488	5
56	8919	·486 2931	·508 0607	·530 2178	7890	7999	·599 2781	·623 2527	4
57	·465 2457	6528	4267	5906	·553 1688	·576 1873	6735	6566	3
58	5996	·487 0126	7929	9634	5488	5748	·600 0691	·624 0607	2
59	9536	3723	·509 1591	·531 3364	9288	9625	4648	4650	1
60	·466 3077	7326	5254	7094	·554 3091	·577 3503	8606	8694	0
′	65°	64°	63°	62°	61°	60°	59°	58°	′

NAT. COTAN.

′	32°	33°	34°	35°	36°	37°	38°	39°	′
0	·624 8694	·649 4076	·674 5085	·700 2075	726 5425	·753 5541	·781 2856	·809 7840	60
1	·625 2739	8212	9318	6411	9871	·754 0102	7542	·810 2658	59
2	6786	·650 2350	·675 3553	·701 0749	·727 4318	4666	·782 2229	7478	58
3	·626 0834	6490	7790	5089	8767	9232	6919	·811 2300	57
4	4884	·651 0631	·676 2028	9430	·728 3218	·755 3799	·783 1611	7124	56
5	8935	4774	6268	·702 3773	7671	8369	6305	·812 1951	55
6	·627 2988	8918	·677 0509	8118	·729 2125	·756 2941	·784 1002	6780	54
7	7042	·652 3064	4752	·703 2464	6582	7514	5700	·813 1611	53
8	·628 1098	7211	8997	6813	·730 1041	·757 2090	·785 0400	6444	52
9	5155	·653 1360	·678 3243	·704 1163	5501	6668	5103	·814 1280	51
10	9214	5511	7492	5515	9963	·758 1248	9808	6118	50
11	·629 3274	9663	·679 1741	9869	·731 4428	5829	·786 4515	·815 0958	49
12	7336	·654 3817	5993	·705 4224	8894	·759 0413	9224	5801	48
13	·630 1399	7972	·680 0246	8581	·732 3362	4999	·787 3935	·816 0646	47
14	5464	·655.2129	4501	·706 2940	7832	9587	8649	5493	46
15	9530	6287	8758	7301	·733 2303	·760 4177	·788 3364	·817 0343	45
16	·631 3598	·656 0447	·681 3016	·707 1664	6777	8769	8082	5195	44
17	7667	4609	7276	6028	·734 1253	·761 3368	·789 2802	·818 0049	43
18	·632 1738	8772	·682 1537	·708 0395	5730	7959	7524	4905	42
19	5810	·657 2937	5801	4763	·735 0210	·762 2557	·790 2248	9764	41
20	9883	7103	·683 0066	9133	4691	7157	6975	·819 4625	40
21	·633 3959	·658 1271	4333	·709 3504	9174	·763 1759	·791 1703	9488	39
22	8035	5441	8601	7878	·736 3660	6363	6434	·820 4354	38
23	·634 2113	9612	·684 2371	·710 2253	8147	·764 0969	·792 1167	9222	37
24	6193	·659 3785	7143	6636	·737 2636	5577	5902	·821 4093	36
25	.635 0274	7960	·685 1416	·711 1009	7127	·765 0188	·793 0640	8965	35
26	4357	·660 2136	5692	5396	·738 1620	4800	5379	·822 3840	34
27	8441	6313	9969	9772	6115	9414	·794 0121	8718	33
28	·636 2527	·661 0492	·686 4247	·712 4157	739 0611	·766 4031	4865	.823 3597	32
29	6614	4673	8528	8545	5110	8649	9611	8479	31
30	·637 0703	8856	·687 2810	·713 2937	9611	·767 3270	·795 4359	·824 3364	30
31	4793	·662 3040	7093	7320	·740 4113	7893	9110	8251	29
32	8885	7225	·688 1379	·714 1712	8618	·768 2517	·796 3862	·825 3140	28
33	·638 2978	·663 1413	5666	6106	·741 3124	7144	8617	8031	27
34	7073	5601	9955	·715 0501	7633	·769 1773	·797 3374	·826 2925	26
35	·639 1169	9792	·689 4246	4898	·742 2143	6404	8134	7821	25
36	5267	·664 3984	8538	9297	6655	·770 1037	·798 2895	·827 2719	24
37	9366	8178	·690 2832	·716 3698	·743 1170	5672	7659	7620	23
38	·640 3467	·665 2373	7128	8100	5686	·771 0309	·799 2425	·828 2523	22
39	7569	6570	·691 1425	·717 2505	·744 0204	4948	7193	7429	21
40	·641 1673	·666 0769	5725	6911	4724	9589	·800 1963	·829 2337	20
41	5779	4969	692 0026	·718 1319	9246	·772 4233	6736	7247	19
42	9886	9171	4323	5729	·745 3770	8878	·801 1511	·830 2160	18
43	·642 3994	·667 3374	8633	·719 0141	8296	·773 3526	6288	7075	17
44	8105	7580	.693 2939	4554	·746 2824	8176	·802 1067	·831 1992	16
45	·643 2216	·668 1786	7247	8970	7354	·774 2827	5849	6912	15
46	6329	5995	·694 1557	·720 3387	·747 1886	7481	·803 0632	·832 1834	14
47	.644 0444	·669 0205	5868	7806	6420	·775 2137	5418	6759	13
48	4560	4417	·695 0131	·721 2227	·748 0956	6795	·804 0206	·833 1686	12
49	8678	8630	4496	6656	5494	·776 1455	4997	6615	11
50	·645 2797	·670 2845	8813	·722 1075	·749 0033	6118	9790	·834 1547	10
51	6918	7061	.696 3131	5502	4575	·777 0782	·805 4584	6481	9
52	·646 1041	·671 1280	7451	9930	9119	5448	9382	·835 1418	8
53	5165	5500	·697 1773	·723 4361	·750 3665	·778 0117	·806 4181	6357	7
54	9290	9721	6097	8793	8212	4788	8983	·836 1298	6
55	·647 3417	·672 3944	·698 0422	·724 3227	·751 2762	9460	·807 3787	6242	5
56	7546	8169	4749	7663	7314	·779 4135	8593	·837 1188	4
57	·648 1676	·673 2396	9078	·725 2101	·752 1867	8812	·808 3401	6136	3
58	5808	6624	·699 3409	6546	6423	·780 3492	8212	·838 1087	2
59	9941	·674 0854	7741	·726 0982	·753 0981	8173	·809 3025	6041	1
60	·649 4076	5085	·700 2075	5425	5541	·781 2856	7840	·839 0996	0
′	57°	56°	55°	54°	53°	52°	51°	50°	′

′	40°	41°	42°	43°	44°	45°	46°	47°	′
0	·839 0996	·869 2867	·900 4040	·932 5151	·965 6888	1·00 00000	1·03 55303	1·07 23687	60
1	5955	7976	9309	·933 0591	·966 2511	05819	61333	29943	59
2	·840 0915	·870 3087	·901 4580	6034	8137	11642	67367	36203	58
3	5878	8200	9854	·934 1479	·967 3767	17469	73404	42467	57
4	·841 0844	·871 3316	·902 5131	6928	9399	23298	79445	48734	56
5	5812	8435	·903 0411	·935 2380	·968 5035	29131	85489	55006	55
6	·842 0782	·872 3556	5693	7834	·969 0674	34968	91538	61282	54
7	5755	8680	·904 0979	·936 3292	6316	40807	97589	67561	53
8	·843 0730	·873 3806	6267	8753	·970 1962	46651	1·04 03645	73845	52
9	5708	8935	·905 1557	·937 4216	7610	52497	09704	80132	51
10	·844 0688	·874 4067	6851	9683	·971 3262	58348	15767	86423	50
11	5670	9201	·906 2147	·938 5153	8917	64201	21833	92718	49
12	·845 0655	·875 4338	7446	·939 0625	·972 4575	70058	27904	99018	48
13	5643	9478	·907 2748	6101	·973 0236	75918	33977	1·08 05321	47
14	·846 0633	·876 4620	8053	·940 1579	5901	81782	40055	11628	46
15	5625	9765	·908 3360	7061	·974 1569	87649	46136	17939	45
16	·847 0620	.877 4912	8671	·941 2545	7240	93520	52221	24254	44
17	5617	.878 0062	·909 3984	8033	·975 2914	99394	58310	30573	43
18	·848 0617	5215	9300	·942 3523	8591	1·01 05272	64402	36896	42
19	5619	·879 0370	·910 4619	9017	·976 4272	11153	70498	43223	41
20	·849 0624	5528	9940	·943 4513	9956	17038	76598	49554	40
21	5631	·880 0688	·911 5265	·944 0013	·977 5643	22925	82702	55889	39
22	·850 0640	5852	·912 0592	5516	·978 1333	28817	88809	62228	38
23	5653	·881 1017	5922	·945 1021	7027	34712	94920	68571	37
24	·851 0667	6186	·913 1255	6530	·979 2724	40610	1·05 01034	74918	36
25	5684	·882 1357	6591	·946 2042	8424	46512	07153	81269	35
26	·852 0704	6531	·914 1929	7556	·980 4127	52418	13275	87624	34
27	5726	·883 1707	7270	·947 3074	9833	58326	19401	93984	33
28	·853 0750	6886	·915 2615	8595	·981 5543	64239	25531	1·09 00347	32
29	5777	·884 2068	7962	·948 4119	·982 1256	70155	31664	06714	31
30	·854 0807	7253	·916 3312	9646	6973	76074	37801	13085	30
31	5839	·885 2440	8665	·949 5176	·983 2692	81997	43942	19460	29
32	·855 0873	7630	·917 4020	·950 0709	8415	87923	50087	25840	28
33	5910	·886 2822	9379	6245	·984 4141	93853	56235	32223	27
34	·856 0950	8017	·918 4740	·951 1784	9871	99786	62388	38610	26
35	5992	·887 3215	·919 0104	7326	·985 5603	1·02 05723	68544	45002	25
36	·857 1037	8415	5471	·952 2871	·986 1339	11664	74704	51397	24
37	6084	·888 3619	·920 0841	8420	7079	17608	80867	57797	23
38	·858 1133	8825	6214	·953 3971	·987 2821	23555	87035	64201	22
39	6185	·889 4033	·921 1590	9526	8567	29506	93206	70609	21
40	·859 1240	9244	6969	·954 5083	·988 4316	35461	99381	77020	20
41	6297	·890 4458	·922 2350	·955 0644	·989 0069	41419	1·06 05560	83436	19
42	·860 1357	9675	7734	6208	5825	47381	11742	89857	18
43	6419	·891 4894	·923 3122	·956 1774	·990 1584	53346	17929	96281	17
44	·861 1484	·892 0116	8512	7344	7346	59315	24119	1·10 02709	16
45	6551	5341	·924 3905	·957 2917	·991 3112	65287	30313	09141	15
46	·862 1621	·893 0569	9301	8494	8881	71263	36511	15578	14
47	6694	5799	·925 4700	·958 4073	·992 4654	77243	42713	22019	13
48	·863 1768	·894 1032	·926 0102	9655	·993 0429	83226	48918	28463	12
49	6846	6268	5506	·959 5241	6208	89212	55128	34912	11
50	·864 1926	·895 1506	·927 0914	·960 0829	·994 1991	95203	61341	41365	10
51	7009	6747	6324	6421	7777	1·03 01196	67558	47823	9
52	·865 2094	·896 1991	·928 1738	·961 2016	·995 3566	07194	73779	54284	8
53	7181	7238	7154	7614	9358	13195	80004	60750	7
54	·866 2272	·897 2487	·929 2573	·962 3215	·996 5154	19199	86233	67219	6
55	7365	7739	7996	8819	·997 0953	25208	92466	73693	5
56	·867 2460	·898 2994	·930 3421	·963 4427	6756	31220	98702	80171	4
57	7558	8251	8849	·964 0037	·998 2562	37235	1·07 04943	86653	3
58	·868 2659	·899 3512	·931 4280	5651	8371	43254	11187	93140	2
59	7762	8775	9714	·965 1268	.999 4184	49277	17435	99630	1
60	·869 2867	·900 4040	·932 5151	6888	1·000 0000	55303	23687	1·11 06125	0
′	49°	48°	47°	46°	45°	44°	43°	42°	′

NAT. COTAN.

′	48°	49°	50°	51°	52°	53°	54°	′
0	1·11 06125	1·15 03684	1·19 17536	1·23 48972	1·27 99416	1·32 70448	1·37 63819	60
1	12624	10445	24579	56319	1·28 07094	78483	72242	59
2	19127	17210	31626	63672	14776	86524	80672	58
3	25635	23979	38679	71030	22465	94571	89108	57
4	32146	30754	45736	78393	30160	1·33 02624	97551	56
5	38662	37532	52799	85762	37860	10684	1·38 06001	55
6	45182	44316	59866	93136	45566	18750	14458	54
7	51706	51104	66938	1·24 00515	53277	26822	22922	53
8	58235	57896	74015	07900	60995	34900	31392	52
9	64768	64693	81097	15290	68718	42984	39869	51
10	71305	71495	88184	22685	76447	51075	48353	50
11	77846	78301	95276	30086	84182	59172	56844	49
12	84391	85112	1·20 02373	37492	91922	67276	65342	48
13	90941	91927	09475	44903	99669	75386	73847	47
14	97495	98747	16581	52320	1·29 07421	83502	82358	46
15	1·12 04053	1·16 05571	23693	59742	15179	91624	90876	45
16	10616	12400	30810	67169	22943	99753	99401	44
17	17183	19284	37932	74602	30713	1·34 07888	1·39 07934	43
18	23754	26073	45058	82040	38488	16029	16473	42
19	30329	32916	52190	89484	46270	24177	25019	41
20	36909	39763	59327	96933	54057	32331	33571	40
21	43493	46615	66468	1·25 04388	61850	40492	42131	39
22	50081	53472	73615	11848	69649	48658	50698	38
23	56674	60334	80767	19313	77454	56832	59272	37
24	63271	67200	87924	26784	85265	65011	67852	36
25	69872	74071	95085	34260	93081	73198	76440	35
26	76478	80947	1·21 02252	41742	1·30 00904	81390	85034	34
27	83088	87827	09424	49229	08733	89589	93636	33
28	89702	94712	16601	56721	16567	97794	1·40 02245	32
29	96321	1·17 01601	23783	64219	24407	1·35 06006	10860	31
30	1·13 02944	08496	30970	71723	32254	14224	19483	30
31	09571	15395	38162	79232	40106	22449	28113	29
32	16203	22298	45359	86747	47964	30680	36749	28
33	22839	29207	52562	94267	55828	38918	45393	27
34	29479	36120	59769	1·26 01792	63699	47162	54044	26
35	36124	43038	66982	09323	71575	55413	62702	25
36	42773	49960	74199	16860	79457	63670	71367	24
37	49427	56888	81422	24402	87345	71934	80039	23
38	56085	63820	88650	31950	95239	80204	88718	22
39	62747	70756	95883	39503	1·31 03140	88481	97405	21
40	69414	77698	1·22 03121	47062	11046	96764	1·41 06098	20
41	76086	84644	10364	54626	18958	1·36 05054	14799	19
42	82761	91595	17613	62196	26876	13350	23506	18
43	89441	98551	24866	69772	34801	21653	32221	17
44	96126	1·18 05512	32125	77353	42731	29963	40943	16
45	1·14 02815	12477	39389	84940	50668	38279	49673	15
46	09508	19447	46658	92532	58610	46602	58409	14
47	16206	26422	53932	1·27 00130	66559	54931	67153	13
48	22908	33402	61211	07733	74513	63267	75904	12
49	29615	40387	68496	15342	82474	71610	84662	11
50	36326	47376	75786	22957	90441	79959	93427	10
51	43041	54370	83081	30578	98414	88315	1·42 02200	9
52	49762	61369	90381	38204	1·32 06393	96678	10979	8
53	56486	68373	97687	45835	14379	1·37 05047	19766	7
54	63215	75382	1·23 04997	53473	22370	13423	28561	6
55	69949	82395	12313	61116	30368	21806	37362	5
56	76687	89414	19634	68765	38371	30195	46171	4
57	83429	96437	26961	76419	46381	38591	54988	3
58	90176	1·19 03465	34292	84079	54397	46994	63811	2
59	96928	10498	41629	91745	62420	55403	72642	1
60	1·15 03684	17536	48972	99416	70448	63819	81480	0
′	41°	40°	39°	38°	37°	36°	35°	′

NAT. COTAN.

′	55°	56°	57°	58°	59°	60°	61°	′
0	1·42 81480	1·48 25610	1·53 98650	1·60 03345	1·66 42795	1·73 20508	1·80 40478	60
1	90326	34916	1·54 08460	13709	53766	32149	52860	59
2	99178	44231	18280	24082	64748	43803	65256	58
3	1·43 08039	53554	28108	34465	75741	55468	77664	57
4	16906	62884	37946	44858	86744	67144	90086	56
5	25781	72223	47792	55260	97758	78833	1·81 02521	55
6	34664	81570	57647	65672	1·67 08782	90533	14969	54
7	43554	90925	67510	76094	19818	1·74 02245	27430	53
8	52451	1·49 00288	77383	86525	30864	13969	39904	52
9	61356	09659	87264	96966	41921	25705	52391	51
10	70268	19039	97155	1·61 07417	52988	37453	64892	50
11	79187	28426	1·55 07054	17878	64067	49213	77405	49
12	88114	37822	16963	28349	75156	60984	89932	48
13	97049	47225	26880	38829	86256	72768	1·82 02473	47
14	1·44 05991	56637	36806	49320	97367	84564	15026	46
15	14940	66058	46741	59820	1·68 08489	96371	27593	45
16	23897	75486	56685	70330	19621	1·75 08191	40173	44
17	32862	84923	66639	80850	30765	20023	52767	43
18	41834	94367	76601	91380	41919	31866	65374	42
19	50814	1·50 03821	86572	1·62 01920	53085	43722	77994	41
20	59801	13282	96552	12469	64261	55590	90628	40
21	68796	22751	1·56 06542	23029	75449	67470	1·83 03275	39
22	77798	32229	16540	33599	86647	79362	15936	38
23	86808	41716	26548	44178	97856	91267	28610	37
24	95825	51210	36564	54768	1·69 09077	1·76 03183	41297	36
25	1·45 04850	60713	46590	65368	20308	15112	53999	35
26	13883	70224	56625	75977	31550	27053	66713	34
27	22923	79743	66669	86597	42804	39007	79442	33
28	31971	89271	76722	97227	54069	50972	92184	32
29	41027	98807	86784	1·63 07867	65344	62950	1·84 04940	31
30	50090	1·51 08352	96856	18517	76631	74940	17709	30
31	59161	17905	1·57 06936	29177	87929	86943	30492	29
32	68240	27466	17026	39847	99238	98958	43289	28
33	77326	37036	27126	50528	1·70 10559	1·77 10985	56099	27
34	86420	46614	37234	61218	21890	23024	68923	26
35	95522	56201	47352	71919	33233	35076	81761	25
36	1·46 04632	65796	57479	82630	44587	47141	94613	24
37	13749	75400	67615	93351	55953	59218	1·85 07479	23
38	22874	85012	77760	1·64 04082	67329	71307	20358	22
39	32007	94632	87915	14824	78717	83409	33252	21
40	41147	1·52 04261	98079	25576	90116	95524	46159	20
41	50296	13899	1·58 08253	36338	1·71 01527	1·78 07651	59080	19
42	59452	23545	18436	47111	12949	19790	72015	18
43	68616	33200	28628	57893	24382	31943	84965	17
44	77788	42863	38830	68687	35827	44107	97928	16
45	86967	52535	49041	79490	47283	56285	1·86 10905	15
46	96155	62215	59261	90304	58751	68475	23896	14
47	1·47 05350	71904	69491	1·65 01128	70230	80678	36902	13
48	14553	81602	79731	11963	81720	92893	49921	12
49	23764	91308	89979	22808	93222	1·79 05121	62955	11
50	32983	1·53 01023	1·59 00238	33663	1·72 04736	17362	76003	10
51	42210	10746	10505	44529	16261	29616	89065	9
52	51445	20479	20783	55405	27797	41883	1·87 02141	8
53	60688	30219	31070	66292	39346	54162	15231	7
54	69938	39969	41366	77189	50905	66454	28336	6
55	79197	49727	51672	88097	62477	78759	41455	5
56	88463	59494	61987	99016	74060	91077	54588	4
57	97738	69270	72312	1·66 09945	85654	1·80 03408	67736	3
58	1·48 07021	79054	82647	20884	97260	15751	80898	2
59	16311	88848	92991	31834	1·73 08878	28108	94074	1
60	25610	98650	1·60 03345	42795	20508	40478	1·88 07265	0
′	34°	33°	32°	31°	30°	29°	28°	′

NAT. COTAN

′	62°	63°	64°	65°	66°	67°	68°	′
0	1·88 07265	1·96 26105	2·05 03038	2·14 45069	2·24 60368	2·35 58524	2·47 50869	60
1	20470	40227	18185	61366	77962	77590	71612	59
2	33690	54364	33349	77683	95580	96683	92386	58
3	46924	68518	48531	94021	2·25 13221	2·36 15801	2·48 13190	57
4	60172	82688	63732	2·15 10378	30885	34946	34023	56
5	73436	96874	78950	26757	48572	54118	54887	55
6	86713	1·97 11077	94187	43156	66283	73316	75781	54
7	1·89 00006	25296	2·06 09442	59575	84016	92540	96706	53
8	13313	39531	24716	76015	2·26 01773	2·37 11791	2·49 17660	52
9	26635	53782	40008	92476	19554	31068	38645	51
10	39971	68050	55318	2·16 08958	37357	50372	59661	50
11	53322	82334	70646	25460	55184	69703	80707	49
12	66688	96635	85994	41983	73035	89060	2·50 01784	48
13	80068	1·98 10952	2·07 01359	58527	90909	2·38 08444	22891	47
14	93464	25286	16743	75091	2·27 08807	27855	44029	46
15	1·90 06874	39636	32146	91677	26729	47293	65198	45
16	20299	54003	47567	2·17 08283	44674	66758	86398	44
17	33738	68387	63007	24911	62643	86250	2·51 07629	43
18	47193	82787	78465	41559	80636	2·39 05769	28890	42
19	60663	97204	93942	58229	98653	25316	50183	41
20	74147	1·99 11637	2·08 09438	74920	2·28 16693	44889	71507	40
21	87647	26087	24953	91631	34758	64490	92863	39
22	1·91 01162	40554	40487	2·18 08364	52846	84118	2·52 14249	38
23	14691	55038	56039	25119	70959	2·40 03774	35667	37
24	28236	69539	71610	41894	89096	23457	57117	36
25	41795	84056	87200	58691	2·29 07257	43168	78598	35
26	55370	98590	2·09 02809	75510	25442	62906	2·53 00111	34
27	68960	2·00 13142	18437	92349	43651	82672	21655	33
28	82565	27710	34085	2·19 09210	61885	2·41 02465	43231	32
29	96186	42295	49751	26093	80143	22286	64839	31
30	1·92 09821	56897	65436	42997	98425	42136	86479	30
31	23472	71516	81140	59923	2·30 16732	62013	2·54 08151	29
32	37138	86153	96864	76871	35064	81918	29855	28
33	50819	2·01 00806	2·10 12607	93840	53420	2·42 01851	51591	27
34	64516	15477	28369	2·20 10831	71801	21812	73359	26
35	78228	30164	44150	27843	90206	41801	95160	25
36	91956	44869	59951	44878	2·31 08637	61819	2·55 16992	24
37	1·93 05699	59592	75771	61934	27092	81864	38858	23
38	19457	74331	91611	79012	45571	2·43 01938	60756	22
39	33231	89088	2·11 07470	96112	64076	22041	82686	21
40	47020	2·02 03862	23348	2·21 13234	82606	42172	2·56 04649	20
41	60825	18654	39246	30379	2·32 01160	62331	26645	19
42	74645	33462	55164	47545	19740	82519	48674	18
43	88481	48289	71101	64733	38345	2·44 02736	70735	17
44	1·94 02333	63133	87057	81944	56975	22982	92830	16
45	16200	77994	2·12 03034	99177	75630	43256	2·57 14957	15
46	30083	92873	19030	2·22 16432	94311	63559	37118	14
47	43981	2·03 07769	35046	33709	2·33 13017	83891	59312	13
48	57896	22683	51082	51009	31748	2·45 04252	81539	12
49	71826	37615	67137	68331	50505	24642	2·58 03800	11
50	85772	52565	83213	85676	69287	45061	26094	10
51	99733	67532	99308	2·23 03043	88095	65510	48421	9
52	1·95 13711	82517	2·13 15423	20433	2·34 06928	85987	70782	8
53	27704	97519	31559	37845	25787	2.46 06494	93177	7
54	41713	2·04 12540	47714	55280	44672	27030	2·59 15606	6
55	55739	27578	63890	72738	63582	47596	38068	5
56	69780	42634	80085	90218	82519	68191	60564	4
57	83837	57708	96301	2·24 07721	2·35 01481	88816	83095	3
58	97910	72800	2·14 12537	25247	20469	2·47 09470	2·60 05659	2
59	1·96 12000	87910	28793	42796	39483	30155	28258	1
60	26105	2·05 03038	45069	60368	58524	50869	50891	0
′	27°	26°	25°	24°	23°	22°	21°	′

NAT. COTAN.

′	69°	70°	71°	72°	73°	74°	75°	′
0	2·60 50891	2·74 74774	2·90 42109	3·07 76835	3·27 08526	3·48 74144	3·73 20508	60
1	73558	99661	69576	3·08 07325	42588	3·49 12470	63980	59
2	96259	2·75 24588	97089	37869	76715	50874	3·74 07546	58
3	2·61 18995	49554	2·91 24649	68468	3·28 10907	89356	51207	57
4	41766	74561	52256	99122	45164	3·50 27916	94963	56
5	64571	99608	79909	3·09 29831	79487	66555	3·75 38815	55
6	87411	2·76 24695	2·92 07610	60596	3·29 13876	3·51 05273	82763	54
7	2·62 10286	49822	35358	91416	48330	44070	3·76 26807	53
8	33196	74990	63152	3·10 22291	82851	82946	70947	52
9	56141	2·77 00199	90995	53223	3·30 17438	3·52 21902	3·77 15185	51
10	79121	25448	2·93 18885	84210	52091	60938	59519	50
11	2·63 02136	50738	46822	3·11 15254	86811	3·53 00054	3·78 03951	49
12	25186	76069	74807	46353	3·31 21598	39251	48481	48
13	48271	2·78 01440	2·94 02840	77509	56452	78528	93109	47
14	71392	26853	30921	3·12 08722	91373	3·54 17886	3·79 37835	46
15	94549	52307	59050	39991	3·32 26362	57325	82661	45
16	2·64 17741	77802	87227	71317	61419	96846	3·80 27585	44
17	40969	2·79 03339	2·95 15453	3·13 02701	96543	3·55 36449	72609	43
18	64232	28917	43727	34141	3·33 31736	76133	3·81 17733	42
19	87531	54537	72050	65639	66997	3·56 15900	62957	41
20	2·65 10867	80198	2·96 00422	97194	3·34 02326	55749	3·82 08281	40
21	34238	2·80 05901	28842	3·14 28807	37724	95681	53707	39
22	57645	31646	57312	60478	73191	3·57 35696	99233	38
23	81089	57433	85831	92207	3·35 08728	75794	3·83 44861	37
24	2·66 04569	83263	2·97 14399	3·15 23994	44333	3·58 15975	90591	36
25	28085	2·81 09134	43016	55840	80008	56241	3·84 36424	35
26	51638	35048	71683	87744	3·36 15753	96590	82358	34
27	75227	61004	2·98 00400	3·16 19706	51568	3·59 37024	3·85 28396	33
28	98853	87003	29167	51728	87453	77543	74537	32
29	2·67 22516	2·82 13045	57983	83808	3·37 23408	3·60 18146	3·86 20782	31
30	46215	39129	86850	3·17 15948	59434	58835	67131	30
31	69951	65256	2·99 15766	48147	95531	99609	3·87 13584	29
32	93725	91426	44734	80406	3·38 31699	3·61 40469	60142	28
33	2·68 17535	2·83 17639	73751	3·18 12724	67938	81415	3·88 06805	27
34	41383	43896	3·00 02820	45102	3·39 04249	3·62 22447	53574	26
35	65267	70196	31939	77540	40631	63566	3·89 00448	25
36	89190	96539	61109	3·19 10039	77085	3·63 04771	47429	24
37	2·69 13149	2·84 22926	90330	42598	3·40 13612	46064	94516	23
38	37147	49356	3·01 19603	75217	50210	87444	3·90 41710	22
39	61181	75831	48926	3·20 07897	86882	3·64 28911	89011	21
40	85254	2·85 02349	78301	40638	3·41 23626	70467	3·91 36420	20
41	2·70 09364	28911	3·02 07728	73440	60443	3·65 12111	83937	19
42	33513	55517	37207	3·21 06304	97333	53844	3·92 31563	18
43	57699	82168	66737	39228	3·42 34297	95665	79297	17
44	81923	2·86 08863	96320	72215	71334	3·66 37575	3·93 27141	16
45	2·71 06186	35602	3·03 25954	3·22 05263	3·43 08446	79575	75094	15
46	30487	62386	55641	38373	45631	3·67 21665	3·94 23157	14
47	54826	89215	85381	71546	82891	63845	71331	13
48	79204	2·87 16088	3·04 15173	3·23 04780	3·44 20226	3·68 06115	3·95 19615	12
49	2·72 03620	43007	45018	38078	57635	48475	68011	11
50	28076	69970	74915	71438	95120	90927	3·96 16518	10
51	52569	96979	3·05 04866	3·24 04860	3·45 32679	3·69 33469	65137	9
52	77102	2·88 24033	34870	38346	70315	76104	3·97 13868	8
53	2·73 01674	51132	64928	71895	3·46 08026	3·70 18830	62712	7
54	26284	78277	95038	3·25 05508	45813	61648	3·98 11669	6
55	50934	2·89 05467	3·06 25203	39184	83676	3·71 04558	60739	5
56	75623	32704	55421	72924	3·47 21616	47561	3·99 09924	4
57	2·74 00352	59986	85694	3·26 06728	59632	90658	59223	3
58	25120	87314	3·07 16020	40596	97726	3·72 33847	4·00 08636	2
59	49927	2·90 14688	46400	74529	3·48 35896	77131	58165	1
60	74774	42109	76835	3·27 08526	74144	3·73 20508	4·01 07809	0
′	20°	19°	18°	17°	16°	15°	14°	′

′	76°	77°	78°	79°	80°	81°	82°	′
0	4·01 07809	4·33 14759	4·70 46301	5·1 445540	5·6 712818	6·3 137515	7·1 153697	60
1	57570	72316	4·71 13686	525557	809446	256601	304190	59
2	4·02 07446	4·34 30018	81256	605813	906394	376126	455308	58
3	57440	87866	4·72 49012	686311	5·7 003663	496092	607056	57
4	4·03 07550	4·35 45861	4·73 16954	767051	101256	616502	759437	56
5	57779	4·36 04003	85083	848035	199173	737359	912456	55
6	4·04 08125	62293	4·74 53401	929264	297416	858665	7·2 066116	54
7	58590	4·37 20731	4·75 21907	5·2 010738	395988	980422	220422	53
8	4·05 09174	79317	90603	092459	494889	6·4 102633	375378	52
9	59877	4·38 38054	4·76 59490	174428	594122	225301	530987	51
10	4·06 10700	96940	4·77 28568	256647	693688	348428	687255	50
11	61643	4·39 55977	97837	339116	793588	472017	844184	49
12	4·07 12707	4·40 15164	4·78 67300	421836	893825	596070	7·3 001780	48
13	63892	74504	4·79 36957	504809	994400	720591	160047	47
14	4·08 15199	4·41 33996	4·80 06808	588035	5·8 095315	845581	318989	46
15	66627	93641	76854	671517	196572	971043	478610	45
16	4·09 18178	4·42 53439	4·81 47096	755255	298172	6·5 096981	638916	44
17	69852	4·43 13392	4·82 17536	839251	400117	223396	799909	43
18	4·10 21649	73500	88174	923505	502410	350293	961595	42
19	73569	4·44 33762	4·83 59010	5·3 008018	605051	477672	7·4 123978	41
20	4·11 25614	94181	4·84 30045	092793	708042	605538	287064	40
21	77784	4·45 54756	4·85 01282	177830	811386	733892	450855	39
22	4·12 30079	4·46 15489	72719	263131	915084	862739	615357	38
23	82499	76379	4·86 44359	348696	5·9 019138	992080	780576	37
24	4·13 35046	4·47 37428	4·87 16201	434527	123550	6·6 121919	946514	36
25	87719	98636	88248	520626	228322	252258	7·5 113178	35
26	4·14 40519	4·48 60004	4·88 60499	606993	333455	383100	280571	34
27	93446	4·49 21532	4·89 32956	693630	438952	514449	448699	33
28	4·15 46501	83221	4·90 05620	780538	544815	646307	617567	32
29	99635	4·50 45072	78491	867718	651045	778677	787179	31
30	4·16 52998	4·51 07085	4·91 51570	955172	757644	911562	957541	30
31	4·17 06440	69261	4·92 24859	5·4 042901	864614	6·7 044966	7·6 128657	29
32	60011	4·52 31601	98358	130906	971957	178891	300533	28
33	4·18 13713	94105	4·93 72068	219188	6·0 079676	313341	473174	27
34	67546	4·53 56773	4·94 45990	307750	187772	448318	646584	26
35	4·19 21510	4·54 19608	4·95 20125	396592	296247	583826	820769	25
36	75606	82608	94474	485715	405103	719867	995735	24
37	4·20 29835	4·55 45776	4·96 69037	575121	514343	856446	7·7 171486	23
38	84196	4·56 09111	4·97 43817	664812	623967	993565	348028	22
39	4·21 38690	72615	4·98 18813	754788	733979	6·8 131227	525366	21
40	93318	4·57 36287	94027	845052	844381	269437	703506	20
41	4·22 48080	4·58 00129	4·99 69459	935604	955174	408196	882453	19
42	4·23 02977	64141	5·00 45111	5·5 026446	6·1 066360	547508	7·8 062212	18
43	58009	4·59 28325	5·01 20984	117579	177943	687378	242790	17
44	4·24 13177	92680	97078	209005	289923	827807	424191	16
45	68482	4·60 57207	5·02 73395	300724	402303	968799	606423	15
46	4·25 23923	4·61 21908	5·03 49935	392740	515085	6·9 110359	789489	14
47	79501	86783	5·04 26700	485052	628272	252489	973396	13
48	4·26 35218	4·62 51832	5·05 03690	577663	741865	395192	7·9 158151	12
49	91072	4·63 17056	80907	670574	855867	538473	343758	11
50	4·27 47066	82457	5·06 58352	763786	970279	682335	530224	10
51	4·28 03199	4·64 48034	5·07 36025	857302	6·2 085106	826781	717555	9
52	59472	4·65 13788	5·08 13928	951121	200347	971806	905756	8
53	4·29 15885	79721	92061	5·6 045247	316007	7·0 117441	8·0 094835	7
54	72440	4·66 45832	5·09 70426	139680	432086	263662	284796	6
55	4·30 29136	4·67 12124	5·10 49024	234421	548588	410482	475647	5
56	85974	78595	5·11 27855	329474	665515	557905	667394	4
57	4·31 42955	4·68 45248	5·12 06921	424838	782868	705934	860042	3
58	4·32 00079	4·69 12083	86224	520516	900651	854573	8·1 053599	2
59	57347	79100	5·13 65763	616509	6·3 018866	7·1 003826	248071	1
60	4·33 14759	4·70 46301	5·14 45540	712818	137515	153697	443464	0
′	13°	12°	11°	10°	9°	8°	7°	′

NAT. COTAN.

′	83°	84°	85°	86°	87°	88°	89°	′
0	8·1 443464	9·5 143645	11.430052	14·300666	19·081137	28·636253	57·289962	60
1	639786	410613	468474	360696	187930	877089	58·261174	59
2	837041	679068	507154	421230	295922	29·122006	59·265872	58
3	8·2 035239	949022	546093	482273	405133	371106	60·305820	57
4	234384	9·6 220486	585294	543833	515584	624499	61·382905	56
5	434485	493475	624761	605916	627296	882299	62·499154	55
6	635547	768000	664495	668529	740291	30·144619	63·656741	54
7	837579	9·7 044075	704500	731679	854591	411580	64·858008	53
8	8·3 040586	321713	744779	795372	970219	683307	66·105473	52
9	244577	600927	785333	859616	20·087199	959928	67·401854	51
10	449558	881732	826167	924417	205553	31·241577	68·750087	50
11	655536	9·8 164140	867282	989784	325308	528392	70·153346	49
12	862519	448166	908682	15·055723	446486	820516	71·615070	48
13	8·4 070515	733823	950370	122242	569115	32·118099	73·138991	47
14	279531	9·9 021125	992349	189349	693220	421295	74·729165	46
15	489573	310088	12.034622	257052	818828	730265	76·390009	45
16	700651	600724	077192	325358	945966	33·045173	78·126342	44
17	912772	893050	120062	394276	21·074664	366194	79·943430	43
18	8·5 125943	10·018708	163236	463814	204949	693509	81·847041	42
19	340172	048283	206716	533981	336851	34·027303	83·843507	41
20	555468	078031	250505	604784	470401	367771	85·939791	40
21	771838	107954	294609	676233	605630	715115	88·143572	39
22	989290	138054	339028	748337	742569	35·069546	90·463336	38
23	8·6 207833	168332	383768	821105	881251	431282	92·908487	37
24	427475	198789	428831	894545	22·021710	800553	95·489475	36
25	648223	229428	474221	968667	163980	36·177596	98·217943	35
26	870088	260249	519942	16·043482	308097	562659	101·10690	34
27	8·7 093077	291255	565997	118998	454096	956001	104·17094	33
28	317198	322447	612390	195225	602015	37·357892	107·42648	32
29	542461	353827	659125	272174	751892	768613	110·89205	31
30	768874	385397	706205	349855	903766	38·188459	114·58865	30
31	996446	417158	753634	428279	23·057677	617738	118·54018	29
32	8·8 225186	449112	801417	507456	213666	39·056771	122·77396	28
33	455103	481261	849557	587396	371777	505895	127·32134	27
34	686206	513607	898058	668112	532052	965460	132·21851	26
35	918505	546151	946924	749614	694537	40·435837	137·50745	25
36	8·9 152009	578895	996160	831915	859277	917412	143·23712	24
37	386726	611841	13·045769	915025	24·026320	41·410588	149·46502	23
38	622668	644992	095757	998957	195714	915790	156·25908	22
39	859843	678348	146127	17·083724	367509	42·433464	163·70019	21
40	9·0 098261	711913	196883	169337	541758	964077	171·88540	20
41	337933	745687	248031	255809	718512	43·508122	180·93220	19
42	578867	779673	299574	343155	897826	44·066113	190·98419	18
43	821074	813872	351518	431385	25·079757	638596	202·21875	17
44	9·1 064564	848288	403867	520516	264361	45·226141	214·85762	16
45	309348	882921	456625	610559	451700	829351	229·18166	15
46	555436	917775	509799	701529	641832	46·448862	245·55198	14
47	802838	952850	563391	793442	834823	47·085343	264·44080	13
48	9·2 051564	988150	617409	886310	26·030736	739501	286·47773	12
49	301627	11·023676	671856	980150	229638	48·412084	312·52137	11
50	553035	059431	726738	18·074977	431600	49·103881	343·77371	10
51	805802	095416	782060	170807	636690	815726	381·97099	9
52	9·3 059936	131635	837827	267654	844984	50·548506	429·71757	8
53	315450	168089	894045	365537	27·056557	51·303157	491·10600	7
54	572355	204780	950719	464471	271486	52·080673	572·95721	6
55	830663	241712	14·007856	564473	489853	882109	687·54887	5
56	9·4 090384	278885	065459	665562	711740	53·708587	859·43630	4
57	351531	316304	123536	767754	937233	54·561300	1145·9153	3
58	614116	353970	182092	871068	28·166422	55·441517	1718·8732	2
59	878149	391885	241134	975523	399397	56·350590	3437·7467	1
60	9.5 143645	430052	300666	19·081137	636253	57·289962	Infinite.	0
′	6°	5°	4°	3°	2°	1°	0°	′

Milli-metres.	English inches.	Milli-metres.	English inches.	Milli-metres.	English inches.	Milli-metres.	English inches.	Milli-metres.	English inches.	Milli-metres.	English inches
501	19·725	551	21·693	601	23·662	651	25·630	701	27·599	751	29·567
502	·764	552	·733	602	·701	652	·670	702	·638	752	·606
503	·803	553	·772	603	·741	653	·709	703	·677	753	·646
504	·843	554	·811	604	·780	654	·748	704	·717	754	·685
505	·882	555	·851	605	·819	655	·788	705	·756	755	·725
506	·921	556	·890	606	·859	656	·827	706	·795	756	·764
507	19·961	557	·930	607	·898	657	·867	707	·835	757	·803
508	20·000	558	21·969	608	·937	658	·906	708	·874	758	·843
509	·040	559	22·009	609	23·977	659	·945	709	·914	759	·882
510	·079	560	·048	610	24·016	660	25·985	710	·953	760	·921
511	·118	561	·087	611	·056	661	26.024	711	27·992	761	29·961
512	·158	562	·126	612	·095	662	·063	712	28·032	762	30·000
513	·197	563	·166	613	·134	663	·103	713	·071	763	·040
514	·236	564	·205	614	·174	664	·142	714	·110	764	·079
515	·276	565	·244	615	·213	665	·181	715	·150	765	·118
516	·315	566	·284	616	·252	666	·221	716	·189	766	·158
517	·354	567	·323	617	·292	667	·260	717	·229	767	·197
518	·394	568	·363	618	·331	668	·300	718	·268	768	·236
519	·433	569	·402	619	·371	669	·339	719	·307	769	·276
520	·473	570	·441	620	·410	670	·378	720	·347	770	·315
521	·512	571	·481	621	·449	671	·418	721	·386	771	·355
522	·551	572	·520	622	·489	672	·457	722	·425	772	·394
523	·591	573	·559	623	·528	673	·496	723	·465	773	·433
524	·630	574	·599	624	·567	674	·536	724	·504	774	·473
525	·670	575	·638	625	·607	675	·575	725	·543	775	·512
526	·709	576	·678	626	·646	676	·615	726	·583	776	·551
527	·748	577	·717	627	·685	677	·654	727	·622	777	·591
528	·788	578	·756	628	·725	678	·693	728	·662	778	·630
529	·827	579	·796	629	·764	679	·733	729	·701	779	·670
530	·867	580	·835	630	·804	680	·772	730	·740	780	·709
531	·906	581	·875	631	·843	681	·811	731	·780	781	·748
532	·945	582	·914	632	·882	682	·851	732	·819	782	·788
533	20·985	583	·953	633	·922	683	·890	733	·858	783	·827
534	21·024	584	22·993	634	·961	684	·930	734	·898	784	·866
535	·063	585	23·032	635	25·000	685	26·969	735	·937	785	·906
536	·103	586	·071	636	·040	686	27·008	736	28·977	786	·945
537	·142	587	·111	637	·079	687	·048	737	29·016	787	30·984
538	·181	588	·150	638	·118	688	·087	738	·055	788	31·024
539	·221	589	·189	639	·158	689	·126	739	·095	789	·063
540	·266	590	·229	640	·197	690	·166	740	·134	790	·103
541	·300	591	·268	641	·237	691	·205	741	·173	PROP'L PARTS.	
542	·339	592	·308	642	·276	692	·245	742	·213	0·1	0·0039
543	·378	593	·347	643	·315	693	·284	743	·252	·2	·0079
544	·417	594	·386	644	·355	694	·323	744	·292	·3	·0118
545	·457	595	·426	645	·394	695	·363	745	·331	·4	·0157
546	·496	596	·465	646	·433	696	·402	746	·370	·5	·0197
547	·536	597	·504	647	·473	697	·441	747	·410	·6	·0236
548	·575	598	·544	648	·512	698	·481	748	·449	·7	·0276
549	·614	599	·583	649	·552	699	·520	749	·488	·8	·0315
550	·654	600	·622	650	·591	700	·559	750	·528	·9	·0354

1 Metre = 39·3707 English inches = 443·296 Paris lines.
1 English foot = 0·304794 metre = 135·114 Paris lines.
1 French foot = 1·0658 English feet = 0·32484 metre.

D. M.	Chords.	D. M.	Chords.	D. M.	Chords.	D. M.	Chords.	D. M.	Chords.
5	·0015	9	·1569	18	·3129	27	·4669	36	·6180
10	·0029	10	·1598	10	·3157	10	·4697	10	·6208
20	·0058	20	·1627	20	·3186	20	·4725	20	·6236
30	·0087	30	·1656	30	·3215	30	·4754	30	·6263
40	·0116	40	·1685	40	·3244	40	·4782	40	·6291
50	·0145	50	·1714	50	·3272	50	·4810	50	·6318
1	·0175	10	·1743	19	·3301	28	·4838	37	·6346
10	·0204	10	·1772	10	·3330	10	·4867	10	·6374
20	·0233	20	·1801	20	·3358	20	·4895	20	·6401
30	·0262	30	·1830	30	·3387	30	·4923	30	·6429
40	·0291	40	·1859	40	·3416	40	·4951	40	·6456
50	·0320	50	·1888	50	·3444	50	·4979	50	·6484
2	·0349	11	·1917	20	·3473	29	·5008	38	·6511
10	·0378	10	·1946	10	·3502	10	·5036	10	·6539
20	·0407	20	·1975	20	·3530	20	·5064	20	·6566
30	·0436	30	·2004	30	·3559	30	·5092	30	·6594
40	·0465	40	·2033	40	·3587	40	·5120	40	·6621
50	·0494	50	·2062	50	·3616	50	·5148	50	·6649
3	·0523	12	·2091	21	·3645	30	·5176	39	·6676
10	·0553	10	·2119	10	·3673	10	·5204	10	·6703
20	·0582	20	·2148	20	·3702	20	·5233	20	·6731
30	·0611	30	·2177	30	·3730	30	·5261	30	·6758
40	·0640	40	·2206	40	·3759	40	·5289	40	·6786
50	·0669	50	·2235	50	·3788	50	·5317	50	·6813
4	·0698	13	·2264	22	·3816	31	·5345	40	·6840
10	·0727	10	·2293	10	·3845	10	·5373	10	·6866
20	·0756	20	·2322	20	·3873	20	·5401	20	·6895
30	·0785	30	·2351	30	·3902	30	·5429	30	·6922
40	·0814	40	·2380	40	·3930	40	·5457	40	·6950
50	·0843	50	·2409	50	·3959	50	·5485	50	·6977
5	·0872	14	·2437	23	·3987	32	·5513	41	·7004
10	·0901	10	·2466	10	·4016	10	·5541	10	·7031
20	·0931	20	·2495	20	·4044	20	·5569	20	·7059
30	·0960	30	·2524	30	·4073	30	·5597	30	·7086
40	·0989	40	·2553	40	·4101	40	·5625	40	·7113
50	·1018	50	·2582	50	·4130	50	·5652	50	·7140
6	·1047	15	·2611	24	·4158	33	·5680	42	·7167
10	·1076	10	·2639	10	·4187	10	·5708	10	·7194
20	·1105	20	·2668	20	·4215	20	·5736	20	·7222
30	·1134	30	·2697	30	·4244	30	·5764	30	·7249
40	·1163	40	·2726	40	·4272	40	·5792	40	·7276
50	·1192	50	·2755	50	·4300	50	·5820	50	·7303
7	·1221	16	·2783	25	·4329	34	·5847	43	·7330
10	·1250	10	·2812	10	·4357	10	·5875	10	·7357
20	·1279	20	·2841	20	·4386	20	·5903	20	·7384
30	·1308	30	·2870	30	·4414	30	·5931	30	·7411
40	·1337	40	·2899	40	·4442	40	·5959	40	·7438
50	·1366	50	·2927	50	·4471	50	·5986	50	·7465
8	·1395	17	·2956	26	·4499	35	·6014	44	·7492
10	·1424	10	·2985	10	·4527	10	·6042	10	·7519
20	·1453	20	·3014	20	·4557	20	·6070	20	·7546
30	·1482	30	·3042	30	·4584	30	·6097	30	·7573
40	·1511	40	·3071	40	·4612	40	·6125	40	·7600
50	·1540	50	·3100	50	·4641	50	·6153	50	·7627

D. M.	Chords.	D. M.	Chords.	D. M.	Chords.	D. M.	Chords.	D. M.	Chords.
45	·7654	54	·9080	63	1·0450	72	1·1756	81	1·2989
10	·7681	10	·9106	10	1·0475	10	1·1779	10	1·3011
20	·7707	20	·9132	20	1·0500	20	1·1803	20	1·3033
30	·7734	30	·9157	30	1·0524	30	1·1826	30	1·3055
40	·7761	40	·9183	40	1·0549	40	1·1850	40	1·3077
50	·7788	50	·9209	50	1·0574	50	1·1873	50	1·3099
46	·7815	55	·9235	64	1·0598	73	1·1896	82	1·3121
10	·7841	10	·9261	10	1·0623	10	1·1920	10	1·3143
20	·7868	20	·9287	20	1·0648	20	1·1943	20	1·3165
30	·7895	30	·9312	30	1·0672	30	1·1966	30	1·3187
40	·7922	40	·9338	40	1·0697	40	1·1990	40	1·3209
50	·7948	50	·9364	50	1·0721	50	1·2013	50	1·3231
47	·7975	56	·9389	65	1·0746	74	1·2036	83	1·3252
10	·8002	10	·9415	10	1·0771	10	1·2060	10	1·3274
20	·8028	20	·9441	20	1·0795	20	1·2083	20	1·3296
30	·8055	30	·9466	30	1·0819	30	1·2106	30	1·3318
40	·8082	40	·9492	40	1·0844	40	1·2129	40	1·3339
50	·8108	50	·9518	50	1·0868	50	1·2152	50	1·3361
48	·8135	57	·9543	66	1·0893	75	1·2175	84	1·3383
10	·8161	10	·9569	10	1·0917	10	1·2198	10	1·3404
20	·8188	20	·9594	20	1·0942	20	1·2221	20	1·3426
30	·8214	30	·9620	30	1·0966	30	1·2244	30	1·3447
40	·8241	40	·9645	40	1·0990	40	1·2267	40	1·3469
50	·8267	50	·9671	50	1·1014	50	1·2290	50	1·3490
49	·8294	58	·9696	67	1·1039	76	1·2313	85	1·3512
10	·8320	10	·9722	10	1·1063	10	1·2336	10	1·3533
20	·8347	20	·9747	20	1·1087	20	1·2359	20	1·3555
30	·8373	30	·9772	30	1·1111	30	1·2382	30	1·3576
40	·8400	40	·9798	40	1·1136	40	1·2405	40	1·3597
50	·8426	50	·9823	50	1·1160	50	1·2428	50	1·3619
50	·8452	59	·9848	68	1·1184	77	1·2450	86	1·3640
10	·8479	10	·9874	10	1·1208	10	1·2473	10	1·3661
20	·8505	20	·9899	20	1·1232	20	1·2496	20	1·3682
30	·8531	30	·9924	30	1·1256	30	1·2518	30	1·3704
40	·8558	40	·9950	40	1·1280	40	1·2541	40	1·3725
50	·8584	50	·9975	50	1·1304	50	1·2564	50	1·3746
51	·8610	60	1·0000	69	1·1328	78	1·2586	87	1·3767
10	·8636	10	1·0025	10	1·1352	10	1·2609	10	1·3788
20	·8663	20	1·0050	20	1·1376	20	1·2632	20	1·3809
30	·8689	30	1·0075	30	1·1400	30	1·2654	30	1·3830
40	·8715	40	1·0101	40	1·1424	40	1·2677	40	1·3851
50	·8741	50	1·0126	50	1·1448	50	1·2699	50	1·3872
52	·8767	61	1·0151	70	1·1472	79	1·2722	88	1·3893
10	·8794	10	1·0176	10	1·1495	10	1·2744	10	1·3914
20	·8820	20	1·0201	20	1·1519	20	1·2766	20	1·3935
30	·8846	30	1·0226	30	1·1543	30	1·2789	30	1·3956
40	·8872	40	1·0251	40	1·1567	40	1·2811	40	1·3977
50	·8898	50	1·0276	50	1·1590	50	1·2833	50	1·3997
53	·8924	62	1·0301	71	1·1614	80	1·2856	89	1·4018
10	·8950	10	1·0326	10	1·1638	10	1·2878	10	1·4039
20	·8976	20	1·0351	20	1·1661	20	1·2900	20	1·4060
30	·9002	30	1·0375	30	1·1685	30	1·2922	30	1·4080
40	·9028	40	1·0400	40	1·1709	40	1·2945	40	1·4101
50	·9054	50	1·0425	50	1·1732	50	1·2967	50	1·4122

Chord of 90 Degrees = 1·4142.

FOR SALE BY

WILLIAM A. BURT,

MOUNT VERNON, MICH.,

To whom all communications (post paid) should be addressed,

WILLIAM J. YOUNG,

MATHEMATICAL AND OPTICAL INSTRUMENT MAKER,

No. 33 North Seventh Street,

McALLISTER & BROTHER,

IMPORTERS AND DEALERS IN

MATHEMATICAL, OPTICAL, AND PHILOSOPHICAL INSTRUMENTS,

No. 194 Chestnut Street, Philadelphia.

www.ingramcontent.com/pod-product-compliance
Lightning Source LLC
LaVergne TN
LVHW011210110826
845150LV00006B/1394

* 9 7 8 1 4 2 5 5 1 7 4 2 7 *